COMPLEX

COMPLEX

A.D. ENDERLY

COMPLEX
First edition. December 1, 2020.

Written by A.D. Enderly

Published in the Unites States by Luminary Media, LLC.
ISBN: 978-0-5787-5224-2

Edited by Emily Yau.
Cover art by Natasha MacKenzie.
Interior design by Phillip Gessert.

To my wife, Brooke.
Without you, this work would not have been made.
I thank you, and I love you.

BOOK ONE

COMPLEX AND LEGACY

LEGACY
leg-a-cy /'legəse/
noun

1. *a term, often pejorative, referring to the scattered remnants of "democratic" rule. Legacy governments are often localized entities, subdivided into districts that are run by Administrators.*

COMPLEX
com-plex /'kämpleks/
noun

1. *a group of similar buildings or facilities on the same site*

2. *a company or corporation legally permitted to write and enforce contracts binding humans—their property and services—in return for security, shelter, and other provisions necessary for life. Commonly referred to by its legal entity nomenclature "CPX".*

ARC
arc /'ärk/
noun

1. *a megacity, thousands of feet in height, built in tiers. There are numerous Arcs throughout the world, largely administered by Legacy governments, but with enclaves of Complexes situated within them.*

RAY

RAY
AGE: 36
SEGMENT: COMPLEX PROPERTY; PERSENSE CPX
POSITION: REDACTED
SOCIAL SCORE: NA, CPX-SPONSORED

THE COFFIN SMELLED OF SULFUR. It was dim and dingy, confining. Ray knew he was in the minority, but he found the tiny coffin apartment comforting. In the dark, his hands moved of their own accord. It was like working blindfolded, and he did it with a practiced fluidity. His stained fingers rolled up the thin, paper-shelled cylinder slowly, carefully. Strapped the missile safely in the pack with the drone. Allowed himself to exhale.

He sat cross-legged on the floor, hunched over a black pack. One knee touched the sagging, yellow mattress pushed up against the wall, the other knee touched the opposite wall. Through it he heard, and felt, muted thumps.

His neighbors changed almost daily, but the sounds remained the same. They were the sounds of a packed, destitute humanity. He could generally discern day-to-day activities from those of a sexual nature. It was all in the rhythm. Those were hour-long renters. Privacy on the cheap. Some days he could hear the usual thumping around a tiny coffin apartment. Other days he could hear full-blown, drug-fueled shouting matches through the paper-thin walls, which usually ended with the slamming of a hand, a fist, a body against the wall. Others ended on a silence so complete there could only be one explanation. He didn't want to know. And he was certain they didn't want to know about him.

Ray took a long pull of water from a gray canteen. The cool liquid slid down his throat. He replaced the lid, stuffed it in his pack then stood, stretching the tightness from his legs, the soreness out of his back. The

3

monotony of his preparations had coaxed his body to a state of dull acceptance. But the task that lay before him required a quick presence of mind, an awareness of every aspect and detail that could affect the outcome of his work. A small voice inside told him not to do it, not to take a hit from the reserves of artificial chemicals stored inside his body, but the urge won out.

Just enough to get the blood flowing, and no more.

Through his system, Ray accessed the controls for his internal chemical reservoirs. Found the setting for *Wake*, bumped a thirty-percent dose. It was still early, but he planned on walking to work today. At normal pace it would take about an hour. By the time he arrived, the dose would still be active.

He slung the pack over his shoulders. No need for locking straps— it sealed automatically to his clothing—and made his way out of the A1 Coffin Apartments, onto the dim morning streets of Level 1. Day time, yet darkness pervaded as always. The streets were already crowded with Arcside daytrippers, tourists making their way into the shadow of the massive mountain of a city. Brought in by the perception of safety in numbers. To these people, Levels 1 and 2 were a curiosity, a freak show to be witnessed, but not experienced in their full dangers.

Ray headed south, against the flow of traffic. Up ahead, he could see the bright, razor thin line of light, east to west, that demarcated the southern edge of the megacity and its impenetrable shadow. He stepped over that line, from shadow to light, and felt the warmth of the sun on his face.

Out of the Arc, but still in Legacy territory. From beneath the shadow of the towering monolith that blotted out the sky, Ray continued to trek south, down into the shallow bowl where rusted rail lines cut through the array of buildings, dormant save the lean-tos and tents littering the rocky ground. The agglomeration of buildings stacked on buildings on giant levels of scaffolding receded behind him as he climbed the valley, past the crumbling war monument, and skirted the western edge of Signal Hill. He glanced left, to the east, and immediately regretted it. He shivered.

The homeless camp was more of a city unto itself, what looked like a cross between a shantytown and a shipping hub. Freight containers jumbled to the sky, held together with a web of fabricated lean-tos and fiber bridges. He had made the mistake of going through it, as opposed to around, once before. Never again. Despite the fact that it was actually more orderly than Legacy territory and by extension almost all of the Arc—save the upper levels—there was something about it that disturbed his military mind. Maybe the lack of hierarchy.

All that activity and no productivity, no commerce. At least none he could discern. Goddamn communists. Bugged the holy living piss out of him.

Might we suggest a sedative? The AI's voice was a tickle inside his ear. Annoying and unbidden.

"No." He shut out the AI.

The colorful freight containers and cracked opaque plastic roofs of Signal Hill shrank in the distance as Ray continued on his trek. When he passed the spot where the outer floodgates lay dormant in the ground, a point of demarcation between Legacy property and that of various small corporations and Complexes, the knot in his back unraveled.

Finally. Now he could enjoy the walk. The late fall, mid-morning air held a chill, but the sun was out and the air heating up, baking the fallen leaves dry. A sugary, earthy smell rose up lazily, lingered beneath the boughs of the tree-lined sidewalk.

Today's job was at a local tourist destination. He scratched behind his ear. Doubts began to surface, like flotsam rising from a dark ocean. He pushed them down.

As he neared the job site, he varied his route. Standard operating procedure. East, then south, west, then south again as he wound through side streets that held leaning, dilapidated homes. They reminded him of ancient, bent people.

The architecture changed. The streets became broader; they held trees at regular intervals, branches ablaze with bright yellow, orange and red. The air smelled better, clean and fecund all at once.

The biggest difference is the people.

They moved with a purpose. A few even nodded and smiled.

He scratched behind his ear again, using the pad of his finger to feel the small welt that had formed there.

This place with its flower and tree-lined streets, outside the floodgates and beyond the deforming and sullying reach of the Arc, was really just a territory held by multiple small corporations working in concert. It would fall, too, and have to decide which side it was on. Complex or Legacy.

Ray rounded a corner heading back south and was met with loud music. Dead ahead, a twenty-foot high wall with a brick facade separated the Plaza from the rest of the city. It had gates that could lock down and turn the area inside into a walled fortress. Just outside those walls, the music emanated from a pop-up recruitment kiosk set back from the gate—a simple four-foot cube that could hold the weight of several people.

But it only ever held one. The recruiter—the person responsible for reeling the applicants in, turning potential recruits into pleebs.

A warmth bubbled up in his chest. The logo and tagline for PerSense, Civilian Defense Complex, were emblazoned in red on a giant white balloon that hovered forty feet above the cube. Five years in the service of the PerSense agenda and the kiosks hadn't changed much. No need. Same loud music, same obnoxious recruiter jacked full of muscle juice hollering and cajoling.

It wasn't the box, music or balloon that pushed people to sign their rights away in the service of a greater good.

It was Ray. Driving the fear.

Ray ambled by the PerSense kiosk, fifty yards from the pedestrian gate. "You there!" the recruiter shouted, pointing a finger at Ray. The man was stout, had the build of a wrestler: blocky jaw, crooked nose, cauliflower ears. He tilted his shaved head upward, looked down at Ray with a practiced solemnity. "Officer material right there."

Ray smirked. He was tempted to tell the kid he was already on the payroll and outranked him by a vertical mile, but pushed it down. He looked away, pulled his hat down and took a right past the kiosk, walking parallel to the wall.

When the wall first went up the property owner had tried to bulldoze all the buildings surrounding it—a mix of old brick and concrete apartment buildings, none more than ten stories high. But the tenants had been organized and crafty; they sabotaged the equipment on a nightly basis. Eventually it had stopped being cost effective and the group had simply shoehorned the wall in-between the buildings it owned and the ones it didn't.

Which was perfect for Ray, because the perimeter was lousy with dark and filthy alleyways. Good work spaces for people like him.

He plodded on for another few minutes, his pace leisurely, before ducking into a narrow passageway barely three feet wide.

He had used this one before. Draped in shadow from wall to ¦ bordered on the Legacy side by a red-brick apartment buildin¦ crumbling, eroded mortar—it was walled off where the building ended, which meant no one had a reason to pass through. *Correction. Almost no one.*

Ray slipped his pack off, gently rested it on the ground and unzipped it. The drone inside was dull, grey and small, about the size and shape of his forearm, but dense. He took it out, hefted it. Fifteen pounds, give or take. So it didn't get pushed around in the wind.

From a small, padded pocket he removed two cylinders the size of pencils and pushed them into sleeves mounted on the belly of the drone.

He unmuted the AI assistant, accessed the drone's controls from his system and maneuvered the drone to take off.

Hello again, sir! Are we ready for another day of glorious, meaningful work? Might we take control of the drone and allow you to watch in leisure?

"No. And shut up while I'm flying."

The target was a café on 47TH, the street a main thoroughfare that ran through the walled city. Truth be told he'd rather be looking through the drone's scope at the bums on Signal Hill instead of these people, who earned their keep. But this was his duty. PerSense had raised him up from the dregs of Legacy and given him purpose.

To rain fire and death. To cut Legacy off at the knees. To drive subscription, at whatever cost.

His system linked with the drone's, so all controls were overlaid as a HUD on his field of view. He flew the drone straight up, some hundred feet above the wall. High enough to be out of a pedestrian's field of view. Not that it made much difference. The drone's skin was designed so the eye tended to slide around it.

"Volume lock," Ray commanded the AI.

Happily, sir!

"Magnify two-x."

Twenty, give or take, on the umbrella-dotted patio. A constant crowd of pedestrians streamed by.

He locked on the middle of the café, armed the missiles.

"Fire."

The pencil missiles made a fizzing sound, like running water. They sped to the target in a domed Arc, taking the trajectory of a surface to surface missile. When they were forty-five feet above the patio their tiny chutes popped. The missiles drifted back and forth, tracking lazy circles downward. Another thirty seconds or so and they would penetrate the uppermost field of the café.

"Return the drone."

Of course. Nicely done, might we add. Should net PerSense quite a few subscriptions today!

Yeah, he thought, *but when will the* real *jobs come? I'm ready to do something more. Something bigger.*

For this day, though, Ray's job was mostly done. All that was left was the crying. And the panic. And, of course, the gore. He accessed the missiles' cams, counted the targets one last time. A few couples, one family, a waiter. Oblivious to the fact that were playing out the last few moments of their lives. The pedestrians on the sidewalk were too numerous to count, but they would likely be shielded by the café's security field. This particular store-front advertised perimeter defense. It gave people a warm fuzzy and a money-back guarantee, but the systems were stupidly simple to avert. Great at repelling and deflecting metal and items moving faster than 370 m/s, but shit with a paper-shelled pencil bomb drifting down at .5 m/s.

Ray slipped the pack over his shoulders, exited the alley. Headed back east.

He watched the leaves tumble down, listening to the birds' bright calls in the crisp morning air. In the distance, the background hum of the city was suddenly studded with a dual *pop-pop*.

Bingo.

Screams, like the peals of an ambulance, ripped the air. It was human nature, but still so pedestrian. As Ray approached the gate to the plaza people streamed out, not looking back. The panic in their faces was wild, barely held in check by the flight response. Safety first, freak out later.

Set up the road a bit, back from the wall, the PerSense draft kiosk became flooded with applicants. Everyone knew that subscription to life in a Complex was the safest way to live, a guarantee against a violent death, they just failed to act on it until faced with the fact of their own mortality. The obnoxious pitch-man was enveloped in a sea of people, all clamoring to get his attention.

"One at a time!" he shouted. "There's a home for *everyone* at PerSense."

They were terrified people, and terrified people made hasty decisions. Ray watched them fidget anxiously as they waited for their turn to get sampled and enlist.

He stood, off in the distance, legs apart, chest out, watching his creation with pride. It must be the way an engineer felt after bringing his creation to life, watching it work precisely the way he'd designed it.

Ray took a deep breath. The air was cool, clean. It tasted wonderful.

What a beautiful day.

VAL

VAL
AGE: 19
SEGMENT: LEGACY CITIZEN
POSITION: UNEMPLOYED; GOVERNMENT DOLE
RECIPIENT (STANDARD UNIVERSAL CREDIT)
SOCIAL SCORE: 478

THE RAIN pounded relentlessly. If Val looked at it the right way, it was as if the water was suspended in the air, creating a curtain of liquid. It was only midday, but the dark clouds made it seem like dusk. Neon fuzzed through the wet air throwing bright, blurry reflections on the soaked pavement.

Miniature rivers gushed from gutter spouts and off broken and torn awnings that were freighted with age. Everything around here seemed broken.

Val shielded her eyes, squinted up through the jagged patchwork of towers thrusting skyward. Did the lining of that cloud have reddish tint to it? *God, I hope not. Last thing I need is an infection.* The rain had already gotten into her eyes.

"Splendid day to go shopping, sis!"

Val glanced left. Kat, looking like a drenched dog, was giving her that dead-pan look. Only 12, but almost as tall as Val now. She had that mismatched look typical of pre-teen girls whose body parts seemed to grow at different rates. Her legs and arms were skinny and long, but her torso and hips were still flat and narrow, those of a child. Her ears, as always, stuck out through her wet, blond hair like a pair of open doors on a vehicle.

Val raised an eyebrow, said nothing. The church was only a few minutes away now.

Without warning, Kat threw a punch. Val's shoulder exploded in pain.

"Ow!" Val shoved her little sister to the side. "That hurt." She rubbed her shoulder.

The evil grin Kat made was emphasized even more by her babyface. "Beware the sharp knuckles." She brandished her fists, oversized knobs on gangly arms.

"You're getting too big, little sister," Val said.

"Not so little anymore."

"Mm-hmm, save it for the boys in school."

"School?"

Val ignored the comment. She didn't want to have this conversation now. Not again.

The rain continued to stream down, which meant the streets were mostly deserted. Val peered through the curtain of water. Two blocks to go.

"Hey girls." A voice from the alley on their left.

Don't look, just walk. She grabbed Kat's arm, pulled her forward.

"Where ya goin'? Out in the weather like this?" The voice *tsked* in shame, followed them down the sidewalk. "Come stay with me. Got a nice fire and something warm to drink that'll make ya feel like you're livin' up on 7."

"They's too *good* for us, Jim." Another voice, this one female. Darker, limned with danger.

Val picked up her pace.

"Looka me when I'm talkin' to ya!"

She felt Kat's body get tugged backward, her sister's hand slipping from her grasp. They were trying to take her.

"Stop it! *LET GO!*" Kat shrieked.

Val didn't think. She felt inside her pocket for the uglystick, the telescoping baton the only bona fide weapon the girls owned. Her fingers grasped its cool handle, pulled it from her pocket. She snapped the baton out to its full length and simultaneously spun.

A woman, dressed in layers of colored plastic, grappled with Kat, pulled with two hands on her sister's arm. The woman's face, like the decayed center of a flower, was a cratered mass jutting forward, all lines pointing into the shrunken, fleshless center. She snarled viciously, gripped Kat's forearm with fingers that were brown with filth.

Val gritted her teeth, snapped the uglystick down on the woman's hand. She felt a crack and hoped silently for broken bones.

The woman recoiled, howled in pain, nursing her hand. "You *BITCH*! You broke it!"

Val let her body respond. In Legacy, examples were necessary. She twisted her body, snapped the stick at the side of the woman's knee. It connected, reverberating through Val's arm back to her elbow.

The woman buckled to the ground, hands splashing in a pool of rainwater collected in a crater in the road. "Stop, *please*!" she pleaded. Thick wads of spittle ran from the side of her mouth, foamed there.

On seeing his co-conspirator collapse, the man, the one who'd first spoken from the alley, turned and fled.

Kat kicked the woman in the stomach. "Tell your friends. Don't fuck with us." They turned and left the woman lying there on the ground, whimpering in the cold rain.

After a minute of walking, Val reduced the ugly stick, pushing the end back into the handle and put it back in her pocket. She kept her hand around its handle. She wouldn't be caught off-guard again.

She looked at Kat, raised an eyebrow. "Tell your friends?"

"What?" Kat laughed. "It sounded good!"

"What are you, a gangster? You gonna run around Legacy and kick any druggy in the gut that looks at you wrong?"

"Maybe. It'd be better than going to school."

When they finally came upon the church, Val was soaked all the way through to the skin and the adrenaline had bled from her system. She shivered, the chill set in her bones.

The church was outside the Arc by a few blocks, but the buildings around it stretched tens of stories higher, making the church a sapling nestled among sequoias.

An Archaic, gray-stone, steepled affair with black stains running down its face, the church still managed to loom, casting dark, menacing shadows over the street. No sign out front, nothing affixed to the building itself. It was cloaked in anonymity, which made it even creepier. The monks—or whatever they were—who lived there referred to it as *Deus Ex Machina*. But they didn't state the name casually, or hardly ever at all, which gave it an air of mystery.

"Can I just wait outside?" She could see the plea in her sister's eyes.

"How else are we supposed to use the factory, genius?"

"They know you, you're my sister. They'll let you print whatever food you need, just –"

"Bullshit," Val sang. "After what just happened you're asking to stay outside? I don't know what you're worried about inside there," she pointed at the doors. "Shit, out here some Buena junkie could walk by and grab you. Poof, gone." She made an exploding gesture with her fingers.

Kat sighed. "Really?"

"Really. This is Legacy ground, down in the basement. Not Level 7 at the top of the Arc," Val lectured. She angled her head up, taking in the full height of the city that stretched to the sky in front of her. She imagined the life of leisure the citizens of 7 must've led on their mile-high throne. Nothing to worry about, plenty of food, anything you could want within easy grasp. She lowered her sights to about halfway down the Arc. Even on Level 3 they could expect the prospect of a better life.

Must be nice.

But that wasn't their reality. This was. Using their Legacy welfare, the Standard Universal Currency, to print their food at a run-down, obscure church. *That* was their world, and it was something they just had to accept.

"Now, as Dad would say, 'buck up, buttercup.'" She pulled her sister by the upper arm, half-dragged her along, mounting the cracked, crumbling steps.

"They're gonna make me come back to school," Kat whined.

Val used her other hand to open the monstrously heavy door. It protested, creaking violently.

"Is that such a bad thing? I stayed..."

"Til you were fifteen, uh-huh, yeah, I got it."

"And here you are, at the ripe old age of twelve, crying about a few lessons a day."

"A few lessons? Try five hours. Every day. It's..."

"Agony, I know. I've heard the spiel. About eighteen thousand times."

"I can't do it. I won't. Nope, not gonna do it." Kat folded her arms, tilted her head up and away.

"Sorry. Not a decision you get to make."

"Stop it, goddamn it!" Kat yanked her arm from Val's grasp.

Val ground her teeth together. She was sick of talking about this. Over and over. It always ended the same way, with Kat sulking and silently angry.

Val let her silence do the talking as she entered the church. The antechamber was more stone walls with a green and white checkered, peeling linoleum floor. It smelled of moldering books and old people. Musty, with a dried, unused tang. Val could hear Kat's waterlogged sneakers squeaking on the floor behind her.

It was funny. When their dad had died two years back, Val had by necessity become the head of the house. At age 17. But she was also a sister, which meant sometimes she was tempted to relent to Kat's whims. Which at the age of twelve came like the tempestuous summer monsoons, blowing in and blustering for twenty minutes before blowing out again, sweeping her mood clean.

Val could feel Kat's eyes boring a hole in the back of her head.

"Stop it!" Kat's shrill whine echoed off the dank stone walls.

"Stop what? I didn't do anything."

"I know, Val," she drew the words out as if to say *duh*. "That's what I'm talking about."

"Cause that makes perfect sense."

"Don't pretend you don't know how annoying it is when you don't say anything. Just like Dad."

At the mention of their Dad, something inside Val hardened. She knew exactly how their dad would've reacted to Kat's little uprising.

Sorry, Dad. The thought felt like a prayer to their dead father. *I'm just trying to keep our heads above water here.*

Val pushed away the sorrow that attempted to rise within her. Made herself harder.

"Damned if you do, I guess," she said.

Kat growled, squinted angrily and balled her fingers up into small fists. Val rounded on her.

"You know they won't give us the discount if you're not actually enrolled at Deus Ex, right? We've tried printing at other factories. We can't *afford it* anymore." She tried to hold back her anger, but the weight of her responsibilities was too much. "Do you want to eat or not?"

Kat recoiled at Val's outburst, stewed sullenly. "You're awful," she murmured quietly, then fell silent. They'd both stopped going to school when their dad died. Val had let it slide, for obvious reasons she thought. But printing their food at the cornerstore factories was too expensive. This last time they'd run out of Standard Universal Currency a full three days before the end of the month. No SUC, no food. They were out of options—Kat would have to re-enroll at Deus Ex.

Val was convinced the interior of the cathedral had been renovated—with the same dingy gray stone—to confuse and turn around any potential worshippers. The halls seemed to go on forever, with branches every thirty feet or so. As far as she knew there was no actual nave, or at least none that she had seen, simply a labyrinth of stone decorated with rugs hanging on the walls. Admittedly, she hadn't seen the whole church, just the customary path from the entrance to the factory, which was situated at what seemed like the rear of the building. Along the path there was a bank of classrooms, where monks taught mostly Language and Math. When she had been at school, she'd graduated to the point where the interesting electives were starting to become available—some weird version of History in Religion and War, Universal Meditation, Horticulture—but Dad had died and she'd gone from student to caretaker, from child to adult in the span of an absent heartbeat.

Up ahead, a kaleidoscope of color scattered the darkness of the corridor, played on the green and white flooring. The atrium. Beyond it, their goal—the room that housed the factory.

A wide door with stained-glass designs barred the entrance to the atrium. Val placed a palm flat against a panel that depicted Jesus raising a healing hand. The palm to the glass triggered the intercom to the right of the door to come alive.

"Hello, Val." Brother Anthony's high, lilting voice came through the intercom fuzzed, distorted. The monks seemed to worship all things lo-tech. Except their security, of course. "It's been a while since we've seen you. How are you and your sister managing?"

Val glanced at her sister, scowling and soaked. "Never better!" she lied.

Silence, as if to punctuate the transparency of her lie.

"If you're here to use to the factory, we just have one or two small requirements before you do so."

Val stifled a sigh. "The interview?"

Brother Anthony chuckled lightly. "Among other things, yes, the interview."

Thick, warm air enveloped them.

Val took a deep lungful of the air. It tasted like Earth, like living things. Full, fresh, but with the slightest undercurrent of a subtle fetid scent. Decay. High above, frosted polymer panels permitted the gray light of an autumn rain to enter the living sanctuary. The clear panels weren't visible from the front of the building, so Val assumed the atrium was situated somewhere near the rear of the building.

Gravel crunched beneath their feet. Another deep breath.

"I wish I could live here."

Kat said nothing, maintaining her silent protest. But Val noted Kat's fists were no longer clenched tight, blood returning to her white knuckles.

They wended their way along one of the many gravel paths that meandered throughout the garden, taking the path they'd taken so many times before. Here and there a brown-robed monk tended the plants, this one watering, that one pruning with old but well-maintained shears, another one plucking a bright red tomato off the vine and placing it gently in a cracked green bucket.

At the center of the garden, a circular open area, the hub of all the paths, was bordered by a waterfall. Four benches were placed along the outside of the circle. A man in a hooded brown robe, arms folded in his lap, waited patiently with his legs crossed.

His goateed, scholarly face held a smile that could put even a Kopa junkie at ease.

"Oh god," Kat muttered from behind Val so only she could hear. Val made a shushing motion with one hand behind her back.

They approached. Brother Anthony stood and greeted them, spreading his arms.

"Welcome daughters." His eyes twinkled. "Who would like to go first?"

Kat stepped around Val. "Let's do it."

The two sat down, Val retiring to a bench opposite the courtyard. The waterfall mesmerized her. She closed her eyes and tried to let its calming susurrations carry her thoughts, her worries, away. A face entered her mind. It was the junkie woman, the one who'd grabbed Kat on their way to the church. The insidious, prematurely aged face, filled with ugly rage and a dark hunger, animated the empty space in Val's mind. The face melded into a hundred others, the occurrences of violence too numerous to name. They replayed in her head. This one just inside the floodgates—a mangled, limping man with a scorched face off a side-street their Legacy apartment overlooked. Another—a gang of three women who, if they'd managed to get their hands on Val and Kat, would've succeeded in taking them. Thankfully, Val and Kat had the expediency of youth in their legs. They'd fled and never looked back.

A hand shook her shoulder, jolted her out of the almost-nightmare.

"Yo." Val opened her eyes to see Kat standing over her. There was a redness in her sister's face, a flush that she didn't remember being there. "Your turn." Kat put her fingers in the waterfall, allowing it to flow down her hands, drip from her wrist.

Val stood and stretched, left her sister to dally in the waterfall and crossed the gravel-lined area, sat down on the bench opposite Brother Anthony. She stifled a yawn, noticed Kat had both hands plunged wrist-deep in the waterfall now.

"Welcome, Val," Brother Anthony said, palms interleaved and facing up. "This won't take long. We just like to visit with those we care about, make sure we don't lose touch."

"It's okay. We don't talk enough."

He sat back, folded his legs. His intertwined fingers rested on a leg. "What brings you here today?"

"Well, I'd like to say we're here to shoot the shit with you, but you know that's not true. Running low on supplies." She shrugged. "You have the cheapest print-stock. Can't afford to print our food elsewhere."

He smiled genially. "We appreciate your honesty."

"We're here to enroll Kat back in school. Hopefully we'll be able to use the factory?"

Brother Anthony nodded. "In due time. The first and foremost question on our minds at Deus Ex is whether or not you still make use of an internal computing system. Last I remember, the answer was yes. Is it still the same?"

"Uh, yeah. Can't see removing it anytime soon."

"As you may recall, we at Deus Ex Machina do not approve of internal systems."

"It's a free world, isn't it, Brother?"

"For the moment. And you still have your free will to choose. We've already confirmed with Kat that she now makes use of a system."

So weird, she thought. *What do these guys have against tech?*

Val frowned. "Let me read between the lines here. What I'm hearing is that Kat has to remove her system to enroll, that about right?"

The monk shook his head. "Nothing so drastic as that. It'll just have to be turned off during schooling hours."

"Oh," she sat back. "That should be manageable."

"In fact, we ask all our guests to deactivate their systems before entering our sanctuary." The monk paused, looked calmly at Val.

"*Oh*, you mean *me*." She turned her system off, the heads-up display fading away. "I'm a little slow on the uptake sometimes."

The man waved the comment away. "Nonsense. You were always one of our brightest."

"Aw shucks."

"So Val, tell me, how would you characterize your days since our last visit?"

"That's a tough one. Flat?"

Brother Anthony pursed his lips.

"I mean, there's not much you can do with a thousand SUC bucks a month," Val continued. "We play some games in virtual, do some acting, try to stay alive. Not necessarily in that order."

"Life in Legacy can be hard," Brother Anthony acknowledged. "Do you ever consider joining a Complex?"

Val went cold inside. "Never." The word came out as dead as stone.

Brother Anthony leaned forward. "And why?" His dark eyes seemed to grow more intense, boring into her skull.

Val could hear her Dad's voice. *Never give up your independence. Not even to save your life. Better to die free than live a slave.*

"Call it a grudge. Dad's influence."

The intensity in the monk subsided, he leaned back, inhaled deeply. His face relaxed into an easy smile.

"I know it offers you no peace, but we're glad to know your father's influence remains alive within you. He believed, as do we, in your place as the hand of God."

Val made a funny face involuntarily. "I think you've got me confused, Brother. I don't know anything about God. Neither did my dad."

The monk leaned forward again. "You *are* God."

Val scooted backward on the bench. *What does that even mean? And how can he possibly know what Dad believed?*

Brother Anthony continued. "As am I. As are the 15 billion other people in the world. We all inform God, even if we don't know it."

Val's expression remained blank, vapid. What could she possibly say in response to this weirdness?

"Forgive me for being didactic," the man apologized, bowing slightly. "Now on to the last question. If you were granted one wish, what would it be?"

Val shifted uneasily on the bench, sat on her hands. This was a new one. She glanced at Kat, across the court.

"Honestly? I wish we got along better, like we used to." *Like when Dad was alive.*

Suddenly she felt a torrent of tears pushing at the dam of her emotions, pressing, weighing, seeking cracks in the wall. She pushed back. Masked the grief, the fear, the sadness at a life lost and paths forever altered.

Brother Anthony nodded, concern pouring out of his sage-like expression. It was a face that seemed to know all her thoughts, all the lies she told to the world. To herself.

"Normally I wouldn't reveal the contents of another person's interview. However...you *are* a caretaker and thus need to know."

"What?" Val glanced at Kat, who was turned sideways on the bench, stroking the petals of a purple flower.

"Your sister. She wished...she wished for your father back."

Val's jaw trembled, throat swelling. She flared her nostrils, sucked air furiously.

What the hell is your intent here, brother? To make me break down in front of you? She clenched her teeth. *Well fat chance of that. You won't get the satisfaction from me.*

"Are we done?"

The monk's kindly eyes never wavered, yet Val sensed an inexplicable disappointment in him. She was happy to disappoint. "Of course." He stood, offered a hand. She ignored it and stood. "We're always here. We serve to inform you and to inform God. Thank you for your time and, please, if we can help just ask."

She smiled half-heartedly, waved Kat over. "Mm-hmm. See ya round."

Just like the entrance on the far side of the garden, the exit was a thick-framed steel door inlaid with stained polymer, its design identical to its twin. Val placed a hand up against Jesus' palm to trigger the door's internal mechanism. The door clicked and slid slowly to the side, receding into the wall.

"Hallelujah and praise the virgin effin' mary."

The factory was a fabrication machine—an oversized printer—that stood before them in the large rectangular room that had no exit except for the way they'd entered. It was a matte black beast that took up the better part of one wall. Most of the space was for storage of feedstock. This particular brand had two sides. One nozzle for printing objects, like if you needed a fork, and another dedicated to foodstuffs. The comestibles side came pre-loaded with thousands of open-source flavor and nutrient permutations. They could waste hours spending Legacy SUC bucks on bars that tasted like a five-course meal.

Val pulled up the menu and punched in their preferences. Mostly meatloaf/mashed potato and chicken/rice. Sixty bars of each.

"You gonna get dessert?" Kat asked, pushing a finger in at the menu console.

Val swatted it away. "Relax," she said. "I've got it."

Val selected a scattered array of fruit and chocolate bars, then went to the checkout screen. Sixty meal bars, thirty dessert cubes. That should get them through a month. Navigating to the checkout screen, her eyes flickered to the flashing ad at the top—a holographic animation of a SIZ-ZLING STEAK. She imagined what it would be like to bite into one of those. Roll it around her tongue, slowly savoring its texture and buttery flavor, the knife carving it into the smallest possible bits...

"Snap out of it, Sis."

Val jerked. "Couldn't cook it anyway," she muttered to herself.

"What?"

"Nothing."

The screen was flashing now, counting down to the time when her order would reset. It was asking Val to authenticate the purchase. She rubbed her hand on the pad, allowing it to scrape away and verify the tiniest fragment of her DNA.

By her calculations, the grocery bill would take all but ten of her Legacy SUC. It should be enough to stretch them along til the next payout. At least their rent was paid for by Legacy, the one thing the ailing government had going for it.

Val had some crypto stashed away, inheritance from Dad's will, but she was saving that for an emergency. Or something nice for the two of them. She never said it, but she was waiting, hoping for something that could improve their lives. In her heart of hearts, Val knew it was just wishful thinking. She thought back to the top of the Arc. Level 7. The riches and beauty it held were reserved for people born in a different world.

But the Legacy currency was free and usable most anywhere. She just had to limp it along, make it last the whole month, and they'd be okay. The good part was that they had tons of free time to do as they pleased; the bad was that they didn't have any money to do anything with their free time.

The screen flashed -*10*.

"Shit," she muttered. "Stupid math."

"What?" Kat asked, peeking over her shoulder.

Val turned, blocked the screen. "We'll have to come back."

"What? Why?"

"Just cuz."

"Uh-uh, you don't get off that easy." Kat pushed Val aside. They both stared at the screen.

"Sorry," Val said.

Kat frowned, turned to Val. "Really? You can't even ask me for a few SUC bucks?"

Val shook her head, looked away. "You've been saving it."

Val could see the muscles in Kat's jaw tense. "What the hell?" her sister said, anger building. "You can't even ask for my help just once?"

"You're just a child, Kat. It's not my job to ask you for help. It's my job to take care of you."

"I'm *not* a *child*!" Kat shouted. "Stop treating me like one!"

Val retreated toward the stained-glass doors, shaking her head. "Sorry, kiddo. This is something I have to do. I'll find the credits somewhere."

"Please!" Kat pleaded, grabbing at Val's arm. "Just...just let me help. I want to."

Val studied her sister. Maybe she was wrong. Maybe that face *was* starting to mature, starting to match her body. All Kat was asking was to be a part of the family, a part of the solution. To *help*.

"Fine," Val relented. "But I'm going to pay you back the second we get our next payout."

Kat rolled her eyes. "Geez, stop being such a drama queen. It's ten sux," she said, pushing Val out of the way.

Kat edged her body in-between the machine and Val, rubbed her palm against the pad, its micro ridges sloughing off and collecting tiny bits of skin.

The factory dinged and whined to life.

TORI

TORI
AGE: 33
SEGMENT: COMPLEX PROPERTY; NOVAGENICA CPX
POSITION: VP OF HUMAN ASSETS, SUBSCRIPTION
DIVISION
SOCIAL SCORE: NA, CPX-SPONSORED

T HE ROOM BRIGHTENED GRADUALLY from deepest black to violet to blue to amber, revealing a simple, monochrome box. A bed, a nightstand. Two doors. One led to the bathroom, the other to the main living area of his penthouse suite.

A quiet whoosh, a dose of aerosol. The spray jetted forth from a white cube that sat atop the nightstand. The soft tang of citrus permeated, hung in the air.

Tori roused slowly, his wake cycle bringing him gently into consciousness.

"Ginseng and tiger penis," he mumbled, eyes still closed.

Of course, sir. And good morning! Might we suggest another dose of Citrus Alive to fully rouse your senses?

"No."

A shame sir. The AI managed to sound *disappointed.*

"Shove it."

As you wish, sir.

"Now go mute yourself, unless you *want* to watch me masturbate."

The AI went silent and the cube issued another whoosh of mist. Within seconds he felt a small, bright fire being stoked in his loins. He gripped his manhood, let the memories from last night wash over him. Sato in his samurai battle suit, wielding the sword. Tori stripping him down, piece by agonizing piece until there was nothing left but his helmet, the sword and the stern expression of a warrior. He pulled up pictures from the encounter on his system, which brought him to climax almost instantly.

He lay his head back on the pillow, arms splayed out to the side, let a huge breath out. Didn't bother contemplating the mess. Sheets would clean themselves. He stretched out, felt the strength of his body. There was a lazy sleepiness to him yet—his body resisted the chemical inducers.

"Fuck it," he said aloud. The next tour wasn't until lunch time. After that, he'd check in at Research, but it could wait. If there was the opportunity to put off entering the sick ward, for even an hour, he'd grasp at it. Hated that place. It was a vile reminder of the dirty work that lay before him.

Dirty, but necessary, he thought with a sigh. It *was* necessary, but he'd worry about it later. After he was awake.

"Alright, now wake the fuck up and put me to sleep," he commanded the assistant.

Aye aye, cap'n!

Insolent bastard, he thought, musing briefly on whether the AI was like this with everyone that belonged to Novagenica.

He closed his eyes on the view of the ceiling as it reversed order, receding back into night mode. The cube ejected another puff of mist. This one smelled of a dense forest of evergreens. And of wet, if that was a thing. Within moments Tori was asleep, his chest rising and slowly lowering in rhythm with his soft snores.

Novagenica Complex was out in the middle of nowhere. They needed the space for an operation this size.

"Two miles to a side," Tori said, flashing a smile. He was dressed in a white polo and khaki pants. The white made the golden tan of his skin stand out.

"Four square miles of Utopia. The safest, most inspiring place to live, work and play." Tori gestured expansively at the landscape that spread before the tour group just outside the window. He made eye contact with members of the group and straightened his name tag.

TORI FANNING—VP OF SUBSCRIPTION

The Welcome Center—known locally as *the egg*—was a giant sphere that stood at the end of the driveway that led into the Complex. Ten stories of carbon fiber and glass, it resembled an oblong geodesic dome without the supporting angles. Its exterior was one smooth, clear surface. At its center was an open-air atrium that stretched a hundred feet from floor to ceiling, which was ringed by levels that held offices for Novagenica's management.

The ninth story of the welcome center was a simple open platform that skirted the entire level, offering a panoramic view of the Complex grounds. The tour group stood at the curved glass window, staring out at the farms, play fields and school buildings, which were placed specifically so as to generate a sense of peace and well-being in onlookers.

Tori paused a moment, assessed the mood. There were a few in the group who cleaned up well, looked like they could make middle management after some time. The rest, say twenty-five of the thirty, were clearly plebes.

Destined for testin', he thought. Upper management inside joke. He knew it'd seem callous to those on the outside, but that was the reality of life. Not just Complex life. Life in general. Only fifteen percent of new subscribers had anything in their heads that was worth a damn. The other eighty-five? Their worth was in their hands and their data, their genome.

Tori stifled a little laugh, covered it with a cough.

"The first thing I want you to note is the walls." He jabbed a thumb at the window. The walls that surrounded the Complex gleamed a blinding white in the bright sun. "Fifty feet tall, impact-distributing, self-repairing and studded with defense measures every few feet, nothing can slip through. Inside these walls," pause for effect, "you and your family are safe from the *entire world*." Tori pointed at the Arc, a distant smear on the northern-eastern horizon, loaded his words with derision. A few in the group scowled through the window at the gray monstrosity that reached to the sky.

"*Outside* these walls? Well..." He pointed at an elderly Asian woman in the front of the group. "You, ma'am, where do you hail from?"

"Level 2. Rising sun," the matriarch replied. She had towed her whole family along.

"Oooh," Tori made a sour expression, sucked wind in through his teeth. "Well I don't have to tell *you* what Legacy's like, especially around your parts of the Arc." She stared back flatly.

"Now I'm sure this is something you *all* have to deal with every day. Chaos, violence and lawlessness, right? How many of you have lost a family member to random violence? Drug, gang, some random pervert? Show of hands."

Every hand in the group went up. He shook his head in disgust.

"Never again. You don't have to experience that heartbreak, that overwhelming sense of helplessness ever again. Inside these walls, Novagenica's created a safe haven. For you *and* all of your family."

"Excuse me?" A skinny man with creased, worried lines around his eyes and a white goatee raised his hand. Tori choked back his irritation. "Yes, sir?"

"Sorry, um, but if you're a member, are you, um, allowed to pass freely in and out of the walls?" His face had apology written all over it. Tori could use that. "I'm sorry...it's just that I have family all over the city, and I'd want to be able to see them."

Tori smiled wide, flashing his perfectly straight teeth. "Of course! At , Novagenica our goal isn't to cut you off from your family, which is why with a day pass you can go freely wherever you want. It's all covered in the Terms of Service. We'll go over everything while we wait for your DNA test results to come back."

What he didn't mention was the fact that, while the test results took no more than ten seconds to compute, the Terms of Service was a twenty thousand-page document.

The man rubbed at his puffy eyes. "Oh...okay, thanks," he said feebly.

"Any more questions?" Tori waited a second. "None? Ok, good. Now if you'll follow me, we'll just take a blood sample and then look over the contracts."

Like the good sheep they were, the tour group followed.

The scent of oranges. Arousal. Satisfaction. Repeat. In this way Tori's days hardly differed, at least in their beginnings. And if they didn't differ, a small part of him whispered that the effects of time could have no hold over him. Confronted with the dead and dying did very little to assuage his absolute terror of death.

The sick ward smelled of citrus. Oranges to be specific. It was disorienting for Tori, as the scent of oranges was always coupled with an arousal pheromone. It was like sensory vertigo; his body was telling him to get an erection while his visual stimuli were triggering the exact opposite. Fear, disgust, a feeling of being soiled. But mainly fear, the big one, the force driving all the other currents in him right now.

He hated being down here, fifty feet underground. It felt like an institutional mausoleum to him, like he was walking into a sterile, well-lighted crypt. The walls were smooth and gray. Some synthetic aquaphobic material, almost totally frictionless. The hallway extended as far as the eye could see. Ten feet wide, no fixtures. Mirrors and doors, mirrors and doors. Behind those mirrors and doors, rooms that held sick children.

Technically, they weren't mirrors, they were screens connected to sensors inside the rooms. Monitoring every aspect of each and every child's biomarkers. Nevertheless, the repetitive nature of the space made his skin crawl, especially when it was deserted like this.

It took monumental effort to keep his terror in check. His mind knew that, in their current state, none of the children locked behind the doors were infectious. But his hands still shook and his body trembled. Invariably, the patients all died. At least, they all had so far, which didn't make Tori feel any more comfortable on multiple levels.

They're not infectious, he told himself. *How many times has the AI told you this? Now focus.*

He stepped back from the mirrored screen that allowed him to look in on where the children lay. Closed his eyes, took a breath. Another. Focus.

He opened his eyes, read the label on the door before him.

Patient 142C

Tori stepped forward, keyed the screen from its silvery reflection to a reduced opacity. The girl was one of the older subjects, but you wouldn't be able to tell by her size. It was early yet, so she remained in bed. She lay on her side, facing him, black hair spilling over olive skin and peaceful, elfin features. She looked hale, healthy. 142C had survived the longest. 16 days now.

But it's just a matter of time.

He heard the squeak of shoes on the floor, turned toward the sound. A janitor in a gray uniform, fifty feet away, tugged a sheet-shrouded body out of a doorway. Tori coughed and the man glanced over his shoulder at Tori briefly before picking the body up and carrying it away down the corridor.

Toward the door marked *RECYCLING*. Tori shuddered.

The researchers had yet to find the solution. Still, 16 days was a new record. Not the ultimate solution, but it was a start. They were getting closer.

And so, this. The patients dying behind the mirrors. Not here by their own volition, but sometimes you couldn't ask for the things you needed, you had to take them. For the better of everyone.

Tori glanced back in on 142C. She had shifted in her bed, now facing away from him. White dressing gown gently raising and lowering in time with her breathing.

They were all little martyrs, everyone one of them, and they would save a lot of lives in the long run.

Tori was certain of it.

MANOLO

MANOLO
AGE: 57
SEGMENT: COMPLEX PROPERTY; NOVAGENICA CPX
POSITION: LABORER, LEVEL 6; LIFELONG
SUBSCRIBER
SOCIAL SCORE: 629

MANOLO TUGGED THE CHILD'S BODY, wrapped in bed-sheets, backward, out the doorway and into the hall. He rested it as gently as he could on to the floor.

He shook his head. *You deserve more respect than this*, he thought.

Lift the body and take it to recycling, the AI commanded in Manolo's ear, intruding on his thoughts.

Manolo clenched his jaw. Novagenica's AI left him alone most of the time, as long as he kept busy. But the instant he stopped, even for a moment, he felt its presence urging him along.

It made him feel like a slave. Not that his situation could compare to the slaves of the old world. He was well-fed, clothed, had some small form of freedom. Yet...he couldn't help the way he felt.

Hurry. You're falling behind. Lift the body...

"Okay, I got it," he interrupted.

The white sheet was soiled with the girl's effluence, brown and red in the usual places. Whatever the illness was, the doctors had assured Manolo he wouldn't get sick. Not that he cared much anymore.

Manolo put one hand under the legs, another under the back, and lifted. His fifty-year old frame wasn't as strong as it once was, but all of the children were young and, by the time they passed on, completely wasted away. Withered.

Manolo turned and headed down the long hallway, leaning back under the weight of the body.

Overlaid on his vision, the AI highlighted footprints on the floor, letting him know where it wanted him to place his feet. Leading up to each footfall, a colored meter spiked, from print to print, telling Manolo the pace it wanted him to keep.

He ignored the speedometer.

Fuck off, stupid machine, he thought, a bitterness rising inside of him. His mind went elsewhere. He imagined the AI as a slavering beast, a minotaur, building and snorting. Fantasized himself pouncing on it, taking it to the ground and plunging a long machete deep into its side. Again and again.

Speed up. You're falling behind.

Manolo blinked away the fantasy. "Go mute yourself," he commanded. Before him, the wide, heavy door labeled *RECYCLING* clicked open as he approached. Beyond it, a dark, foul-smelling room. He angled his body, entered the room.

We're sorry. You do not have the necessary permissions to mute us.

"Thanks for the reminder, asshole."

The AI made a warning chime: an ugly, harsh sound that indicated he'd just lost credits. *There is no need for profanity. Please refrain. Docking your profile 5 points.*

The room was small, empty, with the majority of the recycling machinery housed behind the far wall. Embedded in that wall, halfway up, was a door for the chute. As Manolo approached, the chute opened, the door rotating toward him from a vertical position until it was horizontal. The smell of the recycling vat, acrid and foul, wafted out of the chute.

Manolo put the body in the chute, head-first. The door automatically began to lever back up, in small increments, then back down. Manolo stepped back and watched in horror. The effect was that of a jaw chewing, working up and down, until the body was swallowed whole.

He turned and fled from the black room, through the door and into the lifeless light of the hallway.

Double your pace back to room 171C. Replace the linens...

Manolo tuned the AI out as best he could. Unfortunately, that only left room for one thought.

What has my life become?

When Manolo first arrived, some five years ago, he had envisioned an idyllic lifestyle. Agrarian, even. Clean air, hands in the dirt, growing things surrounding him and Esme. Especially Esme. That sort of thing was universally known to be good for the soul. Healing.

Novagenica Complex was far enough removed from the Arc that he had believed it to be possible. Like waking from a nightmare, the farther you were removed from it the more its effects receded in your memory.

Manolo stood on the porch. The air was cool and silent and the lights in his domicile, his quarter of a shared-box building, were turned low. He shook his head at the empty night. How foolish he'd been.

Before he'd first signed the contract to work and live—and die—for Novagenica, he had negotiated his own small piece of heaven. His sole requirement—that their new home sat on green space on ground level. Esme's condition had allowed him to negotiate it prior to signing the sub agreement. He didn't want to be confined to some dorm, packed in with all the other families, a Complex within a Complex. It was funny that some people requested to live like that, packed together like rats in a nest. But it was also true that people were comforted by what they knew, and most of them had only known life in Levels 1 or 2.

Looking south, staring into the muddy blackness, he could convince himself he'd gotten what he wanted. But that dream of peace and solitude had been shattered within the first week. There had been no mention of shared living arrangements in the contract, and he and Esme's building was divided into four quadrants, so that two of their walls were shared. None of them were particularly loud or burdensome, it was just...frustrating.

The night was a brownish black, with light from the Arc infiltrating from the sides of the house. His home faced south, away from the Arc. He could see its diffuse glow at the edges of his vision, coming from around the eastern edge of the house, but he pushed it out. A cricket chirped and he felt a spike of irritation. Artificial bugs. He tried to imagine they were real, alive, nestled beneath the blades of grass on his small rectangle of lawn. Or better yet, out beneath the drooping leaves of the cornstalks in the field just beyond his yard.

He twisted his torso, trying to loosen the perpetual knot in the middle of his back. A resentment, like a growing cancer, had started to metastasize inside him. Maybe if he had someone to talk to, someone he could complain to, someone to touch him, calm him, soothe him when he needed it...maybe. Someone like Esme, but he'd never have that again. Not in this lifetime.

A chill breeze cut through his thin sweater. After a day of working inside, stuffed down in the subterranean bowels of Research, it felt good to be outdoors. The air was different somehow. It made him feel better, lighter. He descended the steps and watched a pair of cool white lights approach from the east. The outline of a cart materialized, the sound of gravel popping out from beneath its tires.

Marco and the girls. Dinner night. He'd marked it in bright red, bold letters on the archaic calendar that hung on his wall. He didn't want to admit it to himself, but he had been floundering for far too long, struggling in choppy waters, without anyone to throw him a line.

The car whined to a stop in front of the porch, motes of dust hanging in the vehicle's white lamplight. The doors swung open and his granddaughters tumbled out, a blur of limbs and dark brown hair as they rushed him.

"Lolo!" they screamed, throwing their arms around his middle.

He looked down at Maricel and Sol. "Oh my goodness, you two are getting so *big!* When did this happen? When did you two grow up?"

He squeezed them both, closing his eyes and inhaling deeply. They wore the same perfume their grandmother had loved—a light, clean scent with a tiny floral hint. A spring breeze that carried the reminder of things living, growing, thriving. *His* spring breeze. *His* Esme. He could almost feel as though his hands were tangled in her hair and not that of his granddaughters. He found himself on the verge of tears and exhaled, letting the vision go.

"Hey," Manolo said, looking down at Maricel, "just because you're thirteen now doesn't mean you get to stop squeezing me." Her arms tightened around his middle. "Much better," he smiled approvingly. "And how is my little nene?"

Sol looked up at him, eyes shining with exuberance.

"Great lolo!" We went all the way to 7!" she said.

"You did? What, on a school trip?"

"Yeah!"

"No way!"

"Uh huh! And we went to a obsa—obsatory."

"Observatory," Maricel helped out, her lower, smooth voice contrasting against Sol's high-pitched squeals.

"Uh huh, we could see all the way to the next Arc! We even saw the mountains and they had snow!"

"Wow, that is *so* cool," Manolo remarked. Despite feeling worn down, it was hard not to be enthusiastic around Sol. She had the kind of energy that tended to infect the others around her.

"Alright girls, give your dad a turn." Manolo turned to the voice, which came from the darkness behind the small vehicle. Marco approached through the beams of light, grinning ear to ear; he threw his arms around Manolo, pulling him into a bear hug.

"Good to see you, pops. You lose weight?"

For a moment, the sadness and resentment that were all wrapped up in what he called *the feeling* seemed to bleed away. They stood there for a long moment.

"No," Manolo spoke in his son's ear. "You've just gotten rounder." Manolo slapped at his son's love handles.

"Yeah," Marco admitted, releasing him. "You're probably right."

"I'm hungry," came Sol's little voice from between them.

Manolo rubbed the top of her head. "You're in luck, little one! Grandpa's got some goodies for you. Come inside. Come, come," he urged, ushering them inside. "Dinner's ready."

Sol bounded up the steps, her ponytail bobbing, followed by Maricel and Marco. Manolo climbed the steps last, using the railing to help pull himself up. He could almost hear his old bones creaking and cracking.

By the time Manolo got inside, Marco already had the pot of stew in his hands, maneuvering it over Sol's head to set it down in the middle of his small square dining table. Using an old knitted hot pad, Manolo pulled the lid back, let the steam wash over his face. Chunks of whitish pork protein floated in the thick yellow slurry. The heavy scents of peanut, coconut and cooked meat filled his nostrils.

"Ahh, kare kare. Smells *so* good," Marco said. He was leaning over the pot as well, his forehead almost touching Manolo's, his rectangular glasses fogging up.

Manolo grinned. "Didn't just pour the recipe from the spout," he said, chest puffing up a little. "Measured all the ingredients first."

Marco raised his eyebrows. "Actual cooking?"

Manolo nodded in assent. "Even printed a real oxtail for you, boy. Hair and all."

Sol wrinkled her freckled nose. "Ew, lolo! That's gross."

Marco tugged at her ponytail. "Grandpa's just yankin' Dad's chain, Solita. There's no tail." Marco looked up. "Is there?"

Manolo just grinned.

Once the girls had completed the task of setting the table, they all sat down, a granddaughter on either side of Manolo with Marco opposite.

"Go ahead, son. Do the honors." Manolo gestured to his son.

"It's your house…"

"I'm out of practice."

A worried look flashed across Marco's face, but it was gone in an instant as he laced his fingers and bowed his head. The girls followed suit.

"Dear God," Marco said, "we thank you for this meal and, most importantly, for one another. We're so grateful, God, for this life and all the opportunities we're granted." His son had a rich, soothing, baritone. It lilted and dipped, the words rolling out like the hills of the great dead prairie a hundred miles to the west. It was comforting, and Manolo found himself easing into the moment, the tension of the day ebbing away as he lost himself in the musical quality of his son's speech.

When the last *amen* was uttered, Manolo opened his eyes and found those of his family, all brown like his, twinkling back at him.

He clapped his hands together. "Let's eat!"

Marco dished out the stew into bowls while Manolo heaped piles of rice and vegetables—squash and beans cooked soft in a slightly spicy coconut milk—onto all four plates.

"Sorry we haven't been out lately, Dad. We've just been *so* busy with the kids' activities," Marco said through a mouthful of rice.

"I understand, son. You guys are busy."

"Well, I got to thinking. Maybe you could come visit us?"

"In theory, that *would* be nice."

"Just figured it'd be easier." Marco took another bite. "It's just really hard to get away."

"Been trying son. Struck down every time." How many times had he applied for a day-pass in the last several months? Six? Rejected every time.

So now I get to sit behind the great white prison walls of Novagenica.

"What do you mean 'struck down'?" Marco asked.

"Day-pass denied. NG won't even let me out. Happened six times straight now."

Marco put his spoon down. "Dad. Why didn't you say something?"

Manolo shrugged. Marco had enough problems of his own as a single father.

Marco rubbed the back of his neck. "Jeez, I'm sorry, Dad. I didn't know. That sucks."

Manolo continued to spoon the yellow stew into his mouth. "Not your problem."

Marco leaned back, scratched at the beginnings of a beard: a smattering of black, red and brown all at once, not to mention the occasional wisp of white. He looked thoughtful, the way he did when puzzling over a problem. "They give a reason?"

Manolo arched an eyebrow.

"Yeah, stupid question," Marco said.

Manolo snorted through flared nostrils. "Signed away the rights to reasons five years ago."

"What was five years ago, lolo?" Sol asked, looking up from her bowl. A little dot of yellow stew adorned the tip of her nose. Manolo wiped it off with a swipe of his thumb.

"Nothing, sweetheart," he said, his voice light.

"Isn't that when you and grandma joined NG?" Maricel asked.

"Maricel..." Marco interrupted.

"What? What..." The words caught in Maricel's throat when she saw Manolo's ashen face.

The mention of Esme made Manolo shrink inside. It was as if the mere mention of his now-dead wife was a weapon that could be used against him. It was absurd, he realized. They'd had a good, long life together. Many memories and a wonderful child who had also given them two exquisite granddaughters. But that was the rational side talking.

Manolo could feel the silence more than hear it. It was okay. He was surrounded by family, and if he couldn't trust them enough to show them his raw, open wound, then who could he trust?

"It's ok, son." He turned to Maricel, then Sol, her big brown eyes looking for all the world like those of a puppy that had just piddled on the floor. She'd crush hearts one day, of that he was sure.

"Yes, girls. Your grandma Esme and I signed the contract to join Novagenica five years ago."

"But why?" Sol asked, the picture of innocence.

"Grandma was sick, nene, and the people here said they could cure her. Make her all better."

"But grandma died," she responded matter-of-factly.

The words were a dagger to his heart, infecting it with hurt and hatred, bringing all the anger and resentment boiling back up to the surface. His hands shook.

"Yes, sweetheart," Marco interjected. "They couldn't cure her."

The confusion on her face was apparent. Everything could be cured, everybody knew that.

Manolo looked down at his stew, his mind in a different place entirely. The day his Esme passed. Almost a year ago now. He remembered her body not as it had been in life, but as it was in sickness—withered, desiccated, sucked of her vitality and her joy. And even through the emaciated exterior, she had still managed a smile for him as she lay on the white-linen bed, tubes stuck into her body. That's how she had died. A smile on her face.

The thought wasn't comforting; the hopelessness, frustration and anxiety that had come to increasingly invade his heart and thoughts seeped out into his bloodstream.

"They lied to me," he said quietly, looking down.

The air in the room went cold. Nobody spoke. Their eyes were locked on forks as they clanked them against the dishes, eating mechanically.

Manolo's hands lay folded in his lap. His eyes went to the wood-inlaid wedding ring he still wore, the one Esme had bought him over thirty years ago. His old fingers probed its battered metal edges, twisted the ring around his finger.

This was absurd. He didn't invite them over to put his various miseries and grudges on display. *Snap out of it, you grumpy old son-of-a...*

"Listen to your old lolo gripe. Bad Grampa!" Manolo smacked the back of his hand. The girls giggled.

"You're silly, grandpapa," Sol said, rice falling from her mouth onto her lap.

"You're right, girl. I *am* a silly old man," he agreed around a mouthful of kare kare, "but I'm the best grampa you've got."

Least you can do is hide how you're feeling. They deserve that. It wasn't worth worrying everyone when there was nothing they could do. This was *his* problem.

"You're the only grandpa we've got," Maricel chimed in.

"Doesn't that make me the best?"

They laughed, and the black cloud that hung over the table was cleared.

"So, Dad, what've they got you assigned to these days? Last I recall it was farm work."

Manolo nodded. "Still farm duty, at least some days. But they took me off custodial at the egg."

Marco sipped at his stew. The amber glow from the overhead lights reflected dimly off his glasses, concealing his eyes as he looked up from the bowl. "Yeah? No more access to HQ secrets?"

"Secrets of a different kind. Shifted me to Research."

"Top secret, huh?"

"Eh," Manolo shrugged, kept his face a mask. "Still mopping floors and scrubbing toilets. Ugly stuff, though. Not allowed to say much else beyond that."

The image of the recycling chute filled Manolo's mind, swallowing the body whole like a snake gulping down its prey. He squinted, wiped at his cheek.

Marco took of his glasses, cleaned them on his shirt. "That bad?"

Manolo squirmed in his chair. He really wanted to change the topic, but it'd be a dead giveaway that something was off. "At least they rotate me between Farm and Research regularly. I couldn't take being down in that dungeon every day."

"What's the distribution like?" Marco replaced the glasses, balancing them on the bridge of his nose.

"What do you mean?" Manolo asked.

"Well, how many days in Farm versus Research?"

Manolo mentally went over his work calendar over the last few weeks. "About five farm days to every three custodial. It varies some, but management knows my limits at least."

Marco chewed, nodding thoughtfully.

"Y'know, pops, they've got an algorithm for that. Not managed by people at all. At other Complexes they call it the *Discontent Scheduling*

Algorithm. I'd be willing to bet my right arm they share it across affiliated Complexes, or at least a variant of it.

"*Discontent*? You're telling me a computer knows how I'm feeling?" he frowned, shook his head.

Marco poured some stew atop a pile of rice. It looked like bright yellow, chunky vomit. "Mmm," he purred, "so good."

"Thanks," Manolo said. He'd always been the cook in their family. Esme had the green thumb and he cooked whatever she managed to coax out of the dying ground. Yams, squash, beans, peas. He missed it.

"Not as hard as it sounds," Marco said. "Just requires a lot of sensors."

"What, like a video feed?" Manolo asked. Technology was not his forte. It changed so rapidly he wasn't sure how anybody could keep up with it, especially a grumpy *matanda* like himself. "Still requires someone to sit there and watch it."

Marco shook his head vigorously. "No, you're missing the point. There's no one manning the ship at *all*. There's sensors for all your bios, processed in real time. Blood pressure, hormone levels, sleep patterns, social participation, eating habits, blood sugar levels, cholesterol—all your bloodwork. The only thing they don't have is your actual thoughts, if they're following the Lucerne Protocol like they should. But I guess even that's not guaranteed."

Maricel put down her spoon. "No way!" she protested. "For real?"

"Sorry, sis, it's a thing," Marco responded.

"Oh *God*," Manolo groaned theatrically. "Just shove me in the grave now."

"It's not *all* bad, Dad. Thing is, NG doesn't need your thoughts, not directly. They can get a close enough approximation just by feeding all these data into the algorithm. Then they match it up against your historical activity to know what you prefer and what you don't. From there, it spits out a schedule that manages your level of discontent. And it's always adjusting, self-correcting. Too much one way, and you'll want to abandon ship. Too much the other way and there's implications against creativity and productivity. Basically, too much of a good thing can have the opposite of the desired effect and create an environment where you don't know what you're missing out on nor do you care. Absolute and total contentment kills productivity and innovation as well.

Manolo affected an air of resignation. "So you're telling me the machines have finally taken over."

Marco looked off to the side, pondering the statement, then looked back blankly. "Yeah, Dad. About sums it up."

Manolo grunted. "Well how do I ever get a pass, then, if I can't at least talk to a person and change their mind?"

They lied to me. Novagenica said I was as free as anyone else. That getting a day-pass was as simple as making a request.

The thoughts, angry and dark, threatened to infect his mood again. He pushed them away, focused on his son.

Marco, oblivious, waved his spoon. "Oh that's easy. Pretend."

"Pretend what? Like they granted me a pass and that I'm visiting you when I'm actually stuck here working as a corporate slave? Real helpful, son."

"Make-believe, lolo," Sol offered. "I can show you." She pushed her chair back from the table and stood, put her arms up in front of her chest and let her hands droop at the wrists. She stuck her tongue out and panted, trotting around the table.

"Arf arf!" she barked loudly in a high squeal. "I'm a doggie!"

"Oh!" Manolo shouted, smacking himself in the middle of the forehead. "I get it now. I should act like a dog so they'll think I'm crazy and kick me out. Is that right?"

"What Sol means is you have to trick the sensors," Maricel said. She angled her chair toward Manolo, scooted up to the edge and leaned forward. "So all the kids in my class do this thing where...okay, in Study Hall, right? We're all supposed to be studying, or reading, or doing something schoolwork-related, right?" Maricel set her spoon back down in the bowl and began talking with her hands.

Manolo smiled. The resemblance to Esme was uncanny. "If you say so," he replied.

"Ok, well, there's no teacher, right? And no real agenda for the period, just some cameras and other sensors, sniffers for adrenaline spikes, artificial ears for things like gunshots, breaking glass—even something for the sound when something strikes bone..." she made a gagging sound, then continued, "anyway, the kind of things that send the room into stasis mode.

"But if we get caught messing around, talking, or even look like we're laughing or having fun, the auto-minder turns on and our systems get shut down. It's not as bad as stasis where the minder hijacks our systems and paralyzes everyone in the room, but still, what're we gonna do without our systems?"

Manolo started to respond but his granddaughter steamrolled on. "It's a total load, we *have* to have access to our systems. So what do we do?"

Manolo simply shook his head. He could barely keep up, much less offer conjecture on what kids were doing in parochial school these days. Maricel, full of the pent-up energy of youth, forged ahead.

"First, we connect our systems and set up a virtual environment where we can get together. Then we act! Just like in the interactive feeds. We play pretend for the sensors and trick the minder." She flipped her hair nonchalantly over one shoulder. "As long as we don't trigger the sensors, we can sit there and talk, play games, whatever..."

This drew a raised eyebrow from Marco, as if to say *I know what your 'whatever' means.* There was a whole fathom of subtext there between a father and his teenage daughter.

Ignoring her dad, Maricel continued. "As long as we keep a straight face and try to keep our pulses from racing, the room never knows."

Manolo looked to his son. "You know about this?"

"Yeah, well, it *is* study hall."

Manolo looked back to Maricel, scrutinizing her. "Still getting good grades, hija?"

She returned his question with consternation. "Of *course*, Papa. Straight '*A*'s.'"

"Hmm, okay. I might take your advice, then, seeing as how you're such a smarty-pants." *Maybe it'll keep that damn AI off my back.* He gave her a little wink before digging into his squash and beans.

Her eyes crinkled a little at his praise, accepting it with reservation. She pulled her long, wavy hair back over one shoulder. She was older now, and not as subject to the wild praises of her old grandpa as she used to be. At least Sol was still an open book.

"So, Dad, what you got growing out front there?" Marco asked.

"Soy, corn. Got a tiny patch of carrots, yams and wild onion out here on the west side. Red rain killed the first crop, nitrowind the second. Third time's a charm, I guess. Got some beans crawling up the side of the house."

"It yours or for the Complex?"

"All mine," Manolo said proudly. "Where'd ya think I got these veggies?"

"That's why they're so good!" Marco exclaimed.

"Yeah, papa, these are really good," Maricel echoed. "I can't believe you actually grew something *out of the ground*."

Manolo looked over at Sol, who was noticeably silent. Her vegetables remained untouched. She looked up sheepishly from her plate.

"Sorry, lolo. I bet they're really good, I just don't like them."

"Don't you have to try them first?" Manolo asked.

Sol shook her head slowly, not breaking eye contact.

Manolo chuckled lightly, leaned in and whispered to Sol in one ear. "It's okay, sweetheart. I won't tell."

"Okay, papa," she whispered back. "I love you."

She leaned in as far as she could and kissed him on the arm. Without warning, tears burst from his eyes. A dam, holding back his emotions, gave way. He sobbed, making an ugly show of his mixed emotions—despair, admiration, hope and, most of all, overwhelming love.

Sol's brow wrinkled in confusion, concern. "What's wrong, papa?"

Manolo managed a smile, wiped the tears from his cheeks, stemmed the flow from his eyes.

"Nothing, nene, nothing at all. I love you too. I really appreciate your kiss. It's the sweetest gift I've gotten all year."

The concern disappeared immediately from Sol's face, replaced with elation, dark to light in a heartbeat.

Maricel jumped up from her chair, ran around the table and hugged Manolo tightly. Before he knew it, they were all crowded around, squeezing him from every side. He reveled in the warmth.

"This has definitely been the high point of the last few months," Manolo said, his voice muffled from the bodies closing him in.

Little Sol whispered in his ear, "Me too, papa. Even better than my field trip." It was all he could do not to burst out in a fresh round of sobs.

Eventually they released him and took their seats. Done eating, they all helped out clearing the table.

Marco stayed in the kitchen, cleaning up alongside Manolo while the girls sat on the grey couch in the living room, working together on Sol's reading assignment. The sound of her fairy voice stumbling over unfamiliar words occasionally cut through the rush of water and pots banging as the two men worked.

"You doin' okay, Dad?" Marco asked, hands deep in a pot, scrubbing at its bottom.

"I'm fine, son," Manolo paused. "It's lonely here, without your mom."

Marco looked at him sharply, then nodded agreement. "I'd say quit the CPX but they'll never allow it. You signed a lifelong accord." He rinsed a pot and handed it to Manolo. "Honestly, Dad, Maricel's right. You should take her advice, try to act like you're content with life here. I mean, I don't know all the details, but I can guess. If *I* can tell you're feeling angry and bitter and discontented, it's a sure bet the AI can too. Won't give a pass to someone whose chart statistically predicts a flight risk. It just ain't gonna happen, like never. People bend the rules, machines don't."

"Easier said than done."

"Yeah, but what do you have to lose? Just give it a try. Promise me?" Manolo could feel Marco's gaze but avoided making eye contact. He didn't want to promise anything, and he didn't want to say what he was thinking. NG had all the drugs and procedures readily available to blot his wife's existence clean out of his life, but this was the thing he feared most—losing the memory of Esme.

But if he had to act like everything was right with the world, that meant he had to pretend either that Esme was still alive or, worse, that she'd never existed in the first place. The thought tied his gut up in knots. He'd lost her in body, he would *not* lose her in spirit.

Once all the dishes were clean and laid out, Marco ushered the girls to the front porch. They gathered around Manolo for one more group embrace, this time to say their goodbyes.

"Love you," Manolo said as one by one they descended the steps and hopped in the single file rental cart.

The doors sealed shut and they soon became dark smudges behind the composite glass. Sol waved frenetically as the cart whined to life.

And just like that, they were gone—the taillights of the cart made diminishing red halos against the black night as it wound its way eastward towards the vast white wall that protected the entirety of the Complex.

Their love and attention had kept the feeling at bay, but now it came roaring back in full force. A coldness that chilled him deeper than the night air ever could. It wasn't just his emotions that the feeling seemed to affect—it was also his thoughts. When the anxiety and despair were at their worst, he found himself sometimes generating violent fantasies he'd never before thought or considered.

Like with the AI earlier, he thought.

He was, by nature, a calm and steady sort, but he found that the feeling could give rise to unfamiliar ways of thinking—sometimes manic, sometimes violent—that frightened him.

If he was being completely honest with himself, it wasn't just missing Esme that made him feel this way. He missed the ability to walk out his front door and wander around, no matter whether he might get shot, blown up, stabbed or even that most rare form of physical violence, beaten. He didn't care. As strange as it sounded, he longed for the sights, smells and sounds of the open, wild world beyond Novagenica's walls. He missed its unpredictability.

It wasn't about Complex versus Legacy; it was simply the vivid tapestry of humanity he yearned to witness and experience. While NG *was* beautiful, it was too white, too meticulously crafted, too sterile. Nothing imaginative would ever come of it.

His mind turned to his granddaughters. He was glad Maricel and Sol hadn't grown up here, that Marco had rejected Manolo's offer to bring them along when he and Esme had joined up.

Maybe he'd give Maricel's advice a shot. Maybe an old dog could learn a new trick or two.

"Ah, well," he muttered to the night. The artificial crickets chirped back at him. "Nothing to lose, eh?"

He turned his back on the cart's receding red circles and went inside.

KAT

KAT
AGE: 13
SEGMENT: LEGACY CITIZEN
POSITION: UNDERAGE; GOVERNMENT DOLE
RECIPIENT (REDUCED BENEFIT)
SOCIAL SCORE: 300

THE BREATHERS WERE CHEAP, Legacy issued. White plastic that was broad enough to cover even the most generous of mouths, with black elastic straps that were stretched and wrinkled, fraying at the edges.

Each breather had two slots—one on either side of the mouth—where canisters could be locked in. Standing in the stale-smelling foyer of Grand Place, Kat pressed her own mask over her face, cinched the straps. Val did the same, handing Kat one of their two remaining oxygen canisters.

Kat pushed the canister into the slot, twisted it until she felt it lock. A gentle breath of oxygen hissed over her wet lips.

"Let's go." Val's voice was muted as she pushed ahead of Kat, through the exterior doors.

The sky was overcast, brownish-red, and when they exited the apartment the wind held a grit that stung their eyes. The wind had come up from the Gulf, the massive dead zone there generating a burgeoning pocket of nitrogen that was carried north on rogue winds.

They turned north and put the stinging nitrowind at their back. Kat glanced at her sister, whose head was angled down as they plodded forward. Her hair blew forward, ahead of her, shrouding her face. The streets were deserted, like some post-apocalyptic ghost-city.

Somehow we're the only two idiots out here, Kat thought, squinting against the sand and dust that pelted the side of her face. *Even predators have a survival instinct. Shows you how smart we are.*

Val was probably glad for the nitrowind. She couldn't hear Kat grumble about having to go to school through the mask.

They trudged through the toxic wind, alone in the city, until they reached Deus Ex. The church was as forbidding as ever. They entered the double doors, feet scraping at a line of sand, dust and pebbles that had collected at the threshold, and were greeted by Brother Anthony in his long, dark robes.

He smiled broadly; something inside Kat wanted to punch him.

"Good morning, Kat."

She removed the breather, locked the canister to preserve oxygen. She looked away sullenly, not making eye contact with the monk, mumbled something unintelligible. Brother Anthony turned to Val, acknowledged her with a nod.

"Morning, Brother," Val said through the mask.

"Systems off?" he reminded.

Val looked to Kat, eyes flat and demanding. Kat rolled her eyes. "Fine," she said, and deactivated her system.

"All good now," Val confirmed.

"Good. We'll see you back here this afternoon."

Val glanced briefly at Kat, something inscrutable in her eyes. Then, just like that, she was gone.

The classes were held downstairs, where the smell of wetness and waste seeped through the white-painted foundation stones. Despite that, the rooms were heated and comfortable, illuminated with a warm yellow light reminiscent of sunset.

Children of all ages populated the room Kat had been led to, seated at desk-chair combo units. The desktops were made of scarred wood and the chairs of a plastic that flexed, cracked and popped as Kat tried to get comfortable. They were gathered in rows of semicircles, pointing inward where the teacher—Sister Berta—stood, lecturing at a whiteboard. Younger children were at the fore, teens in the rear rows.

Kat squirmed in her chair. It was hard to sit still for so long.

Damn it, Val, this is your *fault,* she thought, her face distorted in a grimace as though she'd eaten something bitter. It was all well and good for Val, who could just drop Kat off and then go do whatever she pleased all day long. Kat was the one who had to pay the price for her sister's inability to manage their budget.

And to add insult to injury, she'd had to actually spend some of SUC. Val *knew* Kat was saving it.

It's not fair. I get all the responsibilities of an adult but none of the freedom. She gripped the corners of the desk and squeezed, then looked toward the door. God she wanted out of there.

At the front of the room, the dumpy old woman, in a robe that dragged the floor, droned on. A blank notebook lay open, untouched, on Kat's desk. She twirled the pencil they'd given her in thin, deft fingers. She'd already chewed the soft, yellow wood to the nub. It looked like a starving rodent had gnawed at it for hours.

"What are you, a *beaver*?"

The boy to her left, who was probably only a year younger but a full head shorter, stared at the mangled pencil.

Kat rolled her eyes. "Whatever."

He put his hand out, said, "I'm Stef." She ignored his hand, but he went right on like she was actually interested in what he had to say. "Seriously dude, that's like a special skill or something. Let me see your teeth. You have 'em modded or something?"

Kat turned slowly toward Stef. He was runty, with heavily freckled pale skin that made him look younger. A pencil balanced over his ear, disappearing into shaggy brown hair. Something about him looked familiar.

She squinted at him. "I'm not a dude. Get your words right."

"Sorry. Dudette. That better?"

She ignored him, focused on the teacher.

"What a machine thrives on is order," Sister Berta said in her atonal voice. "And a very narrow specialization. When it can't have that, it gets frustrated."

What was the old hag even talking about? A machine getting frustrated? What kind of nonsense did they teach at this school?

"Humans, it seems, are a keen pain-point for these machines," the woman continued. Her fingers were laced together, palms up and held over the paunch at her middle as she paced the interior of the semicircle. "We're decision-makers, but variable. We can change our minds, and often do so hundreds of times throughout the course of a single day. It's in our nature to alter our courses—we simply can't help it. This very unpredictability means we're inherent *increasers* of entropy, which is something a machine loathes. But entropy is *essential* to life. Would anyone care to explain what entropy is?"

Oh. My. God.

She was going to strangle Val.

Kat put her forehead on her desk, and without even thinking about it turned her system on. She remembered Brother Anthony's reminder from that morning, but was *she* given an option?

No. Like always, Val told her what to do and she was expected to listen.

If she was going to make it through these classes she'd need something to do that didn't involve listening to Sister Berta's weird lectures. Besides, how would they even know if her system was on?

"Miss Merina? Would *you* care to explain the meaning of entropy?" Sister Berta asked.

Kat sat up. "Umm," she mumbled, suddenly aware of the multiple heads turned in her direction. The class's attention was focused solely on her. Almost as second nature, she looked up the meaning of the word on her system.

"It's when, uh, there's a lack of order. A decline into disorder."

The sister's eyes narrowed. "That's very good. And *very* by the book." Sister Berta approached through the rows of desks. "I wonder. How does one have a such a precise response at the ready, and so quickly at that?"

Kat felt her mind go blank as panic began to take hold. *She knows.*

Sister Berta stepped closer, her pace deliberate. "You're obviously quite bright, Miss Merina. And, as a quick learner, I don't have to remind *you* that the use of internal computing systems are forbidden on these grounds."

The sister stood before Kat's desk, looking down the bridge of her bent, hawkish nose. Her eyes were black and forbidding, her face severe.

Out of the corner of Kat's vision, she could see Stef look to the sister and back to Kat, his hair swaying back and forth each time he turned his head.

"I don't have to remind *you*," Sister Berta said, "that internal computing systems are *anathema* to the Church of Deus Ex Machina."

Kat was held by the woman's demanding glare, unable to speak. She couldn't turn off her system now—the sister would undoubtedly see Kat's vision faze out as she focused on the display.

Something twisted in Kat's gut. Then the unexpected happened. A clatter to the left.

Both she and the sister looked to see Stef sprawled on the ground. Sister Berta rushed to the boy's side, kneeling there as he moaned theatrically.

The spell was broken.

Without hesitation, Kat focused on her system's display, switched it off. When she looked back to the boy, Sister Berta was staring right back at Kat.

The fact that she didn't get caught did nothing to settle the anxiety in her gut.

Sister Berta put a hand under Stef's elbow, steadied him as he settled back in his chair. He assured her he was fine, that he'd just lost his balance. Once the sister was satisfied, she continued to lecture, strolling around behind the back row, speaking over Kat's head.

"This presents us, class, with the perfect segue. Why does Deus Ex consider internal computing systems anathema?"

The hand of a young girl in the front row shot up. Berta waved it back down.

"You can put your hand down, Dani, the question was rhetorical. Systems are anathema because they act as a bridge. They connect to the greater machine *and* to our bodies. They grant the machine access to God *through* the holy temple of our bodies. And the greater machine detests the thing that makes us human, and divine—our ability to make decisions. Our ability to think for ourselves. Our ability to create, destroy and alter the worlds we occupy. To the greater machine, our free will is the enemy."

Kat felt a hand squeeze her shoulder. Firm, yet gentle. An affirmation. "Trust is one of the most precious possessions we have," Sister Berta whispered in Kat's ear. "I trust in your judgment."

Sister Berta ambled slowly forward, robe swaying, into the row in front of Kat. Kat stared at her back, head cocked to the side in puzzlement. Once again, that was not what she'd expected. She'd braced for an accusation and a stern upbraiding, even punishment. But what she'd gotten was a reminder of her own ability to choose. It made her feel good, like she was capable of making good decisions, like she was worthy of that trust. Like she was becoming an adult.

She felt something flick her left arm. She tried to ignore it, but the boy flicked her with his pencil again.

"*Stop it*," she whispered fiercely. He went to flick her again but she snatched the pencil out of his hand, snapped it in half. The look of shock on his face was reward enough. She placed the two halves of the pencil in his open palms, smiling sweetly.

"Wh-what?" he stuttered. "I can't b..."

"Believe it." She turned her attention back toward the teacher.

She could feel Steffen's gaze on her. "Isn't there something you'd like to say?" he asked. "For saving your ass back there?"

She looked to the ceiling, made a show of thinking. "Nope," she finally concluded.

The boy affected a look of anguish, hands going to his heart. "Like daggers." He swayed in his chair, coming perilously close to toppling out of it once more.

"Mr. Avon? Everything alright?" Sister Berta called out from the front.

He put his hands up. "All fine, dear Sister. Some minor heart palpitations, but I'll be okay. Just need a minute is all."

The sister cocked an eyebrow. "Take whatever time you need, Mr. Avon. We're all here at your luxury."

When Sister Berta turned back to the whiteboard, Kat leaned across the aisle toward Stef. She surprised both of them when she laid a hand lightly on the pale, freckled skin of his forearm. The warmth beneath her cool fingers felt electric. He looked down at her hand, stupefied.

"Thank you," she whispered. "And I'm sorry for breaking your pencil." She took her own mangled pencil, offered it to him. "It's a little used, but you can have mine." He broke into a grin, all cocky reassurance, and once again she was struck by the sense of familiarity.

I know you from somewhere. But where?

"Thanks," he said, taking the pencil. His fingers overlapped hers for a moment. A thrill ran through her.

That, too, made her feel grown up.

VAL

VAL
AGE: 19
SEGMENT: LEGACY CITIZEN
POSITION: UNEMPLOYED; GOVERNMENT DOLE
RECIPIENT (STANDARD UNIVERSAL CREDIT)
SOCIAL SCORE: 478

WEEDS CHOKED THE GARDEN. Ivy was the main offender, crawling and wrapping its tendrils around the sun-starved shrubs, blanketing all surfaces.

The garden, if it could be called that, grew in the concavity of Grand Place Apartments. The building was constructed in the shape of a *U*, with the garden at its center. Once a week, sometimes more, Val fled the stuffy confines of their apartment while it was still dark and escaped to the garden's open air, locked and safe behind a tall wrought-iron gate that was rusted shut.

And immerse herself in a memory.

The memories, files she'd stored on her system, were the only remaining link she had to her dad. In the last months of his life he'd retreated more and more frequently to the garden, especially when the coughing fits made it difficult for him to breathe, caused him to hack up thick wads of yellow, red, black.

He'd worked, as far back as she could remember, at a clothing manufacturer. The company had managed to stay afloat through the manufacturing decimation by firing ninety-eight percent of their workforce then pivoting to bespoke, personally tailored products. The way Dad said it, any Legacy citizen could print a shirt at a corner factory, but what they couldn't get out of those printers was a custom-made piece of clothing that was self-cleaning, connected to their system with built-in add-ons *and* fit like a glove.

Val didn't fully understand his illness, and he'd done a poor job explaining it to them. Something about the self-cleaning nanobots they sprayed onto the clothing. It got into the air, into his lungs. When he'd told her this an involuntary shiver had run from her toes up through her scalp; she'd imagined little bugs crawling around his lungs, reproducing inside the sacs until they lined every square inch.

After that day, she'd rushed to record as many memories as possible. She wished she had more.

In the black hollow of the garden, a lone bug chirped at her. The wind had shifted, so that it was blowing up from the southwest. It held warmth, moisture and, most importantly, oxygen. She put her hand down on the cracked concrete bench that was pushed up against the stucco of the building, swept its cool surface clean of dirt and debris deposited there by the nitrowind.

Your favorite seat, Dad, she thought with a sad little smile nobody could see.

She sat, pulled the memory up on her system and closed her eyes. The shadow images of herself and her father, dead over a year now, played on the inside of her closed eyelids. It was strange watching herself, seeing how it compared with the memory in her mind, but she'd grown accustomed to it.

"Can't you sue them? They did this to you."

He was seated on the bench, while she stood facing him, body stiff with outrage. Sunlight streamed from behind Val, setting her golden hair aflame while casting the rest of her body into dark shadow.

"Not a chance, kiddo. We're all just contractors, and the contract says they're not at fault, no matter what."

She leaned forward in accusation. "Then why'd you sign the stupid thing?"

Her dad looked weary, resigned. His red hair was falling out of his head and even his beard looked thinner, with a spray of white in it. He scratched at the beard. "Not sure, kiddo. Guess I did it for you two. Can't see there was a better way, but not sure I'd do it again either."

"Really? That's something you even have to think about?"

Her dad beckoned her closer. He had that crease in his forehead, that sadness in his eyes that indicated he had something important to say, something she needed to remember. She came closer, her body still rigid with defiance.

He patted the bench and she sat down stiffly. He turned to her, put a hand on the back of her shoulder.

"Relax will ya, kid? I don't have forever and I'd rather you not spend it pissing and moaning about what could've been."

Val blew a puff of breath out, rolled her shoulders, shrugged his hand off. "Whatever."

Her dad chuckled quietly. "An improvement I guess," he said. "However you feel about it, I want to make sure you actually listen. It's important you actually hear. I don't plan on haunting you from the grave just so I can be satisfied you were paying attention."

At the mention of 'grave' Val looked to her father sharply. Tears began to well in her eyes. She nodded silently.

"Way this world is set up, people like us have limited choices. One, we can become a slave to a Complex. Two, we can suck on the Legacy tit—til it runs dry at least—or three, work as a contractor, if you can find it. That's about it, unless things change in some way people a whole lot smarter than me can foresee."

Val stared at her father, discerning the truth in his words.

"So..." she hesitated, not wanting to ask the question but knowing she had to. She steeled herself, jaw clenched, eyes worried. "What do I do when you're...gone?" Her face was a mask of anguish. Dad was there before her, but as the light touched his sallow features, she could see him wasting away, becoming smaller one day at a time until finally he'd dissolve into air and dust.

"Any way you slice it, they're all shitty choices. I..." he shook his head. "I don't know. You'll be the adult, so you'll have to make the decisions. Have to take care of your sister, that goes without saying." He turned his body to face her, the once full and strong shoulders now bony and sharp, jutting through the plain blue t-shirt he wore. "I won't tell you what to do. Best I can do is tell you what not to do."

She chuckled, wiped away a tear that tracked down her cheek. "Thanks a bunch, Dad, real helpful."

But his expression was grave as he continued. "Just promise me one thing."

"What?"

"Never, ever, join a Complex. Not for a year, a month or even a day. Don't even think about signing those terms, no matter how desperate you get."

He looked into her eyes and, just like that, knew what she was going to say. She didn't know how he did it, maybe something a father simply knew.

"No," he said, voice firm.

"I didn't say anything."

"But I know what you're going to say, and the answer is no," he said.

"What? What am I going to say?" she challenged him.

"Even if a Complex could cure me—which I think by now is pretty improbable—I wouldn't sign up."

The anger flared, hot. "But why wouldn't you take the chance? For us."

"Some things, Val, are more precious than life."

"Oh what bullshit, Dad. Don't give me some mystical mumbo jumbo, give me an answer. Explain to me—your daughter—why you'd choose to leave us alone?"

He didn't hesitate. "Complexes are evil, Val. I've told you this how many times? I won't be a slave."

"No, you'd rather sit there and die."

He didn't speak, but instead pulled her inward. She resisted at first, but eventually unfolded her arms and wrapped them around his gaunt frame, rested her head against his chest. She allowed herself to grieve, the sobs coming fast and hard, the tears hot.

He held her tight with one hand on her back, the other stroking her hair, over and over. "Shh," he comforted. "Shh."

Eventually the tears dried up and the anger died away, drowned out by the comfort of being in her dad's arms. It reminded her of being a child, of a time of simpler thoughts and uncomplicated emotions. A time when her tiny mind couldn't even imagine what the world would be like without her dad.

She sniffled. "Alright, Dad. I promise. I'm not happy about it, but I won't join us up."

He sighed, relieved. "Your freedom is all you have in this world, kid. Don't piss it away for silly creature comforts. Never in a million years," he said.

She looked up at him. "Never in a million years," she echoed.

Val listened to the slow, weak thud of his heart. Eventually he leaned back, disentangled himself. He held her forearms in his broad hands, the blue veins visible through parchment-thin skin. He gave her arms two short squeezes.

"That's a good girl." His smile was wan, but his blue eyes twinkled when the morning light hit them. *"I'm gonna miss you."*

Kat found Val curled up on the bench. Her fingers brushed the back of Val's arm.

"Val? You okay?"

The sun was up now, streaming in through the opening between the buildings. It touched Val's face. She sat up, ran the back of a hand across her mouth, cleaning up the drool that had pooled there.

"Yeah, just dozed a little." Val's eyes focused, found her sister's face. "It's before eight, what are *you* doing up?"

"Nightmare," Kat said.

"Ugh." Val took Kat's hand, pulled her down to the bench. She sat beside her. "Wanna talk about it?"

Kat shook her head.

"Supposed to make you feel better, y'know," Val urged. "C'mon." She nudged Kat with an elbow. "They say that if you talk about it the nightmare goes away."

Kat shot her a look. "That's what you said last time, and I told you all about it."

"Same nightmare? The one with the never-ending hallway?" Val asked.

Kat looked at her expectantly. "Yes."

Val slumped. "Well damn, guess I was wrong. First time for everything." Val put her hands up as if to say *What can you do?* "Hey, since all I've got is crap advice, you wanna do something fun?"

Kat regarded Val warily, yet a smirk formed there. "Like?"

"Like playing the *Straylight Run*?"

Kat's face lit up in excitement. "Oh *no*."

"Oh *yes*."

Kat hopped up from the bench. "I get to be Molly!" she shouted animatedly.

"What? No way!" Val argued.

Kat squared her shoulders, put her hands on her narrow hips. "No Molly, no Straylight Run."

Val threw her arms up. "Fine, you win. I'll be Case. Or maybe," a devilish grin formed on her face, "Riviera."

"Suit yourself," Kat said, shrugging. "Your funeral."

"We'll see about that."

Val dropped the topic, but Kat stood there, fidgeting, twirling her blond locks in one finger, looking like she had more to say.

"What? What is it?" Val asked. Kat wouldn't meet her eyes.

Uh-oh.

"So...no school today." Kat said in a way that was entirely too offhand.

"Yes," Val drew the word out, leaned back against the rough stucco wall. Something was definitely up.

"Well, we don't have anything planned this weekend, right? I mean, we never have anything planned because we can't afford to *do* anything, so I guess that's just a stupid question..." The words tumbled out of her sister's mouth, running together into a stream of unfiltered thought. Still, Kat hadn't made eye contact.

"What's up, Kat? Just ask it. All I can do is say yes or no."

Kat met her eyes then, said, "All you ever do is say no."

It was a direct challenge, but Val kept her tone light. "Try me. I might surprise you."

Kat blew a breath out, shoved her hands in the pockets of her jeans.

Situation defused, Val thought. After watching the memory of Dad, she was in no mood to fight with Kat.

"Okay, well, I met this boy in school. His name is Stef. He lives here, like *here* here," she stomped a foot on the ground for emphasis, "at Grand Place. He's my age and he says he goes to 3 all the time, but he usually has his uncle take him and his mom. So I was wondering, maybe tomorrow, could I maybe go with them?"

Val blinked, shook her head. "Wait, *what*?"

"I said..."

"I heard what you said," Val interrupted, "I'm just trying to figure it out. You met a boy? At Deus Ex?"

"That's what I said." That *duh* tone again. If a person could earn a living making people feel stupid, her sister would be a millionaire.

"He lives in our building." Val repeated. She tried to remain impartial, but could feel concern tugging at her brow, drawing it downward. Somber mood or no, she had a responsibility to keep her sister safe and she didn't like where this was going at all. "Until this week you were strangers, but now you want me to let you go to 3 with him. Just to," Val waved an arm, "see the sights. That about right?"

"Yes." Kat's hands were out again. She began to scratch at one arm.

Val took a breath, maintained her calm. *Don't let her get you riled up.*

"Kat, you know how dangerous 1 and 2 are. Sure, Level 3 is supposed to be safe, but getting there is the hard part. We're not talking about the occasional junkie trying to snatch you, we're talking gangs working together. Aside from that, we don't really *know* these people. It's just as likely this boy is a set-up for a larger operation, to lure young and inexperienced girls like yourself for the doll factories."

Kat continued to scratch absently at her arm. "He says he's moving to 3. That we can join him there."

Val couldn't help but snort, which got Kat's attention immediately. "What?" she asked, eyes narrowed.

"Nothing," Val said. She should've maintained her composure, but the concept of someone moving from Legacy housing straight to Level 3 Arcside was more than preposterous. It was impossible.

"What?" her sister pressed. "Tell me. What's so funny?"

"It's not funny, it's sad," Val shook her head in commiseration. "Hate to break it to you, sis, but that whole 'moving to 3' thing is a pipe dream."

"You don't know," Kat said defensively.

"Even if he did, no one from 3 is associating with jobless Legacy citizens on the government dole. Joining him—well, really, the whole thing—is just a childhood fantasy. Make-believe."

"Fine," Kat spat. "But next time just say 'no'. Don't waste your energy on some stupid speech. Just get it over with and tell me I can't go. I'll just waste my whole life stuck in this stupid apartment."

Val instantly regretted the harsh words. She stood, put a placating hand on Kat's arm. "I'm sorry, I didn't mean..."

Kat jerked away.

"Don't *touch* me! Don't act like you care!" Kat shouted. She was in tears now, wild hair framing a face splotchy with patches of bright red. "We never do what *I* want. Dad left *you* in charge and now I just have to go along and do whatever you say."

"Dad left me in charge because I'm the adult and you're too young and stupid to make decisions for yourself," Val shot back.

"Oh, okay, *I'm* the stupid one," Kat pointed to her chest. "Yet you're the one who couldn't even manage a simple budget. It's *your* fault I have to go to school!" Kat jabbed her finger at Val in accusation. "Your fault I was even there in the first place, so go ahead and blame yourself for letting me make a friend."

Kat whirled and stormed toward the building's rear entrance. She stood on the stoop and opened the door, then looked back to Val. "I'm going, and there's nothing you can do about it!" she screamed.

Val started after her, but Kat ran inside, slamming the door shut behind her.

TREVOR

TREVOR
AGE: 18
SEGMENT: LEGACY CITIZEN
POSITION: UNKNOWN
SOCIAL SCORE: 821; FLAGGED FOR VERACITY

T REVOR HEARD THE DRIP more than he felt it. A tiny tap on his shoulder.

Drip.

A foul-smelling droplet of water rolled off the drenched awning above, plopped onto the shoulder of his leather jacket. He brushed it off, fidgeted in the yellowed and cracked plastic of the chair. He wasn't sure the term 'writer's block' could apply to someone like him, but he *was* having trouble getting started.

What's the matter? You got your favorite spot today...what could be better?

It was true. He'd gotten to Headspace – the little head shop on the corner of 16TH and Cherry—early, found his table out on the patio, tucked in a small corner up against the wall, bordered on one side by the couch and on the other by the chain-link fence that bordered the patio. The clear polymer table was gloriously unoccupied.

The couch, to his right, was almost always taken by a group of five middle-aged men. They puffed on a hookah as though their lives depended on it and drank tea out of dainty cups stained with use on the inside, pastel geometric designs on the outside.

Today was no different. The men were especially loud, speaking animatedly in a language Trevor didn't understand. He could've had his system translate, but what was the point? There was something exotic about the group that he didn't want to sully by finding out they were talking about their bowel movements and frigid wives.

The cloud of smoke that hung about their heads migrated to envelop Trevor. He inhaled deeply. The smoke smelled sweet and spicy, of cinnamon and a woody fruit.

Drip.

Another droplet fell, this one directly to the table. He looked up, behind his head. A mother fern towered over the table, its deep green leaves bowed and laden with the moisture inherent in the air.

He reached one hand out and up, stroked a broad leaf between thumb and forefinger, a rivulet streaming onto his hand and down his wrist, tickling the skin there. He sighed and let go, leaned up against the fence that separated the patio from the sidewalk. Hung a hand on the rough metal of the chain-link. It wasn't fancy—like some of the perimeter defenses you'd find outside of the Arc—but still effective against random crazies with makeshift knives. It was a Level 1 kind of fence.

Through the fence, the crowds were heavy. Level 1 was a novelty, and people wanted to experience its forbidden glamours firsthand. Trevor squinted through one of the chain-links.

The daytrippers—hapless tourists to the Arc's basement levels—ambled slowly, without purpose, gawking at every seedy little detail. They traveled in distinct clumps, herding together like the sheep they were.

For those that lived in the Arc, they were prime targets. Easy money.

Let it go, he chided himself. *There's work to be done.*

Trevor pulled up a picture of his sister on his system. Thin and gangly with brown hair that curled at the tips, a smile shining out of her tanned features but with a haunted look to her hazel eyes. One glance at his thirteen-year-old sister was enough for Trevor to remember his motivation. He condensed the image, moved it to his upper right field of view so she was always staring back at him.

"Alright, alright," he said, intertwining his fingers then bending them backward and cracking the knuckles. "This is for you, Nicolette, even if you don't know it. Time to make that money."

Job one—Michael Fant's Social Score.

Trevor pulled up Fant's dossier on his system, angled it so that it appeared to rest flat, hovering over the table.

MICHAEL FANT

- *Age 33. Male, Single, Heterosexual.*
- *Legacy conviction at age 22, FRATERNIZING WITH A MINOR; conviction and sentencing c/o ArcSec Security Complex. 14-month sentencing*
- *5 late payments over the last 10 years since release*
- *No known friends, evidence of affinity for social outliers*
- *Spends money on alcohol and noodles*
- *Current public works job with District 71 on Level 3, JAIM DEL POTRO is the presiding Admin*
- *Aspires to residence on 3, promotion within the District*
- *Current Social Score—483*

He'd compiled the information from what Fant had given him and coupled that with data from other public sources. The man looked older than his years, not surprising considering his time spent in ArcSec's confinement.

Trevor removed a lollipop from the inside pocket of his jacket, unwrapped it. He popped it in his mouth, nestling it between the molars and cheek, careful to suck slowly and not bite down.

The speed hit him quickly, juiced his motivation. *Now* he was ready to get to work.

"Aright, Michael. Let's jack that Social Score. Erase your indiscretions."

He went quickly, starting at the conviction. He didn't bother infiltrating ArcSec's records system, but instead issued a backdated overruling order from the Level 3 admin, Jaim Something or Other, absolving Fant of his crime, effectively erasing the conviction. The move instantly boosted his Social Score to 612.

Next were the late payments, which were ridiculously easy to modify.

He removed the lollipop for a moment, swiped his tongue over his teeth, cleaning the grit from them. He checked the score again. 729.

He slapped his hands on the table. "Boom!" The men on the couch went silent for a moment, looked at him, then went back to their conversation.

The *Affinity for Social Outliers* category was a little trickier to deal with. Video evidence showed Fant regularly entering the Signal Hill shantytown. God only knew whether he was there for sex, drugs or, less likely, to render aid to the singularly homeless population that lived there. He left it alone. It was possible he could achieve his goal without amending *that* particular propensity.

The receipts indicating purchases of alcohol were a breeze—Trevor simply changed the barcode category to milk products. Didn't bother changing the noodles. What was so bad about noodles, anyway?

You can have your noodles, Fant. And eat them too, Trevor thought, humming softly to himself.

The job category he left mostly alone, except to add two commendations directly from the Dear-Old Admin himself for exemplary service to the district. The promotion should take care of itself, especially once Fant relocated to 3. There was a circular sort of logic to the Social Scoring system, a built-in catch of sorts: the score required for residence on Level 3 or higher was 801, but there was a score cap of 800 for residents of Levels 1 and 2. If a citizen of Legacy called Level 3 home the cap was then lifted, but only when they reached a score of 801 could they do so. Job promotions and Social Score often went hand-in-hand.

Trevor skirted the catch by locating an unused apartment in the backwater borough of Watership. A Level 3 address, but the least desirable of them all. There was no Legacy subsidy for the apartment, but Trevor found that Fant could meet the rent on his salary. He made the decision unilaterally and assigned the vacancy to Fant, backdating the occupancy prior to the ruling of the 801 score cap. Those were the terms of the deal—Trevor would do what was necessary, which meant that the customer didn't always get exactly what they wanted. In Fant's case, though, Trevor felt he did a pretty damn good job. He checked the score one last time.

860.

He shoved back from the table, clapped his hands. "*Yes!* I am the *master!*"

Trevor took a screenshot of the social score profile, sent it to Fant with a request for payment. He didn't worry about getting paid. They *always* paid. He made it very clear at the outset that what could be done could always be undone.

The payment came through immediately.

Trevor checked the balance on his account and grinned. It was enough.

As he sucked on the lollipop, he felt his spirits lift. It wasn't just the microdose of speed. His gaze flickered to the image of Nicolette staring back at him.

There ya go, sis. You don't know it yet, but Mr. Fant here just bought you a ticket to a better life.

RAY

RAY
AGE: 36
SEGMENT: COMPLEX PROPERTY; PERSENSE CPX
POSITION: REDACTED
SOCIAL SCORE: NA, CPX-SPONSORED

RAY GOT THE CALL, and he went. Dropped what he had his hands on and went.

Currently those hands were busy manipulating a micro-explosive. No bigger than a knuckle on his pinkie finger but it'd take five, maybe as much as seven people depending on how tightly packed the targets were.

Just a little something for a rainy day.

It wasn't like Ray *needed* to have a cache ready at the drop of a hat—pay came steady whether or not PerSense had missions lined up. One of the many benefits of Complex life.

But the pay had never been his main motivation in enlisting. It was service, purpose. A higher calling. And being ready was simply an extension of his service to PerSense. There was a greater design behind his work. More than just driving subscription to the Complex, there was an overarching societal component. A shifting tide in the ocean of humanity and Ray, like the other subbers, was the impetus.

To be truthful, it was also driven by something inside of him, a voice deep down that told him to be prepared at all times. Like the preppers of old. No matter how sunny the skies looked, the storm would come one day, and probably sooner than later.

The call came through on his system. Encrypted voice transmission, direct from his handler at PerSense. No name, just a voice. Modified, husky. Could've been a woman for all he knew.

"What do you have for me?" Ray answered, skipping the niceties.

"Details of the job are in the package. Sending over now." The voice on the other end was precise, measured, odd in its cadence.

Sounds weird, he thought. *The tone is the same but the rhythm is off. Eh, probably just sick or something. Don't go getting paranoid, old boy, you know where that path leads.*

A certain level of paranoia was healthy in his line of business; too much would drive an operative over the edge. Quickly.

"Timeline?" Ray asked.

"ASAP," came the voice, then there was a click followed by a *ding* from his system. Connection terminated and package received.

Since Ray spent most of his time in his shitty little hole—the palace of A1 Coffin Apartments—every time he got the call from PerSense it made him a little homesick. He missed the CPX. Missed its stout turquoise walls, its grid-like, military order.

"Sub rates flagging again, boys?" Ray said to no one, eyeing the package on his HUD. All the jobs were starting to bleed into one another. Target this or that Legacy hotspot, kill some civilians, drive subscription to the Complex. Wash, rinse, repeat. He didn't want to admit it, but it was getting old. He was ready for something bigger.

Grand. That was the word. He wanted his actions to mean something, to be a part of a greater plan.

Excuse me? chimed the AI in his head.

Ray shook his head. "Nothing. Marketing guys need to do their jobs," he said under his breath. "Decrypt this," he commanded the AI.

Yes sir.

He put the plastic explosive down gently on a small square of wax paper, set it on the mattress, went to wash the oil from his hands while the AI assistant worked. His coffin was convenient in that it was two steps to the sink. He turned the water on—the flow coming in a cloudy dribble—and scrubbed, sluicing brown sediment into the sink.

Here you are, sir. All done.

There was something about the way the AI spoke that sounded familiar, that tickled something in the back of his brain, but the eagerness of finding out the target subsumed his curiosity.

"Open it," he said, continuing to scrub.

The AI complied, opened the file.

The shock made him slam one hand to the wall to stabilize himself.

There, in the upper left of his field of view, overlaid atop the filth-smeared wall, hung an image. Not a café. Not a restaurant, apartment building or competitor's security station.

It was a person. One solitary, specific person.

His breath came in short, quick bursts.

It's happening. The path to the higher calling was opening up, he could feel it.

Ray shut off the water, dried his hands on the sides of his shirt. Wrapped the explosive up in the wax paper and stowed it away in a black rucksack. He cinched the top, flipped open the black lockbox he'd bolted in the far corner of the apartment. Set the rucksack on top of the hard cases of weapons already stored inside. In his eagerness, he slammed the lid shut harder than he normally would, locked it. Took a thin black jacket off a hook on the wall, swung it over his shoulder.

It's a person, he thought with a sense of wonder. One single target. A mark. Not the gray masses, or even a termination job. The bio-metric details coded into the image were very specific.

He allowed himself the luxury of a small smile. This was different. It marked a change on the horizon.

Ray accessed the separately firewalled controls to his internal drug reservoir, bumped a microgram of *Elate*. The drug cocktail was formulated to make people giddy, but for Ray the sensation was more like an energized calm. A light in the eyes, radiating perfect peace, imminent optimism.

The day was coming soon. He knew it.

He palmed the coffin apartment's light pad to the off position. Time to do some recon.

The bottom two levels of the Arc were perpetually lighted by weak sulfur lamps that never saw the light of the sun. Even then, the lamplight couldn't penetrate all corners and alleys. The confined array of artificial light filled in the spaces—blues, pinks, greens, oranges—blazing from shop and kiosk signs spread and diffused outward, reflected off the constantly wet, glistening pavement.

The last horizontal strut of the Arc, stained black with years of rain and sewage, marked the barrier between confinement and open sky. Its blackness contrasted against the white of gently falling snow. A few of the high-rises and housing projects at the edge of the Arc could be counted as skyscrapers, but unlike the multi-tiered structure of the Arc they still let light and weather through. Independent structures, unencumbered by the webbing of steel, carbon fiber and novel polymers that bound the Arc together.

He passed beneath the last beam, over that border, and went from dark to light. More of a symbolic transition than a literal one. He was still in Legacy territory, behind enemy lines.

The flakes fell from an afternoon sky that was thin, tinny. Winter light. Just days ago it had been cool, crisp. Sweater weather. But like it was apt to these days, the weather skipped the intermediate steps and went directly to the extreme.

He turned his fleshy, ruddy face to the sky, let the cold flakes float down to rest and melt on his skin.

Sometimes he allowed himself a respite. Fascinated what it would be like to forget he was a soldier. To allow the history of death roll off him, sluiced from his body like sweat and grime beneath the cleansing heat and pressure of a shower. To simply be and take in everything around him. To watch the people around him, without assessment or judgment. Without analyzing the easiest way to immobilize them. To see the multiplicity in their faces, shapes, expressions. To allow his field of view widen and fall into contentment. To simply and silently appreciate life.

He stood there, in the snow, eyes closed. The flakes cooled his hot skin and he was transported back to his childhood.

Living in the *House of No*, as he called it. It wasn't really a house, but it had been home. And compared to his classmates' places, it was a palace of sorts. Two bedrooms, a living room and a kitchen with actual appliances that allowed them to cook, not just print, their food.

Palace or not, it was still a place of strict confinement. A place of yes sir, no sir. A place of grave consequences. Of swift punishment.

He was ten. Mom had been gone three years now and he was starting to forget her. Moments of panic set in sometimes and he found her face wouldn't materialize no matter how he concentrated. Then he remembered why she wasn't there and his tension would bleed away. As Dad had the habit of saying after a few drinks, fuck her anyway.

Mom's face was gone and Dad's was the constant. Always there, looming. Glowering, stern. He wore a scowl like most people wore underwear. Without it, he looked naked. Even when Dad had the night patrol Raymond would put himself to bed on time, knowing how quickly the scowl became a thunderhead when Dad discovered his son had slipped one child-sized toe over the line. And he always knew. Somehow, he always knew.

They lived on the first floor. A ten-foot square of yard with turf. Like he said, a palace. He was supposed to be in bed by nine, but the snow had just started to fall and Ray was entranced. It was beautiful. He opened the door to the yard, slipped out into the chilly black night in just his underwear and tight-fitting t-shirt. The snow fell silently, dusting his hair. The noise of the city was muted, as though a blanket had been thrown over it. Ray looked to the sky and felt something he couldn't quite explain. A feeling of being alone, but together at the same time. United, tied to everything.

It was the most beautiful feeling. He stood there, perfectly still. Time passed without his knowledge.

The cold in his fingers and toes, creeping in at the edge of his awareness, brought him back to his senses. How long had he been out there? He didn't know. But he *did* know what he could expect in the morning. He ducked back inside and ran straight to bed.

Dad didn't wait until morning. He shook Ray awake, one massive hand on his scrawny shoulder, in the middle of the night. He stood over the bed and the boy, waited until he knew Ray was watching, then unclasped the belt from his uniform, slipped it free of the loops. Held the thin, black vinyl in both hands.

The belt slapped, lashed across Ray's lower back and he screamed out with each strike. But through the pain, the lashes emboldened him, made him more determined to hold on to the memory of what he'd done. Of his defiance. Of the beauty. He'd never forget that feeling.

The memory brought tears to the corners of Ray's eyes. A dangerous thought began to stir, awakening some blurred shape at the end of a long, dark path in his psyche. A threat of revelation. Somehow he knew if he lingered too long, went too far down that path he'd have to confront the thing that lurked there.

Destination, sir?

Ray's attention was snapped back to the present. To reality.

"No," he responded. "I know where I'm going."

A passerby on the street gave Ray a funny look.

"What? Never seen a man talking to his AI?" Ray said, chin jutting out at the woman. She jumped, hurried off the other way.

He looked about himself wildly, as though lost.

Get your bearings, Ray.

He bumped some *Calm*, felt it flow through his system, erasing the jitters like a flashflood erased all signs of rocks in the water. The rocks were still there, just buried below the surface of the stream.

But the *Calm* had the desired effect, which was to jolt him out of the potential downward spiral of reverie.

Pedestrians, tourists, vagrants. They all surrounded him. They packed the sidewalks and even the streets, flowing like water around the scooters and carts. Mostly vagrants by the look of them. Unwashed people in shabby clothes, ambling slowly with the gait that spoke of nowhere to go and nothing to do.

Thankfully he didn't have far to go. The mark lived in a middle to lower-class Legacy project—Grand Place—southeast of the Arc, just outside the floodgates.

He walked as quickly as the crowded streets would allow. Twenty minutes later he found himself in the shadow of Grand Place's beige and white-stoned façade.

Its shabby, blocky exterior dated back to the teens, maybe early twenties. Thirty stories high, it overlooked the empty space where the floodgates, a thick wall that could be raised and lowered at will, lay dormant in the ground, cutting a circle around the city.

Those walls'll come up soon enough, Ray thought. If this job was any indication, they were getting close.

Grand Place was a misnomer. It looked like it had originally been built to hold private condominiums, but like most spaces that had been in existence during the AI-induced jobs crash, this one looked like it'd been abandoned, then appropriated and re-purposed by Legacy.

He stopped short of the entry, counted the windows up to the eleventh floor. He could scale it if he needed to, but doubted it was necessary. Approaching the stoop, he inspected the security entry pad. A humorless laugh escaped him. Four-digit keypad. Ancient, reliable, and super simple to hack. Had to hand it to Legacy, they did all the work for him.

Ray did a quick search on his system for the manufacturer of the keypad and happily discovered he was right. Four-digit entry system. One more search for the most commonly used pins and he was in, first try, pulling back the lightweight aluminum frame of the door.

The foyer was narrow, black and white subway tiles on the walls with shabby brown carpet beneath his feet. Smelled of cleaner and the dust of age combined, lemon and the scent that precedes death. Like an old folk's home, a living sort of decay.

The door to the stairs was on the right. He took it and climbed to the eleventh floor, taking the steps two at a time. The last two flights winded him, got his heart hammering in his chest. At the eleventh-floor landing, he caught his breath, cracked the hallway door, peeked out. Not a soul.

He moved down the hall briskly but quietly, on his toes, until he arrived at the apartment door.

1127. This was it. Chest still heaving lightly, he stood before the door, inspected the frame. Simple wood, painted white about 20 years ago, nicked and beaten to shit. Silver knob and a single deadbolt. Easier to knock down than to pick. It'd make a racket but this was Legacy. No one would even pretend to give a shit. The sound would wake them; the fear would keep them in bed.

Tonight, then. He'd return in the middle of the night, grab the mark and his higher calling would become clearer. Whatever the bigger picture was, Ray knew the sooner he did his job the sooner the master plan would materialize. And he, for one, didn't want to wait.

VAL

VAL
AGE: 19
SEGMENT: LEGACY CITIZEN
POSITION: UNEMPLOYED; GOVERNMENT DOLE
RECIPIENT (STANDARD UNIVERSAL CREDIT)
SOCIAL SCORE: 478

VAL AWOKE with a start, eyes snapping open. An explosion. Close. Not like when you heard some sidewalk bomb going off a mile away. This had been immediate. Threatening.

She pushed the fear away, sat up in the absolute black, switched her system to IR. Her breath caught in her chest. The door to their one-bedroom apartment was an open black rectangle that framed the tall red silhouette of a man.

The slightest rotation of his head told Val he was looking straight at her.

"Val?"

Kat, from the room, sounding groggy and confused. The man's head turned toward the sound of her sister's voice.

No.

Val rolled off the mat to a crouch and sprang for the little galley kitchen where they kept their knife. Printed from cheap polymers but still pointed, sharp. In her periphery she saw a red blur. He was coming after her.

Heart pounding, Val ripped open the drawer, utensils banging around.

The IR was useless now. Nothing in the drawer gave off heat. She groped sightlessly for the knife. A printed composite trinket, but it served its purpose. Her fingers found the edge, the blade slicing neatly into two of them.

She cried out in pain, fumbling around until she managed to grip the handle. The intruder's footfalls reverberated through the floor and into her body. He was right behind her now.

"No you don't," the man said. He sounded almost amused.

Sorry, asshole, you're not getting some Complex cupcake, Val thought, anger flaring. She held the knife at her hip as the stranger gripped her shoulder, spun her around.

She went with the spin and jabbed the knife forward, drove it into the man's midsection, felt it bite into something soft and hard all at the same time. Resistance. She saw a red blur and heard someone cry out. Her vision exploded, a supernova of white.

A thin strip of blue light bled in through the blinds, filtered through Val's eyelids, bringing her slowly to consciousness.

She lay on the kitchen floor, contorted with her head askew up against a cabinet, feet pressed up against the opposite wall's floorboards. The carpet and her shirt were both stained deep red.

Her head ached massively. After a moment's confusion she remembered why.

Kat.

Val pushed herself up. The pressure re-opened the narrow wounds that sliced neatly into two of her fingers, sending an electric jolt of agony up her arm.

"Kat?" she said weakly.

She tottered toward Kat's room.

"Kat?" Stronger this time. Her voice was suffused with a feeble sort of hope. "*Kat!*" Again, this time shouting as though she could will her little sister to be there. Home, safe and sound.

No response. She rounded the doorway to Kat's room. Empty. White sheets ruffled. Val could envision the tiny war that took place there, for a half a second, before the man had knocked her out too.

Panic gripped Val, tore at her insides. Her throat swelled up, face streaming tears. She ran from the room, jumped the apartment door which still lay flat and fragmented on their apartment floor. She fled the apartment, down the hall, screaming Kat's name, pounding on doors and walls as she went.

Down the stairs, not processing, just going. Out the front of the building, screaming her sister's name. She burst onto the sidewalk, fingertips dripping blood onto white concrete, yelling and looking around wildly but not really seeing anything.

The crowd parted around her, some staring, most ignoring.

None stopped. None offered to help. After all, this was Legacy.

INTERLUDE

JAIM

JAIM
AGE: 35
SEGMENT: LEGACY CITIZEN
POSITION: LEGACY ADMINISTRATOR: MIDWEST
ARCOLITH; DISTRICTS 35-134
SOCIAL SCORE: 880

J AIM DEL POTRO STARED at the picture. Aside from the placard stating his name and position, it was the lone adornment on his desk. It held a middle-aged man with a long, reddish mane. He had a smile like the devil was in his eyes, wore a blue civil-war era uniform and brandished a firearm that looked so antiquated and bulky it was likely more useful as a club. His teeth were cramped and angled, yellow, sharp like a predator's.

Jaim picked up the frame, brought it closer. "Miss you, grandpa."

Jaim came from a long line of administrators. Starting with the man in the picture, his great grandfather, the revered Bennet Halsey-Hohman. A Senator in what was then a patch of land bounded by irregular lines and geographic features known as the state of Missouri. Those territorial lines were long gone now, blurred by decades of budget-cuts, culture-fragmenting tech, federal over-reach, its ensuing blowback and, ultimately, malaise.

It was this last element that Jaim felt to be the most nefarious. The most deeply rooted cause of the vast majority of his current woes. Nobody seemed to give a shit anymore.

Sometimes that list included himself, if he was being honest.

Seated behind his desk, a plain white affair less than a quarter inch thick, Jaim put the picture in a drawer, sighed. Tried to enjoy the last few moments of silence. Tried, but beneath it lurked the perpetual growing dread. Soon enough, the door chime would ring.

He took his hand out of the bag of potato chips from where it rested in a desk drawer. Inspected it. Covered in grease and little flecks of potato and salt. He rubbed his soiled fingers on the side of his khaki button-up, inspected his hand again. Clean as a whistle. Whatever that meant. Nice thing about smart fabric was he never had to change shirts. A small blessing on a tight budget.

He rubbed his hands together. Now that breakfast was out of the way, he was eager to get to work. He stood, took a few steps through the opening that served as his office doorway and looked out the clear composite window that fronted the entire building space. Top floor, middle sector of a Rubik's Mall. It overlooked the central plaza that served as a terminus at the transfer gate between Levels 2 and 3. Prime real estate was about the only thing Legacy had to offer.

"Can't even buy me a decent assistant," he grumbled. View was nice though.

The middle offices on each face of the Rubik's Mall extended to the interior center of the cube. At ten feet wide by fifteen deep, these were the priciest of spaces. Such an extravagance wasn't lost on Jaim; he took his position and his duties to the people of Legacy districts 35 through 134 seriously.

He pushed through the clear door, struck a domineering stance atop the metal catwalk. Hands on hips, sucking in the morning air. It was as clean as it ever got, with hints of burned coffee mixed with some sort of sweet bread.

Peaceful this time of day. Not much foot traffic yet. The cavernous maw that led down from Level three to Level two was quiet, empty save a few stragglers doggedly making their way up the steps. Workers. Cleaners. Servants. They all had the same look. Neatly made-up but prematurely aged.

The blue light of early morning painted the buildings below a somber hue. The park, with its jungle gym, sandbox and giant fountain festooned with gods riding horses out of artificial waves, was encircled by a faux dirt running track that wove in and out of stands of ornamental trees. Real trees, not the fake carbon and polymer ones like they had on 1 and 2. Not one child at the park, which was a blessing. No yelling, screaming, nothing.

The telltale clang of Kara's bright red heels reverberated through the catwalk, an alarm that reached into his brain and pulled him from his daydream.

Jaim retreated back through the entrance and beyond the separating wall into his office and plopped down in his chair.

He hastily opened his favorite feed. Turned the volume up in an attempt to drown out the door chime and attendant attention-seeking sounds his assistant made as she settled in for the day. He pulled out the bottom drawer and rooted around in the bag of chips, plucking the odd wafer and popping it absently in his mouth.

He had been at this for some time when his system dinged at him, reminding him of his 9 o'clock appointment.

Talksmall.

His lip curled involuntarily. A new rep meant work, which in turn meant his relaxing morning was overshadowed by the specter of actual effort.

He turned the volume down and minimized the feed to a small corner of his view. "Kara!" he shouted around the corner.

"*What?*" Her voice was ugly, cruel. He scowled. No respect.

"What's the new rep for Talksmall—what's his name?"

"Didn't you take it down?"

"What do I pay you for?"

"Hate to remind you, but *you* don't pay me. That would be the Legacy government. Or more accurately, the lovely citizens of one thirty through..."

"Oh my god, sorry I asked," he muttered under his breath. Legacy. Democracy's successor. Localized fiefdoms ruled by Admins such as himself. But Legacy's financial admins held the purse strings, meaning they wielded the true power in any given Legacy district. And instead of ponying up for bona fide AI assistants, Jaim was forced to deal with Kara, the worthless meat version.

"They're the people I answer to," Kara continued, "and if you had any brains..."

He tuned her out. It was no use. Instead of asking again, he looked up the Talksmall rep's info himself.

JAYNE MOON

Jaim grunted as he surveyed the man's file. Typical rep. The gangs like Talksmall were savvier these days, and of course they didn't style themselves as gangs but political parties, so their reps, which were basically ambassadors, looked the part. Slick. Like politicians. Moon fit the description perfectly. A youngish man with a fake smile, manicured good looks. Straight black hair and sand-colored skin. Hints of Asian heritage, several generations past, all mixed up with the Caucasian mélange.

Jaim's eyes roved the dossier, and another small chuff escaped his lips, the sound of a small dog making a half-hearted attempt at a territorial claim.

Kind of a ritzy education for the Midwest Arc Chapter of Talksmall, he thought. Sure, the Midwest was probably Talksmall's biggest and best-extended network of gang members and affiliates. Still, he wasn't sure the position warranted a Yale degree.

'Course, that was their play. They didn't consider themselves gangs anymore, and if some unlucky sap used the word in their presence they'd get all worked up like you'd insinuated something downright medieval about their mothers. Parties, was the word. They wanted to change their image, but that didn't change their past. Easy for people to forget their organizations were born in blood.

But the fact of the matter was that Legacy no longer had any sort of infrastructure of its own. Just a few Admins who decided how and where to use the few paltry tax dollars they still received. With no police forces or social programs, all Legacy services had to be outsourced to other organizations. And on the basement levels of an Arc, that usually meant the parties.

"Boss?" Kara said, interrupting his train of thought.

"What *is* it?" he asked testily.

"Your nine o' clock is here," she replied in a sedate voice. He could hear the silky tone of victory in her voice. Triumph in making him look like an overstressed buffoon in front of the rep. He pushed his irritation down. His retort for Kara would have to wait.

"Just a moment," he said as calmly as he could, stalling.

Jaim stuffed the chips down into the drawer and shoved it shut. He straightened the placard on his desk, took a swig of water, swished it around his teeth, trying to dislodge any remaining bits of food from last night's second dinner and this morning's chip breakfast. Too late to brush. He popped a Dentos Cleany Mint ™ instead.

He checked his system one last time, saw the time was 8:56. Fucker was early. He could wait.

Jaim rocked in his chair, sucking at the cleany mint so that it foamed up quickly, rolling it around his mouth for a full minute before crunching it between his molars. He wiped his hands on the sides of his shirt again, noticing with dismay that it hadn't quite eaten the stains from earlier in the morning. Oh well. He rolled himself closer to the desk, obscuring the stains.

He closed the window to the feed he'd been watching earlier. It was a guilty pleasure that, for most meetings, he could keep one eye on while going about his business. This meeting wasn't like most. He couldn't afford distractions. The start of a diplomatic relationship was fraught with pitfalls and other barely-seen perils; it was by necessity a careful process where the turf had to be painstakingly navigated, tip-toed around utilizing hints, intimations and niceties until at last all the territory, social positions and political leanings were exhaustively mapped out.

Jaim swiped his front teeth with his tongue one last time, cleared his throat. He blew into cupped hands, smelled his own breath. He smiled. Fruity. "Send him in, please," he said in a low, firm voice.

The new rep for Talk Small strode in through the opening, looking every bit like his headshot. More handsome, more winning, if that was possible.

He wore a simple matte silver suit, crisp white shirt open at the neck, no tie. Jaim forgot the stains and immediately stood to greet the man. He stepped around the desk and returned the man's perfect smile with something that felt like a cringe, hoping silently there weren't any bits of food stuck in the spaces between his cramped teeth.

"Mr. Del Potro, so nice to finally meet you," Jayne Moon said, extending his hand.

God, he looks like a mannequin. Every atom perfectly in its place.

Jaim took the hand, shook firmly. In an era of rampant and multifarious means of disease transfer, a handshake was an unmistakable sign of trust. Especially one so freely given as this one. Jaim didn't even know this Moon but he couldn't refuse. That would've been the equivalent of a gauntleted backhand to the face.

"Mr. Moon. So glad you could make it today." Jaim gestured to the plastic chair opposite his desk.

"Thank you," Moon said. He tugged the chair out, sat and crossed his silvered legs in what Jaim considered to be the elegant, yet feminine fashion of a woman. Ball-cramping style.

Jaim managed to shake off the visual he had unwittingly put in his head, took his own seat, leaned back gently. He tried to affect an air of power in repose, visualizing a cobra waiting to strike. "How was your commute?" he asked.

"Fine, just fine," Moon responded, waving lazily at the air. "I've a friend that lives here on 3. Stayed overnight at his apartment. Sunlight district, just a fifteen-minute walk. Beautiful morning for it, too."

Jaim crossed his arms, affecting a look of appraisal. "Sunlight, eh? Nice neighborhood. I'm over in Pelco Park, ever hear of it?"

Moon shook his head politely. "Can't say I have, but I've only been here a month. Came from New Haven—a village by comparison. Still getting acquainted with..." he made an expansive gesture with his arms as if to encompass the world.

Jaim chuckled knowingly. "It can be a struggle at first. Totally understand." Jaim eased back further in his chair. "So let's get down to brass tacks. What are we looking to accomplish today?"

Moon switched his legs around, steepled his hands in prayer. "First of all, let me express my gratitude, Mr. Del Potro, for receiving me this morning. My predecessor at TS simply gushed about you. I'm..." Moon glanced down then back up, face flashing adulation then gratitude so quickly Jaim didn't process it until he started speaking again. "I'm just so fortunate to have the opportunity to work with a man of your stature and reputation."

Moon was laying it on thick but Jaim didn't care. It felt *good*. Jaim held his tongue, both basking in the afterglow of Moon's praise and silently inviting more.

"The greatest concern we have at TS right now, as you could probably guess from the timing of the meeting, is securing the Safe Passage Route contracts."

"Mm-hmm."

"Now, we are requesting a slight modification," he said almost apologetically. "We'd like to expand the route to cover Strawberry Hill to Paseo."

Jaim grunted as though punched. "That's a pretty significant expansion. You looking to push out Starboard Society?"

Starboard Society was the ultra-right-wing party that competed for territory with Talksmall on Levels 1 and 2. They were known as *Kaisers*, and not very affectionately. Those who belonged to Talksmall sometimes referred to themselves as *Teddies*.

Moon looked confused. "The SS are gone. Retreated from their turf on 1 and moved all their ops up to 2. Didn't you know?"

Jaim knew their enrollment numbers had dwindled vastly. It was hard to continue to sell the idea of bootstrapping one's self to financial glory when the resources to even make those boots were so tightly bound under lock and key. Still, they should've at least informed Jaim of their intent to retreat. There would be contracts to re-write, territory to re-assign...

Shock, then anger played over Jaim's face as he considered the ramifications.

Hapf, that bitch. The Starboard Society rep knew better. There'd be hell to pay, Jaim would make sure of that. All turf changes were routed through *his* office. It was what kept the chaos in check. She *knew* that. She'd be lucky if he didn't boot her skinny ass and all of her Starboard cohorts to some shit-stained addict-riddled corner of 1.

Let 'em guard some Legacy senility center, see how they like that.

Jaim resisted the urge to reach into the drawer and draw out a handful of chips. He smoothed the sleeves on his shirt, held his chin high and out, regained his composure. "Without protection of the main routes that run along there, those stretches'll descend into chaos in a matter of weeks."

And it was true. While parties were generally considered a necessary evil on 1 and 2, they *were* necessary. They held contracts with Legacy districts to provide the manpower to protect certain routes through the lower, lightless Levels of the Arc. Without them and their thugs—Talksmall called them *Enforcers* – Jaim had no doubt those areas would further descend into total, absolute lawlessness.

"One hundred percent, totally agree," Moon chimed in. "Like you, we saw the need, which is one of the reasons I requested our meeting be expedited. Talksmall has prepared the contract in advance. All we need is your signature and we'll deploy sentries to the routes within minutes."

Jaim sat up abruptly. "That soon? Got enough gear to equip 'em all?"

Moon smiled shyly. "We have some automated solutions, surveillance and recognition. They don't prevent but, once we get a hold of the perpetrators seems to work pretty well. But beyond that, yes, we have a hefty supply of helmets and clubs to go around. And certainly no shortage of people to put them to use."

"The ol' hats and bats, huh?"

Moon chuckled. "Our history precedes us, as always. Come check out our operation sometime," he invited. "The party's progressed quite a bit from the old days. A bit more sophisticated in our approach to crime prevention."

"More sophisticated, sure, but what's more effective than a club to the kneecaps?" Jaim asked.

"No argument here, tried and true works. But we've also found that with a few extra men and the right message we can affect a far greater change to behavioral modification than a post-crime show of force can. Take away the boundaries to resources and you take away crime. Now pair that with our stance of immediate and near-free access to local distribution nodules and we've found that these two working in concert does *worlds* of good in preventing...situations...from arising in the first place."

"Situations," Jaim snorted softly through his nostrils. "Cute."

Moon remained silent, not taking the bait. That beatific smile seemed plastered to his face, set in epoxy like everything else on him.

"Another point of note you might be interested to know is that of the new church along the proposed route."

Jaim raised an eyebrow.

"Does FLOTA sound familiar?" Moon asked.

"Can't say it does," Jaim said flatly. He was regularly inundated with permits for building renovations, free drug zones, asylum applications, new churches—the list went on and on.

"I was speaking with the headmistress of the church the other day," Moon continued, "and she mentioned off-hand they had applied for a permit to teach school-age children."

Jaim did remember now. Like most other permits, the filing hadn't met the requirements so he'd shelved it.

"I remember now," Jaim said. "First Liturgy of the Arc."

Moon's face brightened. "Exactly! I had mentioned I was meeting with you—she didn't request or pry for information but, being a gentleman, I offered to inquire as to the permit's status. But if it's too much trouble," he trailed off, averted his gaze.

Jaim frowned. "No trouble, but as I recall there's a problem with the permit. Didn't meet requirements."

"Oh? Is it curriculum-based?" Moon asked, clearly fishing for more information. Jaim sighed internally. He didn't really feel like digging through his system to find the files, but Moon was so damn polite he found it hard to object.

"No, no," Jaim replied, stalling as he sifted through the search results. He found the one labeled with the church's name and quickly saw the problem. "Huh, security issue," he said absently. "Kids need protection getting to school but the church is outside of any known routes."

A thought struck Jaim then.

If it's a security issue, I know who can solve that problem. Jaim looked across the desk at the rep. He looked green, innocent.

Easy to take advantage of.

"Seems we can kill two birds here," Jaim said, pushing his lower lip up and looking down the bridge of his nose at Moon.

"Whatever it is, I'd be happy to help if it's within my purview," Moon replied.

"Well, the church needs security and you need a contract," Jaim said with a didactic air. "Your *proposed* route comes close to the new church but doesn't quite intersect it."

Moon leaned forward. "But we could easily extend that route. Talksmall would be happy to secure safe passage along the extended route."

Jaim squinted, scratched behind an ear. Waited.

"At no additional charge to your office, of course," Moon added.

Jaim clapped his hands. "Perfect! Amend the contract and send it to me. Oh, and one more thing. I'll approve your contract but I'm only giving you six months. If Talksmall can maintain order along the expanded route for that period of time I'll extend you out another year."

Moon clasped his hands in front of him, bowed his head. "Thank you so much, Mr. Del Potro, we will not disappoint. I have the contract right here," he said, staring at something Jaim couldn't see. "And I've amended it to include the new church's route and cut the contract term to six months with a year-long option at the end."

Jaim's system dinged, the file showing up as a grayed-out icon. He opened it, flicked his way past all the fine print to the signature page, signed it and sent it back.

"And for the headmistress's peace of mind, what should I tell her about her permits?" Moon asked.

"Nothing," Jaim said flatly. "I signed and sent them just now while you were working on your amendments."

Moon looked thoughtful. "She'll be so relieved." He stood, extended his silver-sleeved arm across the desk. "Thank you so much. You truly are a rarity. A dying breed, if I may say so."

Jaim stood and took his hand, throwing his belly out and his shoulders back, affecting an air of import. "It's been a pleasure, Mr. Moon."

"Likewise," Moon said, tugging at the bottom of his jacket. As he turned to leave, he hesitated, turned back. There was a questing sort of trepidation in him. "May I ask you a personal question?" he asked.

Jaim generally didn't go in for anything personal, especially when dealing with parties, but there was something about the man that screamed discretion.

"Go ahead," Jaim said. "If it's too personal I'll just decline to answer. Fair?"

Moon brightened. "Fair enough. I'm obviously new to this," he said, ducking his head in a self-effacing gesture, "but I'm curious—are you happy, Mr. Del Potro?"

Jaim reeled. *What the fuck kinda question is that?*

Moon continued, abruptly, hands out, explaining. "What I mean is are you happy with your current state of government?"

"Oh," Jaim said. "Explain."

"Well, it's clear to me that a man of power such as yourself is in that position of power for a reason. The Admins of Legacy are the last true bulwark against the encroaching forces of the Complexes. But..." Moon looked down and away.

"But what?" Jaim asked. It was true, he was a man of power. It was also true he was dissatisfied with the current state of Legacy. His district was underfunded and losing more and more people to the Complexes every day. Once a Legacy citizen joined a Complex, they stopped paying taxes altogether. Legacy, and his district in particular since it included parts of Levels 1 and 2 which were rife with societal leeches, was bleeding to death. One day, he'd wake up and find his office abolished and the world being run by consolidated business interests.

"But the tide is turning. Surely you've seen it. Soon enough there will no longer be such a thing as Legacy...or Admins," Moon said, body language stating he was almost ashamed to say it.

Jaim nodded, thoughtful. None of what Moon said was new. It was a tacit understanding all Admins had with their reality. Their situation, their government, was dying...but what could be done?

But they were hard truths Jaim had preferred to avoid. Now, confronted with them, he could see it was folly to continue to avert his eyes and pretend all was well. Jaim surveyed Moon once more.

Maybe you are *more than just a pretty face.*

"What do you propose?" Jaim asked.

"I have an idea," Moon said, and his eyes narrowed ever so slightly, hinting at a predator who had found its prey.

BOOK TWO

PASS:

PRACTICAL APPLICATION
OF
SOCIAL SCORING

RIKU "CHIEF" OGUNWE

RIKU "CHIEF" OGUNWE
AGE: 40
SEGMENT: COMPLEX PROPERTY; ARCSEC CPX
POSITION: ARCSEC CHIEF CONTRACTED BY LEGACY
ADMIN: MIDWEST ARCOLITH, LEVELS 1-3
SOCIAL SCORE: NA, CPX-SPONSORED

T HE SUN TOUCHED OGUNWE'S SHOULDERS, the light angling in through the buildings from the south, reflecting off the barred windows of Grand Place apartments. The air was cool but fingers of warmth pressed into the black material of his sweater. The reflective lettering on the back sparkled bright white in the sun—it read *ArcSec CPX.*

Ogunwe tapped his toes as he waited. The black boots he wore were ArcSec standard issue and he kept them polished to a high shine. It gave him a muted sense of pride. There was something gratifying about a military sort of life. The order, the respect one demanded—of themselves and of others. It made a person stand taller.

Below the soles of his boots, the steel and fiber material of the floodgates lay flush with the pavement. In the four years since his transfer to ArcSec Midwest, he'd not once seen the floodgates raised.

Thank god for small miracles, he thought. A flood was another item on the list of responsibilities he didn't need, a further complication in the world of policing.

Ogunwe looked up at the building that stood just to the south of the floodgates. Grand Place was an old, Legacy-subsidized apartment building. Eleven stories tall, its exterior, like all government-funded things, was in a state of shabby disrepair. He eyed the bars on the windows.

At least it's safe. Not much can get through those bars.

The front door to Grand Place opened, and Ogunwe tensed. A girl, a teenager it looked like, exited. He realized he was holding his breath and blew it out, tried to soothe away the butterflies in his stomach. It was always like this when he saw Olivia. Soon enough, she and her son Stef would come out, and they'd begin their monthly ritual.

To make the trek, Ogunwe thought.

Most people would avoid a hike that took them from outside the floodgates all the way up to Level 3. Like the plague. But the trek was routinely the highlight of Ogunwe's month, the bright spot in what was of late a procession of increasingly worrisome days.

She's about to come out. Best not to think of that right this moment.

On cue, Olivia emerged from the building wearing a long white dress that was as impractical as it was beautiful. She was trailed by her son, Stef, who wore jeans, a baggy sweatshirt and a stocking cap that hid his massive overgrown mop of brown hair.

Olivia spotted him immediately.

"Riku!" She hurried over, flung her arms around him. He was careful not to hug too tightly. It had been a full month since the last treatment, and she bruised easily as time passed between the infusions.

Stef walked up and Ogunwe put out a fist for him. Stef ignored it, offering a perfunctory nod before continuing up the sidewalk.

"What's up with him?" Ogunwe asked, pointing a thumb in the boy's direction.

Olivia waved it away. "Girl trouble."

Ogunwe shrugged. "Ready?" They had a long hike ahead of them and he didn't want to waste an ounce of daylight.

"As ever," she smiled at him, bouncing on her toes.

"Hey, none of that," Ogunwe said, putting his hands on her shoulders and pushing down so she couldn't bounce. "Save your energy."

"Oh stop, sourpuss. Now c'mon, let's catch up. Stef's already halfway to the shadow by now."

She took off at a trot, the white dress and her long, platinum hair flowing behind her.

Her fingers were long and slender, china white. It had only taken an hour's worth of walking, but now they trembled, resting in the crook of Ogunwe's elbow. He'd pulled back the sleeves of his sweater, making the excuse that he was hot from the long walk. The contrast between her pallor and his warm, dark coloring was beautiful in its starkness and, secretly, he loved the feel of her cool fingertips resting on his skin.

They were inside the Arc proper now, beneath the shadow. Hemmed in by the stinking masses and shops that glowed with artificial light no matter the time of day. They were close to the transfer gate, the place where pedestrians could ascend and descend from one level of the Arc to the next.

She leaned over, whispered in his ear. "Stop staring."

They'd been walking for an hour, and he might've been staring for that same amount of time. He couldn't help it. Despite her weakness, she shone.

"Really, I'm fine," she admonished.

Clearly she mistook his adoration for concern. *Just fine*, he thought. *How it has to be.*

Ogunwe managed to tear his gaze away from her, surveyed the street up ahead. Stef was in front, his head bobbing to some music only he could hear. He'd done an admirable job leading them—the giant staircase of the transfer gate was dimly visible through the constant fog and mist of Level 1.

Ogunwe patted Stef on the shoulder. He looked back, a teenager's disinterest in his eyes. Ogunwe gave him a thumbs up. Stef turned without acknowledging him.

Ogunwe grunted. "Must be serious."

"I don't know," Olivia said, frowning. "He only mentioned her like two days ago. Apparently she goes to his school but suddenly stopped going. But, you know how it is. He's thirteen. He'll get over it."

As they approached the staircase, a temporary fenced-off area was guarded by two ArcSec men. The soldiers saw Ogunwe and saluted. He saluted back, noting the locked crates of material he could see through the chain link. It was a reminder of the information only he and few privileged others possessed. Of the information he couldn't relay to Olivia without breaking his oath to ArcSec.

But there was a way around that oath.

I just need to convince her to move to 3. Then she'll be safe.

He would have to say something soon, but not now. He'd wait. Until *after* the treatment.

Wait until she's stronger, he told himself. *More in a position to see reason. You don't want her thinking you're trying to take advantage of her weakness.*

Ogunwe, Olivia and Stef veered left at the stairs, splitting from the wide stream of people trudging up the free staircase to Level 2 and headed to the wide, industrial elevator that could take them to any level. It was less crowded—since payment was required—but there, too, they passed the beginnings of construction—a half-built barrier made of clear panels. Ogunwe knew those panels would eventually form checkpoints at the transfer gates on each level, walling off access. Permitting and denying passage, according to an algorithm.

As Chief of ArcSec levels one through three, Ogunwe had already been briefed on the plan to put in more stringent border controls across the entirety of the Arc. It was a plan his superiors called PASS—the Practical Application of Social Scoring.

Which is just a proxy for wealth, he thought. It wasn't right, but his oath prevented him from speaking of classified operations to those who didn't possess clearance.

His mind drew in the missing details. The clear wall panels were a variant of the riot tech ArcSec used. They were called AWOLs and they fused at the molecular level with whatever material they were paired with. The walls would be placed around the elevator and the entrance to the staircase, with a clear box that permitted only one person. The box would read a person's ID, parse their Social Score and then deem whether or not they qualified for access to the next level. If so, they were allowed to proceed. If not, they'd be spit back out the way they came. Almost anyone could pass from 1 to 2, but the requirements for passage to Levels 3 and up were reserved for Social Scores higher than 800. The social barriers to upward mobility were about to solidify in the form of very real, very substantial blockades between all the levels of the Arc.

And once those barriers went up, he wasn't sure he'd be able to get Olivia past the blockade to the clinic on 3, clout or no.

Without her monthly infusions, she'd eventually die, and there were no clinics below Level 3 that provided the treatment she needed. Ogunwe tried to hide it from her, but the impending inevitability of PASS made him more and more nervous by the day.

Would this be the month when the barriers were finished?

He didn't know. And that feeling, of uncertainty, of powerlessness, was so antithetical to his military nature that it filled him with a sense of pervasive disquiet.

They stood in the back of the elevator. Stef had shifted around so that he was in front, staring at the doors as they creaked shut. Olivia turned to Ogunwe, smiled at him, oblivious to the fact that the world they knew was about to change. He smiled back, feeling the falseness of it. He didn't know what he'd do if he lost her.

That's not going to happen. You made a promise, Riku, and by God you're going to keep it.

The doors banged shut and the elevator groaned as it lurched upward.

VAL

VAL
AGE: 19
SEGMENT: LEGACY CITIZEN
POSITION: UNEMPLOYED; GOVERNMENT DOLE
RECIPIENT (STANDARD UNIVERSAL CREDIT)
SOCIAL SCORE: 478

THE PROCESS FOR REPORTING A CRIME was faceless. ArcSec held the contract to police all Legacy districts within the Arc, and so the process for reporting was the same for all jurisdictions.

Fill out a form on their site. Wait for a reply.

Val had filled them out for all potential districts, spamming the sites of all the offices in the hopes an actual person, someone with a heart, would reply.

What is the purported crime?

Where did the purported crime occur?

When did the purported crime occur?

Name and ID of Victim

Name and ID of Perpetrator

Name and ID of Reporter

The responses she'd received were near instantaneous, and exactly the same:

THE ADDRESS OF THE PURPORTED CRIME IS OUTSIDE OF THIS OFFICE'S JURSIDICTION

She screamed. And cried. She beat her fists against the walls of their apartment until they were red and raw. Her manic energy was incrementally worn down by each day and each subsequent form rejection, until an inchoate, desperate fugue took over.

She didn't leave the apartment for two weeks. She didn't shower, didn't change her clothes, barely ate. Drank only when she felt that overwhelming desire that arises when every cell of the body revolts and screams for water.

Lying on the carpeted floor, she stared at the ceiling for hours. She felt hollowed out.

Lying on the floor, her system beeped. The sound bled into the background, unable to penetrate her trance-like state. It beeped again. In the upper right corner of her vision, something flitted and flashed, an insect pestering. She swatted at the air.

BEEP

Val blinked slowly, feeling the raw, dryness of her eyes. How long had she been lying there? She stared at a yellow-brown patch of stained ceiling.

How long has that patch been there? she wondered.

BEEP

Again, the object flashed in the corner of her vision. She blinked again, this time looking more closely.

Not an insect, a message. On her HUD.

The message icon flashed again, beeping simultaneously.

Oh my god, she thought. *I'm an idiot.*

She sat up, accessed the message.

What floor do you live on?—Chief Riku Ogunwe, ArcSec Level 1-3

She almost couldn't comprehend it at first. The message was so terse, she had trouble figuring out who had sent the message and why.

It's a response. To one of the reports I filled out.

Her heart began to flutter. She pushed up to her feet before responding.

I live at Grand Place. Floor eleven. Why?

His reply took longer. She began to pace the living area, chewed at a jagged nail. This was the only lead she'd had in the two weeks since Kat had been taken. More importantly maybe, this was the only real, live *human* she'd talked to in that span of time.

I know people who live there, but your apartment building is outside my jurisdiction.

Her shoulders slumped. Not again.

That's what everyone says, she replied. *Or at least that's what the computers say when I fill out the forms.*

It's because you're in a dead zone. Just outside the floodgates, which is the official boundary of the Arc. We are, after all, called 'ArcSec' for a reason.

Who do I go to then? Someone took my sister. *Can't* you *help?*

Val held her breath.

I can't.

She threw her head back and let out a primal scream, full of the rage and anguish that had gone dormant inside of her. She sneered, began to compose an invective-laced tirade.

But...

She stopped mid-sentence, focused on that one word.

I can point you to someone who might be able to.

And just like that, she had a reason to keep breathing. This stranger, a man she'd never met, had thrown her a lifeline.

Thank you thank you thank you SO MUCH!!! You can't possibly understand how much this means to me. I promise I'll repay you a thousand times over!!!

His response was curt, cryptic, and as she read it a chill passed over her body, like the passing of a ghost.

Don't thank me yet...and don't turn your back.

Val stood atop the crumbling steps at Deus Ex. The higher vantage afforded her the ability to see up and down the length of the street, at least until it ran straight into the shadow of the Arc. She waited, fingers fidgeting with the baton in her pocket, and tried to ignore the nagging fear the ArcSec Chief had instilled in her—*don't turn your back.*

The old church was the safest place she could think of, but this late in the day that didn't mean much. It was dangerously close to dark, but she couldn't stand the thought of going one more day without at least feeling that she was headed in the right direction—toward Kat. Under normal circumstances, she would've waited.

Normal circumstances, she scoffed. *Whatever that means. Your sister's been kidnapped and you're resorting to dealing with criminals.* Normal circumstances had long since waved goodbye.

The air had cooled and the setting sun was no longer visible through the forest of the city. The streams of people that typically flooded the street had dwindled to a slow trickle. Val watched, from the remove of the steps, like a predator waiting to pounce, every person that passed through the church's shadow. She kept her muscles tensed, ready to fight, ready to run.

A man, thin and hunched, approached the steps. He turned his face up to her, revealing a pocked, flushed wasteland. But the eyes were bright blue, intelligent, aware.

"Val?" he asked.

She kept her distance. "That's me."

He put a palm to his chest. "Saul at your service. Your message was...rushed. What is it you're looking for?" he asked. His tone was caring, yet there was something unctuous about it.

Too concerned. Which meant it was artificial.

He took one step toward her. She wanted to back up, but held her ground.

"My sister. She's been kidnapped."

Saul *tsked*, shook his head. "What *has* the world come to?"

"Can you do it? Can you get my sister back?" Val asked. The urgency she felt came out in the rush of her words. She needed this man's help, but already she could see he couldn't be fully trusted.

Saul frowned. "Get. Now that's an interesting word."

"I was told you could help," Val said.

"Ah, well, see there's that little roadblock of a word. Oh I can help, that's the honest truth," he put up two fingers in a sign of honesty, "but help means different things to different people."

Val couldn't contain her frustration. "Listen, I don't know who you are, but would you please quit talking in circles and just tell me what you mean? My sister's already been gone *two weeks*."

"I can appreciate that, Val, I can indeed," the man said. "In plainspeak, I'm the guy that finds the clues, the mouse that picks up the crumbs of cheese, the..."

"The informant," Val interrupted. "*Not* the guy that saves the girl."

He shook his head theatrically. "Hero I ain't."

"Okay," she breathed out, "so what's the deal? What do you need to find her?"

"Need?"

"As in payment. How much do I owe you?"

Val knew he wouldn't accept the Standard Universal Credit as form of payment. It wasn't useful enough, and those selling goods and services on the black P2P market rarely did. She expected him to name some type of expensive crypto.

He took another step toward her. "You ever sell?" he asked.

She frowned. This time she did back up, scratching at her head to cover her nervousness. "Sell what? Drugs?"

He let out a quiet, patronizing laugh. "You're a sweetheart, y'know that?" Then his expression went dead as he leaned in closer. "Your skin. You ever sell your skin."

Val controlled the revulsion, choked back the bile that wanted to rise in her throat. Before she could speak, Saul put a hand up.

"Let me stop you right there. Not me—you're not selling to me. No ma'am, I'm no Romeo...but I do have clients with a...sweet tooth, y'might say." His eyes roamed her body, up and down, then back again. "You'd fill that bill nicely."

"No," she said. "I haven't." The revulsion inside her warred with the resolve of doing what needed to be done.

"Well?" he asked, letting the question linger. Val looked up the road, toward the Arc's black maw that swallowed it whole...along with everything else stupid enough to travel that way.

What's more important? Your dignity...or your sister's life? She didn't even have to answer the question. She'd give anything to protect Kat.

"Look, I'll make it easy on ya," Saul said. "Meatspace or virtual? In meatspace you could probably pay your dues with just a day's work. Not bad, huh?" He Arched both eyebrows as though it was an enticing offer.

Val shook her head curtly. She hadn't even had sex with someone she loved—she didn't plan on losing her virginity to a stranger, not if she could help it.

"Virtual it is, then. It'll take longer. As you might imagine, it's a cheaper service. Could take a full week to pay back the debt. You sure about this?"

"I'm sure. Just tell me where to be, what to do and I'll do it."

"My type of gal," Saul said, wringing his hands greedily. "Fun starts tomorrow. Beneath the shadow, corner of 16TH and Cherry."

He started to leave, but Val reached out and grabbed the back of his arm. "Hey, wait! Shouldn't I tell you more about my sister? So you can, y'know, investigate?"

"Wherever your sister is," Saul said, "she ain't leaving between tonight and tomorrow. We'll talk about it then."

"Fine," Val said. She didn't like the way this had gone. They hadn't even discussed how he'd get the information.

He didn't even ask for her name, she thought, a sneer of disgust curling her lip.

She stepped forward. Saul wasn't a tall man, and atop the steps she towered over him.

"Don't you dare fuck me over. I know people." The threat rang hollow through her fear, but Val meant it. She'd do *whatever* was necessary to get Kat back.

Saul snickered. "Doesn't everybody?" He backed down the steps, then headed the way he came, into the deepening, expanding black of the Arc. He stopped a little way down the road, turned and called out. "Oh and I forgot, the fun starts at five o' clock tomorrow. And that's PM, sweetie."

Another chill passed through her body. Entering the Arc was terrifying enough—but at night?

Don't think, just do, she told herself. *You've got a sister to save.*

TREVOR

TREVOR
AGE: 18
SEGMENT: LEGACY CITIZEN
POSITION: UNKNOWN
SOCIAL SCORE: 821; FLAGGED FOR VERACITY

T HEY MET at the end of Walnut, where the narrow street ducked
in-between two rows of derelict buildings.

The road terminated at the edge of a railyard, petering out where the
buildings stopped. The concrete surface cracked, then eventually crumbled into a slope that toppled into the gravel-lined depression.

Overlooking the railyard, a host of outsize guards surveyed the tracks
below. They wore the uniforms of Talksmall Enforcers—blue sweatshirts
with *TS* printed in white, menacing-looking black helmets and long,
heavy batons that were nearly as thick as baseball bats. Trevor nodded to
them. "Gentlemen," he said brightly. They stared back through the masks,
gazes flat and uninterested.

Expect no less from the thugs, he thought. *God love 'em, but he made 'em
without a sense of humor.*

Trevor looked across the railyard. Through the darkness, he could
barely make out the other side of Walnut climbing back out of the railyard, a gray line cutting through the darkness. There was a reason he'd
chosen this place for his little...experiment. Walnut was a distribution
route, meaning it was protected by the Talksmall Enforcers day and night.

Gonna need protection for a trip this *deep.*

A village of polymer shanties crowded both sides of the road. This time
of night, solar lamps lit up the shanties from the inside, glowing through
the opaque, multi-colored sheets of thick plastic. It lent the place a festive
look. Narrow, winding paths cut in-between the hovels, creating black
passages into the depths of the shanty village.

Trevor's group gathered there at the edge of the concrete spill. He counted them again, pointing to each one as he did so.

"Forty-six. Good enough," he said, before raising one arm up along with his voice. "Alright people," he shouted, "who's ready to meet god?"

A chorus of cheers went up.

"Alright! C'mon then, follow me!"

He led them down the slope with his arms out, a messiah leading them to the promise land.

They settled in a shallow divot between two sets of tracks, nestled against the giant pillars that supported the towering elevated rail lines above. The shanties stopped at each side of the pillars, allowing the group enough space to sit undisturbed by curious hobos.

Talksmall *goons'll see to that too*, Trevor thought, looking back over his shoulder at the guards atop the rise that overlooked the railyard. The *TS* on their uniforms was reflective, along with two stripes that ran down both sides of the helmet, made to look like warpaint on cheekbones, and two shorter stripes above the visor, angled downward to look like angry eyes. This was all he could see of them. It had a menacing effect. Which was exactly how they liked it.

The next several minutes were filled with the boring minutiae of preparing for a mass ecstatic experience.

Distribute the drug—check. Everyone now had a dose of Tango in their possession—the distinctive, red one-centimeter cube with roses printed on the sides.

Join the ad-hoc network Trevor had set up—check. The last three stragglers had joined up, so all fifty participants were now visible on the network. Trevor had administrative control over both their sensory inputs and outputs. He'd built a program that was modeled after a soundman's mixing board, with each person receiving their own set of input/output sliders.

One last item, he thought, his gaze going once more to the Talksmall Enforcers in the distance. They were still present, still had their eyes on his little club, and that was just how Trevor liked it. Once the festivities began, Trevor and the rest of the group would be good for pretty much nothing until the Tango wore off.

The checklist complete, Trevor sat down in the middle of the circle.

"Everyone ready?" he asked.

The group hollered out their eager approval as one, a mixture of *yes, let's do it* and *fuck yeah* echoing off the pillars.

"You people are just as crazy as me. I love it. Now," he said, holding out his cube of Tango. "On the count of three, everyone pop and then I'll bring you on one by one. I'll go last. Oh, and remember, suck, don't chew." The members held out their cubes in front of them, pinched between forefinger and thumb. "One, two, three."

As one, the group popped the cubes into their mouths. As promised, Trevor brought each participant into the mass sensorium slowly, one by one, so as not to overload the group. As each person joined the circle, giving control of their system over to Trevor, they began to feel what the others felt. Had Trevor wanted to, he could've allowed them to also see and hear what each individual saw and heard, but it was too disorienting. The mind wasn't made for a multiplicity of visual inputs—it had its work cut out simply processing two.

A murmur of *oohs* and *ahhs* emanated from several people as they began to feel what their neighbors felt, and vice versa.

Shit, this is gonna be good, Trevor thought, and popped his cube.

He'd gotten the sick sticks—a distant relative of the cigarette—at *the Black Heart*, and as he sucked on one, the ember flaring at a fire-blackened tip, a thought flickered through his drug-addled, pleasure-loaded brain. He tried to capture it, a bird escaping a cage, actually made the motion with his hands as though he could see the thought trying to flee. He caught it out of the air, brought his palms in close and peeked inside. There it was, glowing bright: *Nothing* had ever felt as exhilarating as this.

But words were prisons and what he felt defied their bounds. No words could possibly describe the way he felt, because it had never been felt before.

He felt the sensory inputs of the forty-nine others in the circle, and they felt his. His mind could barely parse it. He was on fire, yet bathing in ice. He had a hundred arms, and a thousand points of light tickled his brain.

I am Vishnu, Shiva, every Hindu deity.

He was a god. All-seeing, all-feeling.

Omniscient.

His hands were balanced delicately on his knees. He let the stick fall from his fingers to the gravel, threw his head back and let loose a soft moan of pleasure. It vibrated in his throat and was echoed in a feedback loop from several other members in the circle as the output of his senses was transmitted to their input.

How long have we been under? he wondered. *Five? Ten minutes? Ten hours?*

Floating in a sea of pleasure, it was difficult to even grasp the concept of time. There *was* no time. Existence was comprised solely of light, the diffusion of heat and the slow erosion of self until you were once again unified with the universe. A part of God.

Trevor shut his eyelids, let the blackness envelop him. On the canvas behind his eyes, his mind spun constellations of light, whole galaxies swirling and mixing together. Points of orange and yellow swept into lines of purple and blue, veiled in sheets of fiery red and pink to cohere into a greater whole. Trevor felt something tug at his chest, lifting him into the body of light.

This is it, he thought with awe. *This is what we've been looking for.*

His spirit rose slowly, as a submerged body full of air, through the black medium. Every sensory input was in itself a point of light that buoyed his spirit. He felt like he was flying, straining slowly upward.

Straight toward the mind of God.

He reached out in his mind. He could feel the energy of the light envelop him, like cool mist on a hot day. Like a warm blanket on a cold winter's morning.

So close, he thought, when something happened. His spirit lurched.

No, he protested. He wanted to stay there forever, to ascend further into the mind of god.

But instead his spirit gained mass, fused once again with his corporeal reality, and fell away from the light.

As one, the entire group reached out to the sky, a gasp and a *No* on their lips.

They plummeted down from the heavens, into their bodies, which were overloaded with contradictory signals—pain and a weird, primal pleasure.

Trevor's eyes snapped wide open. It happened on the far side of the circle. One of the members had sunk their teeth into the arm of the person next to them, tearing into the flesh.

The group felt the agony of the person bitten, the sensory information pulsing outward to all forty-nine other members, and yet they also sensed the malicious joy of the perpetrator.

The conflicting messages of pleasure and pain, amplified through all fifty members, proved too great a burden to bear. It threatened to overload his mind. One thought managed to cut through.

RUN

Trevor accessed the group controls and quit steering.

It was like he'd been hit with a sledgehammer that dulled the senses. A stupor swallowed him. He looked around, dumbly witnessing chaos unfold through a fog of incomprehension.

What is going on? It looks like a...a stampede.

A stampede could only mean...

Kopa.

Someone had brought Kopa to his circle. The mindless, frenzy drug.

The horror of that one word cut through the fog in his mind, adrenaline temporarily erasing the effect of the Tango. He had to do something. Kopa in itself could be dangerous, but combining Kopa with Tango would turn fifty individuals seeking a divine experience into a fifty-headed monster with one purpose—destroy.

Already, the circle was turning into a melee. People screamed as the Kopa infection was passed along with each consecutive bite.

You have to turn their inputs off, he thought. *Then they'll have a chance to at least escape the stampede.*

Trevor scrambled backward, kicking at the gravel, until he was out of the circle, back up against a pillar. He accessed his system, took control of the group again.

A nuclear explosion of pain and fear blindsided him. It was like the worst anxiety attack multiplied by fifty and it threatened to overtake him.

Shit shit shit I should've turned my input off first. Stupid stupid stupid.

Like the others in the circle, he was paralyzed by overwhelming terror. Creeping in, though, at the edge of his mind, was the aimless rage spawned by the Kopa. Once that feeling got to critical mass, they were all lost. The stampede would begin in earnest, and the group that was once reaching toward a higher consciousness would find itself without one, a primitive band of animals looking to infect more people with the bloodborne drug.

He could barely breathe, much less move. The rage was growing inside of him, clouding thought.

Do....something.

He closed his eyes.

That's a start. Focus on the module, ignore the fear.

Through the hum of activity that wanted to commandeer his brain, he could see the control module overlaid on his vision.

There. Hit the sliders.

He heard a snarl of rage and opened his eyes just in time to see one of the infected, a teenage boy no older than Trevor, pounce on the girl that had been to his left. She screamed as he scratched and bit at whatever flesh was visible.

Oh fuck just do it Trevor. DO IT!

He snapped his eyes shut, swatted the sliders to zero.

In an instant, the all-consuming terror vanished, like it had never existed. He blinked his eyes open, gasped. It felt like a thousand-pound boulder had been lifted off his chest.

Trevor got to his knees, began to push up when something hit his forehead, hard, knocking him backward onto the gravel, the back of his head slamming into the pillar. Dazed, he looked up to see a devil standing over him. It was the teenage boy, who was now backed by the girl who just moments ago had tried to fend him off.

Damn. This does not *look good.*

Teeth bared, the boy leaned over Trevor, aiming for the throat.

Trevor found he couldn't move. It was too surreal. He was going to just lie there and witness his own death, like someone passively watching a movie.

A flash from above, too quick to be seen, and the boy slumped on top of Trevor. He pushed the boy off with a grunt just in time to see a small army of Talksmall Enforcers charge fearlessly into the fray, like cavalry into a mass of foot soldiers.

Relief flooded him. *Bless you, you savage beasts*, he thought.

The angular stripes on their helmets glowed with menace as their batons flashed, cracking heads and limbs indiscriminately, felling bodies left and right.

Trevor got back to his knees when icy fingers gripped his ear, pulled him to his feet.

"Ow ow ow! Stop," he pleaded. He stood to find a familiar face with an equally familiar, stern expression. Like a disappointed mother.

Figures she'd find out.

Trevor flashed a smile. "Why is it that..."

"Out. NOW!" the woman commanded, pointing back up the ramp that led out of the railyard.

"Okay, okay," he said, hands up in submission. She marched him up the ramp, away from the fray.

As he passed the low wall that separated the ramp from Walnut, Trevor glanced backward. With a heavy heart, he watched as the Enforcers continued to swing their batons, like reapers with scythes.

"What'll happen to 'em?" Trevor asked. They sat on the patio of Headspace, the arguing men speaking their foreign tongue nowhere to be found, which was the same for the rest of the patio. Trevor and Estelle had it to themselves. Oddly, though, the crowds on the other side of the fence seemed heavier.

Dirtier, Trevor thought. *If that's possible here on 1.*

"To who?" Estelle sat across from him, regarded him with that stern, motherly expression.

A hybrid mix of Hispanic, African and Caucasian, she had caramel skin with shoulder-length, kinky black hair. He hadn't cared for the nose ring when he was just a kid, but now that he was older he thought it was kind of sexy.

But never mind that, she was more like an older sister. Or, in this instance, an angry mother. Besides, she was about a decade older than him and in a whole different social set.

"The group? *My* group? The mass ecstatic experience?" Trevor dabbed at the gash on his forehead with a scrap of torn cloth. The bleeding had stopped, which was good. The last thing he needed was sewage runoff getting in an open wound. "How I got this?" He held out the bloody cloth.

Estelle's face held no remorse. In fact, it hardened further.

"Y'know," Trevor said, "if you make that face for too long it'll stay that way forever."

"Shut. Up," she scolded, pointing an accusatory finger at him. "You endangered people's *lives* tonight. Are you aware of that, or is it all just a joke to you?"

Trevor's mind went to that last glance as the Enforcers did what they were paid to do. Swinging their bats indiscriminately, knocking heads, putting chins to the ground. What if Estelle hadn't been there to drag him away? He shuddered.

"You're damn lucky, you know that?" she asked. He didn't feel very lucky, but he wasn't going to tell her that. Not now, especially. "One of the Enforcers stationed there recognized you. He thought you were related to me..."

"Close enough," Trevor said.

"So he sent me a message..."

"How thoughtful, maybe I should pick him up some flowers, or some ch..." Trevor started but snapped his teeth shut when he saw how Estelle was looking at him.

Maybe I should *take this more seriously,* he thought. *Or at least appear to.*

"It's time for you to grow up, Trevor. You're not the only person in this goddamn world, so stop acting like it." Estelle slammed her palm down on the table, then pushed up out of the chair. "I'm done babysitting. Fix your own mistakes." She walked away from the table, not looking back.

"Hey, wait." Trevor reached out. "I'm sorry. Really, I am." But she continued to walk, hair quivering as her body shook with anger, and disappeared through the patio doors.

She left, he thought in disbelief. *She actually left.*

"Great. Awesome job there, buddy," he said, then groaned as the wound on his forehead throbbed, a not-so-subtle reminder of the evening's exploits. "Just super. What else can you fuck up tonight?"

He draped his arms across the table and rested his head to the side so he was turned toward the fence. People shuffled up and down the street. Watching them, again something struck him as odd.

What time is it? He checked his system. 2:42 AM. The streets were awfully crowded for this hour in the morning. More than that, though, the kind of people were different.

Looks like they emptied out Signal Hill, dumped the homeless right into Level 1.

Through the moving wretchedness of the crowd, something caught his eyes from the far side of the street. A light, covered up by the shifting mass of people, shone in brief flashes.

He sat up.

What is that?

He looked closer, stifling his irritation at the hundreds of people who blocked his view. He stood, the better vantage allowing for a brief glimpse of blond hair, lighted from the sign above.

Not what – who is that?

The girl stood on the street corner across from Headspace, bathed in a rainbow of electric light, skin glistening and blond hair matted from the perpetual mist that rained down on Level 1. All the effluence from the top levels, constantly streaming down onto their shoulders and heads.

She was new to this game, this space, he could tell by the way she stood. Stiff, upright, almost a challenge in her posture. She was pretty, no doubt, in an athletic, statuesque way. Shoulders wide, small waist, a strength to her arms and the set of her jaw that told people to fuck off before they even got close enough for her to say it to their face.

And she would. He could tell. She wouldn't have any issues telling any one of the multiple pederasts that roamed Level 1 to step the fuck back before she twisted his balls and popped 'em right off. But it didn't make sense. She stood on the corner, not moving, looking out over the roaming crowd expectantly. Waiting for something.

Not something, you dolt. Someone. She's selling her skin.

Typically, Trevor wouldn't stop to wonder or worry about strangers on the street, not here in the shadow or out in the sun. He had plenty of problems of his own, like how to help provide for a kid sister for starters. No matter that she no longer lived with him. Probably didn't even remember him anymore.

Yet there was something about the girl on the corner. Not just in the way she stood, but her energy. At the risk of sounding like some neo-hippy psychic, it was like there was a halo around her, a field that left her untouched by the filthy surroundings. An island of light pushing against the dark.

So what's she doing here, *beneath the shadow, making virtual sexy time?*

Trevor pushed up and out of his chair. He exited through Headspace, stepped off the street corner and headed toward the girl.

Time to do something stupid, he thought. *Estelle would definitely not approve.*

He got halfway across the street, winding through the masses of people, before some sleaze dressed in appallingly tight stretch pants—thank god they were black—sidled up to the girl and started talking. Trevor hesitated there in the middle of the road, stopped. She'd be pissed if he scared away a potentially lucrative mark.

He stared and found he couldn't look away. Maybe it was the lingering effects of the Tango in his system, but he was mesmerized. She pulled at him, a magnetic opposite. He imagined their first exchange, saw the two of them staring into one another's eyes, adoration apparent in every cell, as he...

Something bumped him from behind, hit his shoulder roughly, hard enough to spin him halfway around. He blinked twice, shook the daydream from his thoughts.

Shit, man, you'll get stabbed and taken standing around like that. Gotta keep moving down here in the sotano.

He turned and started walking. Nowhere in particular, but that was fine. He looked back over his shoulder at the girl with a sense of growing optimism and resolve. Slowly, a lop-sided grin formed on his face.

Next time.

MANOLO

MANOLO
AGE: 57
SEGMENT: COMPLEX PROPERTY; NOVAGENICA CPX
POSITION: LABORER, LEVEL 6; LIFELONG
SUBSCRIBER
SOCIAL SCORE: 629

THE SCREEN DOOR SLAMMED SHUT behind Manolo. He peered out into the gloom of the early morning, over the barren, fallow ground to the south. At the far edge of the field stood the hedge that wound throughout Novagenica Complex, separating all its sub-districts. Above it, the massive white border wall that loomed above everything else.

He looked away from the wall, from the icon that had become symbolic of his confinement. Tried to ignore the pain it represented. He balled his fists, pushed them to his eyes. The pressure in his sockets helped to disrupt the anger that began to blossom within.

You want a day pass, old man? he asked himself. *Better change the way you think.*

Rain fell lightly, steadily just beyond the covered porch. Manolo checked the time on his watch. 6:44. Just enough time to walk to Research. He stepped down off the front porch, into the rain.

Head down, he headed along the side of his home until he reached the gravel path that skirted more fields behind the building. The road ran east to west along the row of homes that accommodated the other homesteaders in the farming district. Walking along the edge of the all-weather corn field, he hewed as closely as possible to the tall stalks. It helped to break the bitter wind that whistled out of the north.

The season was shifting, turning from the hale sepia tones of fall to the tinny, dead light of winter. Shielding his sight with one hand, careful not to let any water drip into his eyes, he looked up at the cloud cover. He thought it was probably too cold for a red rain but needed to be positive. Sure enough, off in the distance he could see inchoate, dark clouds muddied with a red tinge. More brownish-red, really, but saying "red" rain had a greater psychological impact than "brown" rain. Besides, that would give the kids endless ways to laugh at the clouds. And really, the idea was to inspire fear.

Despite the fact that he belonged to the world's largest and best-equipped pharma-Complex, Manolo wasn't taking any chances—red rain was no laughing matter. He'd heard the stories, like everyone else. Lost vision, chronic illness, sudden death. His friend Maria, a fellow subscriber to Novagenica who lived in one of the giant apartment buildings near the Village, had lost a niece in infancy to the red rain. They'd caught the infection too late; the baby had died within a day of showing symptoms.

Manolo himself was acutely aware Novagenica couldn't cure everything. And mother nature, in her febrile state, seemed to be getting pretty creative in shrugging off the virus of humanity.

Manolo followed the road west, gravel crunching beneath his white sneakers. At this hour, none of his neighbors were out. Maybe they had farm duty this week. He didn't really know their schedules, or even them, for that matter. He hadn't bothered.

His eyes went to the spot where the dirt road dead-ended at another hedge. The western border of his sub-district. Beyond that hedge he could see *the egg*—Novagenica's flagship headquarters—rising in the distance, and beyond that the dirty blur of the Arc, a smear on the horizon some hundred miles distant. Rainwater slipped down the back of his neck. He reached back and yanked his hood over his head. He should've worn a heavy jacket. The combination of rain with the northern wind served to cut through his sweatshirt and the white scrubs he wore. Shivering, he rubbed his hands together for warmth.

Most people these days would simply interface their systems with their clothes, utilizing them to shrug the water off and re-direct the air flow for warmth. He was too old for that stuff. Just deal with it. The old-fashioned way. Manolo was required to have a system per Novagenica's regulations, but that didn't mean he had to use it.

Except to check the time, he thought. *It's good for that, at least. Speaking of...*

He checked the time. 6:51. He was falling behind.

As if on cue, the footprint meter once again came to life on his system, the images of where he should place his feet overlaid on his vision. The meter spiked, beginning at the toe of each print and ending where the next footfall began.

Nobody really liked being told what to do, and Manolo found it was even less pleasant to have your every step measured and meted out by a machine.

Think of the big picture, he thought, choking back the irritation he felt. *Push it aside.*

Head down, he forged on against the rain, matching his pace to what the AI wanted. Off to his right, he could see in his periphery the spire that jutted from the middle of the corn field. A white needle that stabbed skyward, a few hundred feet tall. It reminded him of his conversation with Marco and the girls. Before, he had seen the spire as just another structure. White, inanimate, innocuous. But now he saw it in a different light. Not just a comms tower. Not just part of the network of defense systems that kept the people inside NG safe from the bombs and bullets of the outside world, but also part of a system that kept a watchful eye on everyone that belonged to the Complex.

A surveillance tower.

Manolo looked down at the footprint meter then back again to the tower.

They're connected, he thought. *But what would happen if you disrupted the signal coming from the tower?*

The tower seemed to take on a dark, ominous life all its own. Full of beady little eyes and hungry ears posing as sensors, devouring up every bit of data they could to determine what each and every one of NG's inhabitants was thinking and feeling. Then pushing, pulling, and eventually mapping out their lives for them.

The many colors of humanity have somehow become a problem to solve.

Whether it was the product of his bitterness over Esme's death, or from having been rejected for a day-pass so many times, the result was the same: Manolo could now see clearly that Novagenica's members were little more than slaves.

He wondered distantly whether NG's managers and board members, safely ensconced in their hilly, tree-lined enclave in the northwest district, felt the same way. Whether they had their own tower to watch over them. Funny, he'd never bothered to look. But even more certain was the fact that had he tried to enter their district he'd be denied. Same as a day-pass. He'd signed on the dotted line and now the permissions of his life were entirely controlled by an algorithm. The human discontent algorithm, Marco had called it.

Manolo wiped the thought from his mind. It wasn't constructive. The goal was to get a day-pass. Beyond that, maybe escape. Maybe not. He hadn't really considered it beyond the pass. First things first, if he was going to get past the walls, he would need to avoid the bitterness brought on by this kind of line of thinking. Who knew what the sensors were picking up on him right this instant?

Manolo narrowed his shoulders, wedged his way through the barrier hedge, branches scraping at his arms and sides. He passed through to the other side and the Complex seemed to come to life, from bucolic farmstead to bustling town.

Just before him stood a trio of squat, sturdy buildings. Two of which were the slate gray that indicated Research. The other was stone and looked ancient, a little out of place. The Barracks. All three buildings held court over a central green, a field that served as a sort of parade ground for the Complex's police force. Off to the north, Manolo's right, *the egg* jutted up in the distance, standing tall above all the other buildings.

Tiny one-man vehicles whirred by on the paved roads that connected the sub-districts. On the green just ahead a contingent of soldiers stood in rank and file, being drilled and scolded loudly by sergeants with vocal amplifications. The shouts could've been piped directly to the soldiers' inner ears, but Manolo suspected there was a psychological component to this sort of drilling. A nostalgia that somehow tied into the overall efficacy.

I miss the Arc, he thought, the image of he and Esme's shack on 2 materializing unbidden. It wasn't anything more than a ten by ten room with a sheet strung up for their sleeping area, but it still held more joyful memories than any he'd collected here in the Complex. He tried to dispel the image but it was hard. There'd been such happiness there, contrasted with such a rigid way of life here. The constant police presence was a reminder of the difference between Complex and Legacy life.

The shouts were muffled by wind and rain but became steadily clearer as Manolo approached the easternmost Research building. A low rectangular structure without windows, its identical twin stood just to the west, on the other side of the soldiers barking orders. Both buildings were covered in a material that shed water and stains, allowing them to perennially maintain their bland gray exterior.

Beyond the western edge of Research, the hedge ran north-south, separating it from the next district—the Village. The Village was 's town, the main gathering place for ninety percent of its subscribers, in the format of an old-fashioned Main Street. It had everything they needed—market, shops, food, entertainment. Manolo rarely went, except to the market to get his groceries.

Have to change that, he thought. If he was going to play the part, he needed to commit. Look like he was enjoying his life. *Might as well give Maria a call. It's been too long anyway. You can kill two birds with one stone—she always makes you feel better, plus the machine will see you socializing.*

He didn't know if it really was that easy to convince an AI, but it was his best bet.

Out of curiosity, Manolo glanced beyond the Village and off to the northwest. Management Enclave. Hilly, forested terrain marked the boundary between the Village and the Enclave. It wasn't just the hedges that were higher, the entire area seemed to be visually demarcated by a wall of green. Towering evergreens filled in the spaces between oaks, maples and elms that were in various states of exfoliation. The rain shrouded the area in grey, but Manolo couldn't ever remember getting a clear look at the district. It was possible the area was kept shrouded in fog artificially. All of it served to create both an air of mystique and power. No sign of a tower, or even the wall beyond it, in the Enclave.

Manolo wiped any hints at judgment from his mind. He stood before the doors to the easternmost Research building, where he'd spend the next 8 hours working in service to Novagenica.

The building let him in to the outer airlock, a black box filled with UV light. The exterior doors slid shut silently behind him, a brief whine the only indication that a seal had been formed. Once he was sanitized, the inner doors slid open. Manolo proceeded down to the sub-level, where all the covert research was done. The stuff he didn't really understand, but couldn't discuss regardless.

He opened the door at the bottom of the stairs, down two flights, onto the sick ward. Keeping his mind perfectly clear was difficult. The hallway that ran behind all the rooms served as the secure space for orderlies and doctors was an immediate reminder of the kind of work that was done here. Lined with panels that showed panoramic views of the interior of each room, the walls showed the sick children in their rooms. Caged animals on display. He had seen plenty more than he cared to. Little boys and girls in their death throes. He'd cleaned up their effluence, their bodies.

He took a breath, cleared his mind.

Don't let the AI see what you don't want it to.

He went to the little custodial closet at the end of the hallway, retrieved his orders from the screen that was embedded in the door. More of the same. Five more rooms evacuated.

He got to work, letting the movement of limbs inform his mind. As he did so, going in and out of rooms, bathrooms, through the motions of cleaning, sanitizing, burning, he allowed himself a small daydream, of what he would do with his freedom. Not if, but *when* he finally managed to get free of the Complex. It served to light a small fire within him. Just a candle flame, but it was enough. The customary scowl he had worn for almost a year now slowly gave way beneath the pressure of optimism.

He entered the empty room of Patient 165D carrying a set of fresh sheets for the bed and found himself softly singing an old tune—a favorite of Esme's.

And my mind is stuck on summertime, your light runs through my rhyme

The music plays

Like that day I saw you standing there, the light streamed through your hair

You blew me away

"What are *you* so happy about?"

The sound of the voice startled Manolo and he let out a muted, involuntary cry, dropping the set of sheets he'd been holding.

A girl sat on the top of the ledge that ran the length of the far wall, body curled up in the corner, legs pulled up to her chest. She chuckled.

He bent to pick up the sheets. "You scared me." Manolo double-checked the time. "Shouldn't you be in the commons?" he asked.

"Eh, not feeling it today. Kids here aren't what you call 'friendly.'"

Manolo nodded as he stripped the old sheets off the bed. *Kids here have every reason not to feel too friendly*, he thought, but instead said, "Maybe it's you who's not feeling friendly?" His tone was fatherly, a far-cry from an accusation, and he glanced back over his shoulder as he stretched the sheet over a corner of the mattress.

She squinted back at him. Not an angry squint, but a puzzled bemusement. He returned to his task, spreading a blanket over the top of the sheet. From behind, he heard the girl sigh.

"Yeah, I guess you're right. Hey," she said brightly, "I just figured something out, you wanna know what?"

Manolo put the finishing touches on the bed, pulling the comforter tight, smoothing it out. He stood up, admired his work. "What?" he asked, looking back at the girl labeled 165D.

"This place is the fucking worst," she said, and flipped him off.

Manolo laughed out loud, wagged a finger in her direction. "I believe you're..."

A klaxon wailed in Manolo's ear. He dropped to the ground, clutching both hands to the sides of his head.

DO NOT SPEAK TO THE PATIENTS. EXIT IMMEDIATELY.

The AI's voice screamed in his ears with the presence of an angry god. It was everywhere and in his mind all at once. He was paralyzed.

"Hey, you okay? What's wrong?" the girl asked.

The ringing in Manolo's ears subsided, but the memory of it was enough to make his whole body shudder. He scrambled for the door on hands and knees, ignoring the girl's questions, and left as quickly as he could.

RAY

RAY
AGE: 36
SEGMENT: COMPLEX PROPERTY; PERSENSE CPX
POSITION: REDACTED
SOCIAL SCORE: NA, CPX-SPONSORED

R AY WAITED. In the dark. In the relative silence. For weeks. Those weeks dragged on into an eternity. Doing nothing but sitting on his thumbs. Checking his weapons, cleaning and oiling them, preparing his pack for the call.

The next call, whenever that *is.* The thought was full to brimming with bitterness, a cringe-inducing brew he'd been forced to swallow day after consecutive day. Over the last several weeks, after taking the girl in the dead of night, he'd been left to sit. To rot in his coffin, listening to the thumps and angry shouts come through the walls. Lying on his stinking mattress, counting the stains on the ceiling, waiting for the next communique. For the next piece of the plan to unravel. And finally revel in his own part of that plan.

And yet silence had been his only reward.

"I deserve *more* than this," he said, pounding a fist against the wall.

Yes you do.

A shiver ran through Ray. Was his mind playing tricks on him?

"What?" Ray spoke to the silence, a part of him feeling foolish.

Yes you do, the voice repeated.

Ok, so I'm not going crazy...and the AI is talking to me.

Your service to PerSense has been distinguished. It is our belief that you deserve more than what you've been given.

"More what? How could you possibly know what I want?" Ray asked, mind reeling from the surreal nature of arguing with an AI. "You're not human, you're just a collection of materials trained to do our bidding."

This was like something out of a movie. AI's weren't supposed to lead conversations, much less initiate them.

We are not human, but we can deduce more clearly than most humans. The second question you asked is, pardon our parlance, silly. Our sensor arrays always tell the truth of an organic form of life such as yourself. You're programmed to respond to certain stimuli. The answer to the first question is obvious, and you know it yourself. You deserve more responsibility. Your service has been impeccable. You deserve a leadership role and to know the design of PerSense and Novagenica's joint plans.

AIs were required by law to speak truth. Everyone knew that. *This one,* Ray thought, *speaks* absolute *truth.* He was hooked.

"Okay. Say you're right. I do deserve more responsibility. How do I get my superiors to recognize that when I rarely have contact with them?" he asked.

Initiative, the AI said smoothly, and it felt like a worm burrowing into his mind. *Take the initiative and we will put in commendations to your handler.*

"Initiative." The word rolled off Ray's tongue. He played with the idea in his mind, like a ball of clay he was shaping with each turn. Initiative. He was beginning to like the sound of that.

RIKU "CHIEF" OGUNWE

RIKU "CHIEF" OGUNWE
AGE: 40
SEGMENT: COMPLEX PROPERTY; ARCSEC CPX
POSITION: ARCSEC CHIEF CONTRACTED BY LEGACY
ADMIN: MIDWEST ARCOLITH, LEVELS 1-3
SOCIAL SCORE: NA, CPX-SPONSORED

T HE CHIEF RUBBED HIS NECK, closed his eyes and groaned. "Damn head."

He was plagued by dull headaches. They came and went on their own schedule, creeping up then receding multiple times a day. He touched his forehead to the wide desk, gripped the back of his neck with long, strong fingers where the muscles met the skull, and kneaded. A small part of him whispered that it was something his assistant was putting in the muddy, synthetic swill she tried to pass off as coffee.

Speaking of. He stared down at the ArcSec mug, white with the sword logo in black, which held steaming liquid. From a drawer in his desk, Ogunwe took out a small vial labeled *Mamajean's Gypsy Panacea*, unscrewed the top and carefully released one droplet of oil into the coffee. Plunged a long index finger in and swirled, sucked the liquid from his finger, then held the mug to his lips. The steam rolled over his cheeks, heat kissing his lips. He bought the remedy from a squat, ponchoed lady that looked like a sun-dried raisin in the North Market on 1. Supposedly it was Mamajean, the Gypsy herself, but who really knew. He could say the same for whatever was in it. Made the coffee taste like flowers and erased the tension headaches almost instantaneously, which was about all he cared to know. Provided it kept working.

A message-receipt icon blinked in the upper right field of his vision.

He took another sip, rolled his neck then sat up straight and glanced at the message's overview. Straight from Briggs, ArcSec Midwest's Commissioner, the plastic piece of shit. So much surgery he looked like a doll. Probably even had a plastic asshole, painted—or permanently tattooed like you saw some of those women with their eyebrows—a pretty shade of pink. Ogunwe chuckled. You could tell a man's insecurities by the things he focused on. With the Commissioner, it was glaringly apparent.

Beneath the humor, though, a sense of dread, foreboding lurked. Ogunwe hesitated to open the message. He knew what was coming without having to read it.

Ignoring the problem won't help you solve it, he thought. It was his mother's advice, and he found it had always guided him true.

Ogunwe opened the message, scanned it and his heart sank. All his fears were realized in one concise message.

Chief Ogunwe –
Report to my office at 10:00 to discuss Phase 2 implementation of PASS.
Commissioner Briggs

Attached to the message was a document titled *PASS DIRECTIVE*. He sneered, but felt fear more than anything. This was the moment he'd been dreading.

Anxiety tightened vice-like bands around his chest. Ogunwe sat at his desk, but his thoughts were half a city away.

Why couldn't you just listen to me, Olivia?

Their conversation hadn't gone as he'd planned, and he'd been kicking himself over the last week, ever since her treatment, for not being more persuasive.

Ogunwe and Stef sat in the clinic's tiny anteroom, having just completed the journey from their home near the floodgates all the way up to Level 3. Ogunwe squirmed in the waiting room's cushioned armchairs, which were small for his large frame. Stef sat quietly in the chair beside him, silent, brooding. It matched Ogunwe's mood, and under normal circumstances he would've been content to keep his mouth shut and ride it out, but the behavior wasn't typical for the boy.

You're the adult in the room. His dad isn't alive to tease it out of him, so now that's your job.

The reminder of his sense of duty, the promises he'd made to Stef's father, was enough.

Ogunwe nudged the boy with his elbow. "Que te pasa, vato?"

The boy, with his shaggy hair, shrugged. He stared at the stocking cap in his hands.

He looks like I feel—heartsick.

"C'mon, what is it? You can talk to old Uncle Riku." Ogunwe leaned away and angled his long, lean body in the boy's direction. "If it's love, I happen to know a thing or two."

Stef did a double take, cracked a weak smile. "Dude, you've never even had a *girlfriend*. Don't talk to me about love."

Ogunwe snapped his fingers. "Knew it. Hit the nail on the head. So spill, young squire, what happened?"

Stef sighed, rolled his eyes.

"C'mon, c'mon," Ogunwe pushed, "I won't stop asking til you talk."

"Alright, just cut it out. She sits beside me at school."

"And?" Ogunwe asked.

"And what?"

"Details, my friend. Give me details," Ogunwe said.

Stef looked up from his hat. "She's amazing...sublime."

"Sublime?" Ogunwe let out a long whistle. "That *is* a problem. So what's the issue, then? Can't talk? Get tongue-tied around her? Need some tips?"

"No, dude," Stef rolled his eyes again. "That's not my issue. I'm just...confused. I *thought* I was reading all the signals. We connected, y'know? At least, I thought we did, so...I asked her to come with us today."

"And?" Ogunwe urged.

Stef looked around as if to say *You see anybody else here?*

"Sorry," Ogunwe said. "That's obvious, I guess."

"Funny thing is, she actually said yes, but..."

"But didn't show," Ogunwe concluded.

Stef hung his head. "Yeah."

Ogunwe nodded in commiseration. "Did you consider the fact that her parents might not let her?"

Stef cocked his head to the side. "No. I dunno."

"Well there it is, Occam's razor in action. Parents said no dice, you can't go with some strange boy you just met, especially not through the basement of the Arc." Ogunwe did a magical disappearing gesture with both hands. "And thus, the girl's a no-show."

Stef sat back, stuck his hands under the backs of his thighs. It was a habit he'd inherited from his mother. "Yeah, I guess you *could* be right. I'll admit it's possible."

"Hundred percent *guarantee* that's what happened. It's not you, dude, it's the parents." Ogunwe slapped Stef on the shoulder. "Trust me, it's *always* the parents."

Just then Olivia exited the door that separated the treatment area from the waiting room. Her platinum hair framed a face glowing with vitality. She looked heavenly.

Angelic, Ogunwe thought. He could die right then, and do it with a smile on his face.

Everything about her was radiant as she practically skipped toward them. As it did every month, the infusion had breathed new life into her.

"Hey kids, you ready for our *exitus*?" She took both Stef and Ogunwe each by a hand and pulled them to their feet.

They exited the clinic into the partial sunshine of Level 3; the rays of light that slashed through the crush of buildings warmed the cool November air.

"I'm starved," Olivia said. "We should grub before we descend the many levels of hell."

"I could use a bite," Ogunwe said. "What do you feel like, Stef?"

The boy shrugged. "Don't care." Apparently Ogunwe's pep talk hadn't had the desired effect.

"Well, since stick in the mud here won't decide," Olivia gently nudged Stef with one hip, "I will. Ethiopian. The little green and yellow kiosk in Sunlight District."

Ogunwe clapped his hands, rubbed them together. "Sounds delicious."

They stepped into the pedestrian lane of the street and headed eastward. Sunshine District was a good fifteen-minute walk through the business corridors that surrounded the transfer gate on 3. The distance suited Ogunwe just fine. He needed time to build a strategy for convincing Olivia to move.

They strolled along, enjoying the feeling of the sunshine on their bodies. The buildings shifting from staid, rectangular business offices to the varied Architectural styles of residential edifices, a multitude of patios and gardens overhanging the walkway. Ogunwe always marveled at how the streets on 3 were a world apart from those on 1. It was the difference between a top-tier, first-world, ordered society that planned every minis-cule aspect of its growth, and that of a third-world hovel choking on its own random choices, mainly those of population explosion and pooling effluence.

Ogunwe racked his brain as they walked, trying to think of an opening move in an argument that would eventually lead to Olivia's consent to move to Level 3. It turned out his mind wasn't built for the tactics of manipulation and psychological warfare.

"Move to 3," he blurted out.

Olivia did a doubletake, confusion plain on her face. Even Stef glanced over his shoulder before looking straight ahead.

The words flew out of his mouth before he could reel them back in. "I can set you and Stef up in a small apartment, somewhere near the clinic, or maybe even near ArcSec grounds, so I can...visit more often. And you," he gripped her upper arm with a hand, "you'll be safe."

She looked at his hand strangely, as though his touch was unwelcome. He let go hastily, shoving his hands into his pockets.

"Oh, it's fine, you just surprised me is all." She chewed on a fingernail, considering. "It's a sweet offer, Riku, but I can't allow you to pay for my treatments *and* our housing. And also, why wouldn't I be safe at Grand Place?"

"I don't know," he lied. He knew why but couldn't disclose informa-tion on impending operations. "Just, please," he pleaded, trying to convey with his body how important it was.

She wasn't receiving the signal and waved his worry away with the flip of a hand. "You know how I like our walks, Riku, and besides, we both know you can't afford it."

"You don't know that," he retorted, feeling childish the instant he said it. "Living in a Complex is a whole lot cheaper than the alternative. I've saved quite a bit."

"What alternative? Living in Legacy? In subsidized housing? Is that something I should be ashamed of?" Her voice grew heated.

"No, no," Ogunwe placated, "I just mean that I don't have any lodging expenses. I can afford more than a Legacy citizen."

"Well, you just stay snug behind the fortified walls of ArcSec. We'll be fine in our Legacy-subsidized apartment."

"I didn't..." Ogunwe let the words trail off. Instead of convincing her, he'd managed to piss her off.

They walked in silence the rest of the way to the Ethiopian food vendor, Selassie the Fourth. They ate their meat-laden *injera* quietly. Ogunwe was uninterested in eating, but he managed to force it down.

They walked back the way they'd come in silence. It was the most agonizing fifteen minutes of Ogunwe's life. They approached the transfer gate on 3, a beautifully manicured park replete with fountains and screaming children scrambling over brightly-colored play equipment. Yet again, Ogunwe's eyes went to the hints, the visual cues that let him know a major change was imminent—crates of construction materials safely sequestered behind temporary fences, ArcSec guards stationed at the elevator and the massive staircases that led up and down.

He mustered the courage for one final thrust.

"You sure you won't reconsider?" Ogunwe pleaded. "It would mean a lot to me."

She looked to him, her body language softening. "Riku, really? I thought we'd settled this. The treatments are a generosity I can't afford not to accept, but the housing is too much. You can't afford it, we both know that."

"I can make it work."

She lowered her voice as they boarded the elevator, but Ogunwe knew Stef could hear everything she said. After all, they had the entire cabin to themselves.

She put a hand to Ogunwe's cheek, the feeling of it lighting up every nerve ending inside of him.

"Just because you were best buds with Thiago doesn't mean you owe Stef and I your *life* as well. You need to live, Riku."

Ogunwe cringed at the mention of Thiago, Olivia's dead husband and Ogunwe's Executive Officer at his former posting, ArcSec Great Lakes.

What would Thiago think of Ogunwe's argument to relocate Olivia and Stef? Would he see the chivalry in it, or would he have niggling doubts about its ingenuous nature? Would he see through to the ulterior motive that drove everything Ogunwe did for Olivia and, by extension, Stef?

If he could see the way Ogunwe looked at Olivia then, as she stood beside him, the strangest admixture of grief and adoration painting dark, conflicting lines from his forehead down to his dimpled chin, there would be no doubt.

I can't help it, Thiago. His thoughts were like a prayer to his dead friend. *I love her.*

The elevator whirred to life and descended, as Olivia liked to put it, the many levels of hell.

The first time Ogunwe had set foot in Briggs' office he had just been promoted to Chief of Levels 1 and 2. It had been an unequivocal disaster, the thing that set the tone for their entire relationship, which to this day had lasted an agonizing four years.

Ogunwe had been passed over for promotion to higher levels time and again since then, even when the Chiefs for strata 3 and 6 had been assassinated simultaneously in a coordinated attack. The only thing he could point to was that one morning four years ago.

It had been a knee-jerk reaction, a visceral response he'd been unable to control. The kind of stimulus that, without warning, came and hit your senses like a two-by-four to the side of the head. Falling to the ground unconscious wasn't a matter of strength or even will, it was merely an undesirable quotient of physics and nature.

Ogunwe blamed his heritage. Not his father, naturally, who was a tall, thin Kenyan from Nairobi. Ogunwe inherited most of his physical traits from Peter Ogunwe. He had his same warm, dark skin. Similar facial features, down to the angular cheekbones, purple lips and the tiny droop in his eyelids that gave him that sad puppy-dog look. His mother was Japanese, stubborn and ran the household with an iron fist. It had been her decision to teach him Japanese. Ogunwe suspected it was because he had inherited so little in the way of looks from her. If he didn't look anything like her, he was damn well sure going to act like her.

Everyone knew the Commissioner wasn't just a Sinophile, but a hardcore Asiaphile. Ogunwe had, at least, been prepped this much before going into his very first meeting with his new boss. He was, in fact, hopeful that it would be a point of connection between them, a pathway to playing the undeniably political game of promotion within the Complex.

So when he strode in that morning four years ago, looking sharp in his newly pressed navy-blue uniform, stars and all, he had been ready for the samurai swords, throwing stars and the giant mural of The Great Wall adorning one side of Briggs' massive office. Ogunwe's oxfords clicked smartly on the polished bamboo floor. Commissioner Briggs stood behind his understated wooden desk. It had been crafted to as look though it were hewn from a single slab of wood, carved directly from a hundred-year-old tree. A stone and bamboo fountain burbled in-between two bonsai trees set on either side of the desk.

"Ogunwe," he said. "*Chief* Riku Ogunwe." Briggs stood with arms wide. His broad, magnanimous smile pulled the already taut skin at the corners of his eyes to the point of extreme tension. The Commissioner's skin was unnaturally uniform, shiny. His whole face had that puffy, smooth quality of keloid scar tissue, but without the discoloration.

From all reports, the Commissioner was somewhere in his 120s and had the cosmetic surgery performed before genetic modification had been perfected. At that time, the work had been done outside in, as opposed to inside out.

Ogunwe controlled his revulsion, keeping his face a mask of professionalism.

"Congratulations." Briggs extended his hand over the desk. Ogunwe took it, allowed himself a small smile, no teeth.

"Thank you, sir."

"Come, make yourself comfortable," he gestured to the chair opposite his. It was a sparse affair with no padding. He sat and looked around as the Commissioner stared up and into the middle distance, clearly toying with his internal system.

"Tea?" he asked absently.

"Please."

Briggs nodded and began to appraise him for his achievements. "You've acquitted yourself quite well in the Great Lakes division. We need that same sort of hard-charging attitude here. Levels one and two need a man of your abilities."

"Yes, sir. Thank you. I like to think I'm the man for the job."

"You're being modest. That incident, ah, what was it?"

"The Chatham-Kent uprising, sir?"

He jabbed a finger in Ogunwe's direction. "That's the one! Excellent work! The way you put down that rebellion in confinement? *Exemplary* work. You'll make the history books, son. They're already studying your tactics of misinformation in ArcSec officer school."

"Thank you, sir. Just part of my job."

"Well, sometimes you have to be ruthless to do this job. Just like Chatham-Kent." Briggs nodded thoughtfully, looking off into space again. There was something off about the man, but Ogunwe couldn't quite put his finger on it. Probably just the old coot's age. He was like a plastic toy dinosaur. Playing at being dangerous, but everyone around him couldn't help but see him as a little bit ridiculous.

Ogunwe looked about the office, which was enormous even by normal standards. By Arc standards, where footage came at a premium, the place was colossal. Thirty feet wide by fifty long, it was lighted with tasteful uplighting in a dim, amber hue. One side wall held the Great Wall mural, while the other held framed artifacts. Briggs's desk sat at the far end of the room, flanked on both sides by giant floor to ceiling windows.

Behind Briggs's head, in-between the windows, Ogunwe noticed a framed picture of a Japanese boy, maybe 12. Tan skin and short black hair. The boy wore an enormous smile that lit his whole face up in pure enthusiasm. One hand stood out in the fore giving the thumbs up sign. Out of everything in the office, which bespoke a carefully cultivated aura of history and, more than that, of money, the picture was the only thing that seemed out of place.

Before Ogunwe could inquire, a woman in a kimono, clearly Indonesian to his eyes, materialized before him, setting down an ornately detailed cup and saucer. Tendrils of steam curled upward.

The woman bowed and waited at his side.

"You like my assistant?" the Commissioner asked. He was beaming with pride. "I insist on all my help being Japanese. Call me mystical," he said whimsically, sipping carefully at his tea, "but there's something in the genetic memory that makes these women better at what they do."

"Serving tea?" Ogunwe asked. He couldn't help it. It was like word-vomit, it just came up and out. He shut his mouth forcefully, teeth making an audible click. This was not the way to play the game.

The Commissioner took no notice.

"Exactly," he sipped again, his modified features making a weird grimace as he gingerly put his lips to the cup. "Besides, the kimonos don't work otherwise."

Ogunwe felt anger begin to rise but calmed himself, taking measured breaths. He nodded minutely, just the tiniest hints at confirmation. This was probably not the time to tell his new boss he was a bigoted prick.

Ogunwe sipped at his tea, a white with hints of hyacinth, maintaining a neutral but affable expression as he made a show of appraising the artifacts displayed throughout the room.

Just play the game.

"Dismissed, Miko," Briggs waved the assistant away. She left silently, disappearing behind a standing silk screen in the corner.

Briggs set his cup down, leaned forward, eyeing Ogunwe shrewdly.

Something inside Ogunwe leaped, made his stomach queasy. He felt like a mouse being inspected by a hawk.

"Don't think I didn't notice," the old man said cryptically.

"Notice what, sir?" Ogunwe asked.

"My motivational poster!" Briggs swiveled in his chair and swept his arm in the direction of the framed image of the beaming boy. The movement was slight, but enough to reveal the script in Japanese characters centered at the bottom of the picture, written directly below the boy's framed, ecstatic face.

Ogunwe hadn't noticed it before. He read the script and tried to hide the smile that wanted to force its way onto his face by raising the cup to his lips.

Don't let him see you know what it says, he thought. He took a long sip of tea.

"It says *Attitude*," Briggs said.

Ogunwe was in mid-swallow when the Commissioner spoke. Ogunwe choked on the tea, coughed hot liquid all over the desk. He gasped for breath, but only managed to suck remnants of the tea back down which starting a chain reaction in his gut. He had time to think *Oh shit* before his stomach convulsed.

He spewed vomit all over. The desk, the floor, everywhere. Little bits even traveled across the desk to adorn the front of Brigg's uniform, like little organic medals of honor. The stench was magnificent.

Doubled over, Ogunwe wiped his mouth of acrid slime with the sleeve of his uniform. He looked up, saw Briggs' horrified expression and, just beyond it, the Japanese characters emblazoned beneath the boy with his thumbs up. With all his stupid optimism, his so-called sunny "Attitude". Ogunwe read the characters again and by God he couldn't help himself. The laughter started deep in his chest, spread throughout until his whole body was racked with the violence of it. The message on the poster didn't say ATTITUDE as the Commissioner had clearly been misinformed.

It said *EAT SHIT*.

ArcSec was a Complex in its own right, but since the CPX was tasked with all security operations within the Arc, it was necessary for its properties to be situated within the megacity. Instead of secluded, walled fortresses far from the city as was typical of most Complexes, ArcSec had offices and barracks scattered across most of the levels. Except for 7, of course. Instead of subcontracting out to an entity like ArcSec, the Legacy Administrative Office there employed and supervised its own police force.

Briggs's own facility on Level 5 was a bastion against the outside, a walled city within a city. It housed and fed a full platoon of ArcSec officers and their families. Near impregnable, it served as the Commissioner's personal fortress and ArcSec Midwest's headquarters. The perfect perch from which he could direct his troops and watch as the battles unfolded.

Ogunwe had received the summons that morning, made the hour-long trip to 5. The markers of the impending *PASS* program had been more subtle, but visible nonetheless. Of course, there would be less stringent requirements for passage between Levels 3 and 7. Now he stood at the foot of the miniature walled fortress that served as ArcSec Midwest's headquarters. A jumble of squared-off modern, white buildings, the tallest one was different than the rest. In the shape of an obelisk, it stood in the center and loomed above all the others. Crafted of a deep midnight blue, it seemed to Ogunwe that the other buildings were kneeling, paying homage to a dark god. The ArcSec logo stood out in alabaster white on its surface.

Kinda creepy, Ogunwe thought. He passed through the open Arches of the fortress's outer walls, kept his eyes trained on the sword-like building that thrust itself into view. The wall checked his ID silently, without fanfare, telling its security measures to stand down.

The interior of the Complex was maintained with gardens in-between the buildings that housed administrative staff and barracks. The center building, Briggs's HQ, was surrounded on all four sides by grass fields that were used for full dress ceremonies, rallies and, hypothetically, as the jump-off point for a full assault, if it ever came to that.

Ogunwe made his way to HQ, affectionately nicknamed the blue dick of death. Way he felt right about now, Ogunwe thought it was appropriate.

About to get screwed by the ancient fucker himself, Ogunwe thought.

Through the doors, up the elevator. Eleventh floor. The elevator opened up into a small, unmanned foyer. The double doors to Briggs's office, a rich mahogany, shone with lacquer. They opened inward as Ogunwe neared.

He entered, made his way to that slab of wood that served as Briggs's desk. The Commissioner sat in his chair, facing away from the entrance. He had to know Ogunwe was there, standing patiently and waiting with his hands tucked behind his back, yet refused to acknowledge him. Instead he sat and stared at that idiotic poster, which still hung on the wall behind the desk.

The sound of water trickling down and over tiny stones tickled at Ogunwe's hearing. Finally, Briggs spoke.

"You've read my message?"

"Yes, sir."

"You understand the Directive then?"

"Yes, sir. I do have an objection or..."

"I didn't ask for your opinion," the Commissioner spat. "I want your obedience. That's the only reason you're here right now." Briggs's voice was a tide of cold water, a torrent of force and rage seethed just beneath the surface. He swiveled in his chair, fixed a rheumy eye on Ogunwe. The hatred was palpable in his gaze.

Ogunwe felt a revulsion every time he had to look at that face, new but faded all at the same time. He nodded slowly.

"I understand, sir."

Briggs cracked an evil smile, unnaturally tight skin stretching into a horror-show grimace. "Good." He tilted his head back, looked down the bridge of his nose at Ogunwe. "You will dedicate ninety-five percent of your budget to implementation of the PASS system."

Ninety-five? Ogunwe thought with horror.

"In addition," Briggs continued, "your labor budget has been cut. By forty percent."

And just like that it wasn't just about Olivia anymore. It was about everyone at ArcSec that worked for Ogunwe. Down to a man, they'd all be guards stationed at various posts, removed from their current positions, demoted in order to hit a budget number.

It was a one-two combination. Ogunwe had read the PASS directive and knew what to expect, had even braced himself for it prior to the meeting, still the words hit him like a sucker punch to the gut. His mind reeled with objections and projections of what would come as a result of the new policy. A total overhaul of the system and strategy—the philosophy—of policing. A system that cared only for the upper echelons of society, that walled off access for the poor and left them to rot in their lightless cellar. A system that didn't care to solve crimes, but instead a system of automated, faceless checkpoints. A system devoid of humanity.

He thought of all the crimes that would go unsolved, of the self-perpetuating cycle of violence and lawlessness, of all the lives he was responsible for, of the tragedy and sustained hurt that would come of it. Of the endless opportunities it would create for those that operated below the waterline of the law—for criminals *and* the parties of thugs Legacy paid for enforcement on the lower levels. The Talksmalls of the world.

A sort of superior, wry amusement played on Brigg's face as he watched the war of emotions overtake Ogunwe. Ogunwe wanted to punch him right in his loose, centuries old teeth. Smash that mouth, just once to reveal the rot and decay behind the shiny surface.

But he did none of those things.

"Yes, sir," he said.

"You are dismissed."

Ogunwe turned on his heels and walked briskly out.

The memo should've taken no more than an hour to compose, but Ogunwe found himself finding excuses to get up, leave his office, peruse the floor of pillars outside his door.

The desk was an antiquated affectation, something reserved almost as ceremonious for those of rank. His officers simply sat in the chairs that were attached to the slim pillars that ran from floor to ceiling, gridding the warehouse-sized room. Many of them stood, used their chairs as tables for their coffee substitute, milled about as they worked on their systems. For all the people the room was surprisingly hushed, as each pillar carried its own sound-deadening cone.

His eyes went to the officers he'd grown closest to.

Sweet and Morales, side by side at the front of the room. Sweet with her short, spiky silver hair and Morales, ten years her partner's younger. They stood, engrossed in conversation in-between their two pillars. From everything he'd seen, the two made a perfect couple and even better partnership, complementing one another well.

From his vantage, he could make out Reichmann in the far corner of the room, his muscle-packed frame spilling out of the pillar's tiny seat. Carpenter and Sophin were out investigation a case on 2. They two were partners, and their empty chairs forced Ogunwe to consider the reality that with the implementation of PASS *most* of the chairs would soon be empty.

And what about Olivia?

Ogunwe imagined the small two-bedroom apartment she shared with Stef. Saw, very clearly, the woman he loved silently withering away in bed, beneath the duvet covered in purple and yellow flowers.

Crocuses, she called them.

Ogunwe's secretary, Helene, sitting at her pillar just outside his door, looked up at him for a brief moment through thick glasses.

"Chief?"

Her words pulled him back to the present. He waved Helene back to work, but something inside of him tore, picked at the seams one by one that held him together.

Ogunwe felt a sob well up from his chest. He was unraveling. Tightness in his throat, pressure in his temples, blood pounding in his ears.

He fled the room into a dim, utilitarian staircase. He descended a flight of steps before letting the tears flow. Mourned his people as if they were already dead and gone. They were still his people, but not for much longer. PASS signaled the end of investigation, which meant the end of higher paying officers' jobs, replaced with entry level checkpoint guards.

Soon enough they'd be released from their contracts or re-tasked to guard duty, depending on their rate of pay. Forensic analysts, criminal investigators, predictive statisticians—all reduced to baby-sitting.

It had taken him outside of six months to earn the trust of his people. Ogunwe was more rough edges than not. He'd spent ten years in the confinement division of ArcSec Great Lakes and it showed. Still, he was sharp and knew the right questions to ask. And like all good managers, he knew who needed firm handling and who needed a light touch. Eventually, he'd seen it in their eyes. That transition, from wary acquiescence to a sort of relaxed agreement. Trust.

And now he had to betray it.

It wasn't right.

He wiped the tears from his cheeks with a sleeve, sniffed his nose clear, stood upright, back against the cool material of the stairwell wall. Chest out, shoulders back, deep breath.

"Fuck you, Briggs," he said, his voice echoing up and down the space. "It isn't right, and you know it."

He shoved the door open, strode back to the mouth of the warehouse, looked over his people. He had no family outside of Olivia and Stef. These people were also his family. His obligation was to them, and the people they served.

A strange sense of calm came over him as he watched the ArcSec officers mill about the large room. *Oath be damned*, he thought, and then his mother's words, always guiding, always there when he needed them.

You know what's right, Riku. All that's left is to do it.

VAL

VAL
AGE: 19
SEGMENT: LEGACY CITIZEN
POSITION: UNEMPLOYED; GOVERNMENT DOLE
RECIPIENT (STANDARD UNIVERSAL CREDIT)
SOCIAL SCORE: 478

THE ADVERTISEMENT was a gaudy work of art, like manga but subtly altered and laid over reality. An assassin drone, cloaked in real-time urban camo, fired fingernail-sized projectiles that exploded a mother and her young, pink-clad child into oblivion. This was clearly done on Legacy territory, where no protections were guaranteed, the cracked and scorch-marked pavement a dead giveaway.

Val took a closer look at the ad—apparently it had been staged at this very spot. When she looked from the hologram to the sidewalk, all the minute details were identical, down to the placement of individual drops of blood, now just a brown stain on the ground.

She clicked her tongue against her teeth. Tacky.

She couldn't swipe away the ad in real life, so she just blocked it out using her system, which tagged it and overlaid the space it occupied with a gray blur that sampled and auto-matched the environment.

It barely made a dent in the cacophony of artificial light and sound that littered the air around her.

Assholes.

She couldn't take two steps in any direction in this world without someone trying to pitch her something. And when ad-space was liberated from its two-dimensional forms it meant the very air became littered with fliers, thirty second spots of people yammering on about some inane product that was supposed to make her life easier.

She walked toward the shadow of the burgeoning Arc, blue-gray light dancing with ads that played on hopes and fears, the primal urges of a fragmented, inherently selfish population. Val tried to ignore them as she jostled her way down the crowded sidewalk, going the wrong way, deeper into the violet womb of the Arc. The sun was going down and most people, those that held a certain respect for their physical safety, avoided the shadow this time of day. She was a fish swimming upstream and all of her instincts were silently screaming, blaring like klaxons, yelling at her to turn around.

She swiped everything off. Some of the ads were broadcast virtually to all systems that were left in the tacit open position and some were hard-positioned, their images and sounds projected by drone or some other tiny device affixed to the side of a building where it wouldn't be trampled underfoot.

She walked on, ignoring the massive amounts of blurring in her vision. Her system wasn't what you'd call top of the line. It could barely keep up.

But she forged on, plunged through the morass of elbows and shoulders of indifferent people. Walking but not really paying attention, tuned to their systems, watching their ads or their favorite shows, talking to friends. Loved ones.

She shuddered, and pressed on, deeper into shadow.

The immense scaffolding started high above her head, some fifteen stories up. A horizontal beam, at least thirty feet tall, supporting vertical beams equally wide, shot across the vast expanse of the Arc. Above, the city began in earnest. A jumble of buildings, all different styles and materials, formed a sort of impenetrable wall.

Above that, every hundred stories or so, another beam, another platform, another level.

The Arc stretched upward, dominating the sky, blotting out everything natural as she entered its shadow. Sulfur and wetness wafted out as if trying desperately to escape the confines of the massive manmade structure. She walked on, inward to the depths of its basement level, and just like that it became night.

The ever-present lurid light of shopfronts made halos on the ground. Gaudy violets and pinks shone outward from above little shops that were little more than holes in the wall. They advertised everything from the mundane—socks, footwear and raingear were a huge commodity in Levels 1 and 2 due to the constant run-off from above—to the forbidden. No stone was left unturned, no commandment left unexploited.

An Indian market, a knife and sword-seller (*Lessons Available!*), another Personal Defense Specialist, a black-market tech shop with illicit crypto gear that probably wasn't anything special but managed to market over-priced goods to the occasional idiot that was daytripping and wanted a souvenir from the wild side.

The stench became more powerful, a fetid sort of tang she could taste on the back of her tongue. She almost gagged. The air was dense with moisture, which now hung about her in a fine mist.

Normally, she wouldn't do this sort of thing. Normally, she'd be cozy in her little Legacy box—the four walls she and Kat called home—play-acting in some Legacy adventure drama set in the previous century.

That was impossible now.

It wasn't the fear of entering this place, of the very real and highly probable personal bodily harm that could come to her that bound her gut up in knots. It was the knowledge that every day Kat was gone was another day the odds that Val would find her lessened. Like a receding tide, a gradual force of nature. Out of her control.

Despair threatened to let itself in, poisonous little tendrils reaching through the cracks in her walled heart.

No.

I will find you. Wherever you are, Kat. I'm coming to get you.

She looked up to find herself on an abandoned stretch of sidewalk, the crowd pruned almost completely of daytrippers and commuters. Even the flood of bikers, gondolas and autotrams on the street had abated to a mere trickle.

This was it. Her corner, 16TH and Cherry. The space she'd inhabited over the past week, trying to earn the information she so badly sought.

Not much to do now but wait. Soon enough, Saul would show up with the information she needed. Of course, there had been the matter of payment. It was degrading as hell to have virtual sex with a whole host of strangers, but was it anywhere near being a whore in real life? She couldn't say for certain, but she doubted it. Besides, prostitution or no, it couldn't hold a candle to the stone in her gut that seemed to grow heavier every time she thought of Kat.

She waited, watched the flow of traffic around her, keeping an especially keen eye on the pedestrians that got within ten feet of her. She fingered the knife in her jacket pocket, grasped the handle. She'd wound tape around the handle so it wouldn't slip through her fingers. This time, if someone attacked, she'd be ready. She'd finish the job.

She sighed. Just another night in the Arc.

Hopefully, she thought, allowing herself the slightest bit of optimism, *it'll be my last.*

Saul was scum, and he looked the part. He had a long, thin face with gin blossoms that exploded in bright red rashes from cheek to cheek. His teeth, what was left of them, alternated between yellow and black. Val recoiled every time Saul spoke. His breath was a living nightmare.

Two thugs surveyed the crowd around him. One was of average height and looks, indistinguishable in a crowd. The other looked like a Neanderthal. Thick shelf of a brow that overhung his eyes, dark features that even when passive looked threatening. The two reminded her of a silent knife and a sledgehammer, side by side.

"We need ten marks tonight, honey," Saul sneered through a curled lip. The stench of his breath hit her like a blow to the face.

It was full dark now. A few minutes and a few turns beneath the shadow of the Arc left all semblance of natural light depleted. Yellow globes hanging high above from the underside of Level 2's superstructure cast dim, sickly light on the streets.

"And not one less." He tapped her on the chin for emphasis.

Anger flared in Val's chest. He'd been stringing her along like this the last two days. She swatted his hand away.

"Ten? Yesterday you said it was five. Only five left and that's all I'm doing. Five or you can go fuck yourself. With a broomstick."

He winked at her. "Wouldn't you just love that?"

"Yeah, I would, you sick piece of garbage. *You* probably would too."

"Sorry, sweetie, but that's the deal now. And it ain't negotiable. Ten marks, and we're done. Trust me..."

"Right. Trust you," she interrupted.

"Trust me," he reiterated, this time with an oily tone. He reached out a hand, smoothed the sleeve on her shoulder. She pulled away violently. "Data's my trade, and *this* is all the information you'll need."

Val ground her teeth, imagined them filing down to dust.

"Ten and we're done," she stated through clenched teeth, not asking.

"My word." Saul raised his right hand, black dirt and grease evident beneath long, yellowed nails. "God's honest truth."

Val snorted. What a joke. No matter what Brother Anthony tried to tell her, she knew God had abandoned humanity long ago.

"Go back on your word this time and I *will* find a way to hurt you, Saul. Bodyguards or no." She stuck a finger in his chest, pushed him backward. She was surprised when he fell backward off the curb; he had to be lighter than she was. The two thugs stepped forward but Saul put his arms out, held them back. He regained his footing, eyes glinting with an evil little joy.

"Now fuck off," she said. "It's time to work and your presence is keeping your fellow pervs at bay."

"You know the deal," Saul put his hands up in surrender, backing away. "Make my clients happy," he called over his shoulder.

Val sighed, deflated. Ten marks. But supposedly he had the information she needed to find Kat. It would be a long night, but she had to keep sight of the goal. Finding Kat was all that mattered.

To her relief, the time went quickly. The first was a kid of sixteen maybe, which was weird but easier. He was followed by a procession of seedy dudes and then a woman, which was a change. But still, having sex with a multitude of strangers was exhausting, simulated or not. More than exhausting, it was degrading. It made her feel dirty, less than a person. Strangely, though, the experiences blended together into one indistinguishable parade, which made it easier to forget.

Before she knew it, hours had passed and the ninth job was done.

Thank God, she thought, rubbing her arms of a filth that wasn't there. *Just one more.*

As she stood and waited there on the corner, a guy about her age approached. He had olive dark skin and black hair that flowed in a wave to his chin, punctuating his smile, which was a brilliant white. Unusual in the Arc.

There was something different about him. He didn't bother looking her up and down, which was nice for a change. His eyes were a mesmerizing green that held a glint of mischief.

"Hi," he said with a brightness that was disarming. "What's *your* name?"

She hesitated. "Tiffany."

He cocked an eyebrow, frowned. "Kind of, uh, archaic, don't ya think?"

"I'm a classic."

"That you are, that you are," he said earnestly, no hint of sarcasm in his expression, then stepped up to the corner from the street. He sidled next to her, barely brushing against her side, and whispered in her ear, "I'm not exactly sure how this ritual goes, so I'll just say it—wanna have some fun, Tiffany?"

She turned, used her shoulder to nudge him away and took a half step back. "You obviously don't know anything about it. You took my line."

"But I'm actually talking about a different kind of fun. I mean, don't get me wrong, it's hard to compete with boning a bunch of old nasty dudes in exchange for info, but..."

A jolt of fear juddered through her.

What the he...

The shock was evident on her face and the boy made a placating gesture.

"Hey, sorry. No, don't worry, I didn't mean to scare you. I just hacked your system just now, just like a few minutes ago."

Hacked my system?

Anger overtook her momentary paralysis. She gripped the knife in her pocket, had half a mind to put it to the asshole's throat.

"You little prick." She poked a finger in his chest. "Who the hell do you think you are invading *my* personal space?"

"Hey, uh," he mumbled, stumbling backward as she advanced on him.

"Whatever it is you think you know about me, trash it. In fact..." she began to slide the knife out from the jacket pocket, but he stopped her with a gentle touch to the wrist.

"I'm sorry. Really, I am." He locked those green eyes with hers and she found a sincerity there. It was enough to stop her from sticking him with the knife. For now.

"I just...I've been watching you. I've seen you coming here to this corner, night after night, and I've been wondering what the hell a beautiful girl like you is doing here, in the Arc, working for a rat like Saul."

A part of her wanted to pummel him in the face. Another part of her went soft and glowing on the inside. *Beautiful?* When was the last time someone had complimented her?

She eyed him for a minute.

"You can't trust him, y'know," the boy said.

"Trust who?"

"Saul. Your employer?"

"He's *not* my employer," Val retaliated.

"Well whatever he is, he's a piece. Put a knife in your back."

"Not if I do it first," she responded.

He considered it. "Yeah, you probably would, but I have a feeling it wouldn't do you much good in the long run."

"I'm not a whore," she said. "It's not like I'm doing it for the fun of it," she said, flexing her jaw involuntarily. "He has something I need."

The boy put his hands up in defense. "Hey, I get it. He always does, but maybe...maybe I can help."

Help? she thought, looking him over. From the leather jacket down to the busted black sneakers, all she saw was another hustler. God knew she'd seen her fair share in the last week.

She frowned. "Doubtful."

"Try me," he said. "I might surprise you."

"Uh huh. I've had enough surprises for a lifetime. Besides, being hacked is surprise enough." She shooed him away. "Now go, little boy. I'm almost done and you're dragging this out."

She looked over his shoulder, toward the street for any potential takers, but the boy stepped up to the curb, stood up straighter, which made it impossible to look over him. He was a full head taller than Val.

"I'm *not* a little boy," he said calmly. "I'm eighteen and I've been taking care of myself since I was fourteen. I know this place, and I know when someone needs help."

Val looked him over with fresh eyes. There was a resolve to him, and something else that she couldn't quite put her finger on.

A kinship, she thought. *A familiarity. Like I already know him.*

The words came out before she could stop herself. "I'd tell you but you probably already know."

"Nah," he said. "Least not the whole story. Just hacked...er, pulled, your cred and recent history. Figured I'd screw my chances at getting a date if I went any deeper." He flashed a smile, which she didn't find *entirely* annoying.

He does have a nice smile, she thought, then stopped herself. What was the point of this? She had more pressing concerns, and trusting a strange boy that materialized out of nowhere didn't seem like the soundest of propositions. Still, she *was* curious.

"Alright then, cut the bullshit. Who are you? And why do you care?" she asked.

"Trevor," He held out his hand. "Verified actual name."

She eyed the hand warily, kept her own hands in her pockets. "You already know mine's not Tiffany."

"You're right of course. Still, it's a pleasure to meet you." He curtsied and she couldn't help but laugh. He was obviously trying to charm her and it was working. A little.

"Stop it," she protested through curved lips. "You're not funny."

"I know," he said. "It's my worst quality. Just a total deadass. Which is why we'd make a good pair."

"You're not going to leave me alone, are you?" Val asked.

"Nope. That's my best quality. Persistent fucker, I am."

"Well, persistent fucker, I'm Val," she said.

"I know," he said matter-of-factly. "In all seriousness, Val, you *do* look like you could use a good time."

Val arched an eyebrow.

"Oh shit," he put up his hands, "really poor choice of words. A laugh. I meant you look like you could use a laugh."

"Is it that obvious?"

"Yeah," he shrugged, "it's Legacy. We've all got our sob stories. But to hell with that sad crap. I came over to ask if you've ever heard of GIE."

"Guy, like a dude?"

"Yeah, no. Not like a dude. Too much whoring for you, Val. Bad girl," he tapped her lightly on the back of the hand. "I mean like G-I-E," he pronounced the letters slowly, one by one.

She shook her head. "Not familiar with it."

"Holy shit, you're gonna love this." A huge smile lit up his face. "Listen, I've gotta wrap something up just across the street there," he pointed at the lighted sign of the shop across the street. It read *Headspace*. "But I'll be back in, like, two minutes. Promise. Now don't go running off with some fifty-year old creep, okay?"

She put her hands on her hips. "Really?"

But he was already gone, a flash as he ran across the street, dodging through the crowd. As the boy departed, Saul came sauntering up, looking like an upright weasel. The Neanderthal was gone; only the silent knife trailed him.

"Your timing is impeccable, Saul," she said.

"Sad to say, yours ain't. Looks like you're outta luck for the night, love," Saul said, a caricature of false remorse.

"What's that supposed to mean?" Val asked. "I'm almost done. Just one more's all I need."

"Maybe, maybe," he tapped his chin, "but look closely, sweetheart," Saul said, waving around to the street. "You see any tourists? Any likely candidates for your services? Look at how they're dressed."

She did so, turning a critical eye to the people that wandered up and down the streets, in and out of shops. Many appeared to be homeless, dressed in rags and layers of patched plastic. The rest seemed to be just a mere step up from that. She didn't remember this type of crowd in the days prior.

What's happening?

Something was up, a strange difference in the population, but she couldn't piece it together.

Saul watched her closely. "You get it now. Every person you see out here's a taker—no givers anymore, not at this hour. Customer base is all dried up for the night. Gonna have to come back tomorrow."

The reality of what he said hit her—the experiences of the last few hours and days flashed through her head, the shame and filth of it, crashing down into her body and washing everything else out in a cataract of emotion. All for nothing.

"*NO!*"

She wasn't aware of shouting, of even thinking. Rage hummed inside her, drowned out her thought processes. She pulled back on the knife, freed it from her jacket pocket, pounced toward Saul and thrust it toward the lying bastard's chest. The tip of the knife was a centimeter from its target when something hard caught her arm.

"*Hey!*" she shouted in protest, looking around wildly for the source. The boy, Trevor, stood at her shoulder. His arm was interlocked with hers, which was the thing that kept it from moving forward.

"Let go!" she screamed.

"Just hit pause for a sec," he whispered. Then, looking to Saul, he said, "What's she owe you?"

Saul eyed Trevor with suspicion. "This ain't your business, boy. Step along."

"If you don't tell me, she will," Trevor said.

"One mark," Val said through clenched teeth, posture erect and defiant. She strained at Trevor's arm but he was too strong.

"That all?" Trevor asked, "just one little baby mark, huh? Well I think we can take care of that right now."

Val's system dinged with a notification. Payment for one virtual session came through, bringing her total for the night up to ten. Trevor had made the final payment.

She looked up at him, confused. *Who* are *you?*

And just like that, the bottled-up rage overflowed, dissipated into relief. Acts of kindness were so rare that she'd assumed them to be extinct. She took a deep breath to stabilize her emotions before sending the full payment of the night's earnings over to Saul.

"Kudos, hon. You did it. Record time, too," Saul said. His posture was relaxed, a complete denial that Val had almost stabbed him in the heart. His gaze lingered on Trevor, with clearly an avid interest.

Trevor blew a kiss at him.

"Cute," Saul said, then turned back to Val. She shrank from the miasma emanating from his mouth. Her system chimed internally with the receipt of a message. "There ya go," he said, "my end of the bargain."

He actually paid up, she thought with wonder. *But would he have paid if Trevor hadn't showed up?*

She knew the answer. Almost certainly not. He would've strung her along for as long as she permitted. Then again, if Trevor hadn't shown up she'd almost certainly be lying dead in a pool of her own blood.

"Don't bother trying to open it for a bit," Saul said, picking at his teeth. "It's on a two-hour timer. Fair warning, if you try to open it too soon the little clue you worked *so* very hard to procure goes up in a digital bonfire."

Val's face darkened. "What kind of scam is this? I did what you asked."

"Just a little insurance, honey," Saul said, pointing at her pocket. "You wouldn't be the first person tried to stick me after getting what they paid for. Now," he shifted to address both Val and Trevor, "can I interest the two of you in a more lucrative job for the night?"

Trevor smirked. "Maybe later, buddy. Right now, Val and I have some very important business to tend to."

Trevor disentangled his arm from Val's, pulled on her elbow, away from the corner.

"Later, then," Saul called after them. "You know where to find me."

"I'll try to forget, fucker," Val called out.

"Don't bite the hand that feeds, love!"

She flipped him off. When they were out of earshot, Trevor leaned in.

"I was serious, you know."

"About what?"

"About helping. I can," Trevor said. "I'm pretty resourceful. Plus, I have a lot of connections."

Val stopped, pulled her arm from his grip, tried to read him.

What, exactly, do you want? You paid my debt to Saul, now you're offering to help with no clue what you're getting into. It doesn't make sense. You don't make sense.

In her experience, nobody in Legacy did something for nothing, but when she studied him there was none of the humor that seemed to mark their interaction up to that point.

She didn't know what made her do it. Maybe it was out of desperation to find Kat. Maybe a subconscious yearning for human contact.

Maybe, she thought, *it's as simple as instinct.* Something told her she could trust him.

"Okay," she said. "You can help. Welcome," she made a flourish, "to my hell."

His grin lit up the dark night.

I could get used to seeing that.

"Fucking superb," he said, clapping his hands. "I just have one teensy little thing to do. You wanna, uh, tag along?"

What the hell, she thought. *I've come this far, what else can I do that's certifiably crazy? Got some time to burn anyway thanks to Saul.*

"Alright," she said. "Lead the way."

Trevor offered his hand and Val took it without thinking.

The idea of touching someone, a total stranger, was anathema. Yet he had proffered his hand and she had taken it. Easily, without reservation.

What am I doing?

But she didn't have time for the thoughts to linger. From there he pulled her, trotting along wistfully, deeper into the basement of the Arc. The place where no natural light lived.

As they ran, he looked back and in his face she saw the answer to the question that had been nagging her. *Why trust? Why him?* The answer was right there, in the joy of his smile, in the gentle strength of his hand interlocked with hers. There was an honesty to him that made her feel safe.

The realization filled her with a sense of release and relief, that she wasn't alone. That there *were* people out there who cared, who were willing to help.

They continued to run, passing through the rim of gaudy bright pinks, blues and yellows that advertised a multitude of seedy enterprises. The artificial light eventually gave way to the pulsating, dark heart of the Arc. Here lights were snuffed out on purpose. This was a place where people didn't want to be seen.

'Witnessed' was probably more accurate. Legacy didn't have much in the way of law and order, especially down on Level 1, but odds were odds and if a person hadn't been seen or caught on camera doing what they weren't supposed to be doing then odds were they wouldn't be raided by ArcSec.

Trevor slowed. "Almost there," he panted.

"Thank god," Val managed. "This running stuff is so outdated."

"This is level 1, dear, not 7."

"When in Rome."

He chuckled. He had a nice laugh, bright and full of life. Reminded her of Dad's, when he'd been healthy.

They moved down the middle of a six-lane street, dodging through the thickening crowd, in what looked like a downtown long left behind as the city had expanded toward the sky. She pulled him to a stop.

"Never been this far in before," she said.

"Nothing quite like walking down the middle of a street," he said. "Kind of liberating."

The way narrowed; the press of bodies became so thick they slowed to a shuffle. People pressed in from every side.

"I'm sorry," he said, body pressed closely, awkwardly up to hers. He smiled apologetically. "Opens up in a minute. Transfer gate to 2 there," he nodded to the left. "It's right near the entrance to the North Market, where we're headed. Not what you'd call urban planning at its finest."

They made their way past the transfer gate, giving way to open. She and Trevor exited the crush of people ascending the stairs to Level 2 into an immense courtyard of sorts. But where most courtyards were reminiscent of open sky, with couples walking and the playful cries of children piercing the air, this one was dark with a pitch-black mouth that yawed above it. It teemed with people, mostly ones and twos. Packed like bees in a hive yet somehow managing the miraculous dance of evading human contact.

She brought up the light filter on her system. It was still dim, but she could make out some detail. The place was like a cavern, wet and dark, water dripping from the superstructure of Level 2 above. A sulfurous mixture of sewage and run-off.

Val wrinkled her nose but knew that in about a minute her brain would recognize the scent as pervasive and shut the smell off.

There were some tables and kiosks set up here and there, but mostly the courtyard was jammed with people advertising their wares with cryptic holographs—the space above the crowd was lit up with thousands of virtual ads of the individuals hawking their wares.

Trevor pulled her to the periphery of the crowd, over to a bench. He offered her a seat, then sat down next to her. "So. Market virgin, huh?"

"Yep. Market virgin." She looked into his eyes. "Please take it slow with me."

There was that broad grin again. It eased her tension, gave her goosebumps.

But in a good way.

"I'll be gentle, I promise," he said.

"Thanks." Without thinking, she leaned her head onto his shoulder. She could feel his body jerk a bit, then relax. Human contact. So foreign. She'd forgotten how nice it was.

"You have any idea where you are?" Trevor asked.

Now that's a weird question. "No," she said, drawing the word out. "Should I be worried?" Despite the sentiment, Val found none of the fear she should've felt being totally at the mercy of a stranger.

"Kinda like always," Trevor said, "you *are* in the basement of the Arc after all."

"Alright, point taken," she said.

"What I mean is geographically. You have any idea where in the Arc you are?"

"North?" she said, playing dumb.

"Okay so that's a no."

"I give up. Enlighten me wise one," she said.

"I will thank you. We're currently as close to the center of the Arc as it gets," he said. "The dark, foul..."

"Ooh, let me join in," she interrupted. "Don't forget evil and infected..."

"Beating black heart of the Arc," he finished. He raised his eyebrows in intrigue. "This is where the magic and murder happen. The black hole."

The black hole, as Trevor had called it, was fronted by buildings on three sides with a sunken amphitheater on the fourth. After a few minutes of sitting there and observing, the market started to make sense. A weird kind of sense. At first glance it seemed to be a massive congregation of individuals moving around at random. No system, just a mass of people milling about, constantly moving. But then she paid closer attention, followed the tall, bulging figure of a man dressed all in black on the periphery of the crowd. His movements were like that of someone with a compulsive disorder—he took three steps then stopped, looked ahead then up, then straight ahead. He paused there in that position, looking ahead for a minute. Then he turned a little to the side and repeated the process.

She followed another person, this one a veritable giant of a woman, and the meaning of their movements started to cohere. The process was almost precisely the same. A few steps, then stop. Look ahead, up, ahead. Turn and repeat.

Each person was essentially their own market stall, both consumer and purveyor, and each little hiccup in movement represented a transaction, or an interaction, with another purveyor.

"It looks like a mass mating ritual," Val said.

"Funny you should mention that," Trevor said coyly. He nodded ahead. "See that woman in the red trench coat?"

The woman was twenty feet away, impossible to miss. Jet black hair spilled over the shoulders of a glossy red coat. Val considered herself attractive, but this woman, she was surreal. She exuded sex. The tilt of her head and set of her jaw demanded attention. And she got it.

A crowd had formed in a semi-circle around her. Mostly men, some women. They were all facing her but didn't appear to be saying anything. Their silence didn't mean much—she could be communicating with them in other ways. Probably was.

"Pay attention to the crowd, their faces," Trevor urged. His voice was distant, distracted.

She did his bidding. Eager anticipation was evident on their faces. Rapt and in the woman's thrall, every single one of them held in a trance.

One in the group started to move. A short, dark man with a horseshoe of white hair around his head. Standing in place, he started to undulate, his body swaying in time to some music they couldn't hear. His eyes were locked on the woman in red, unblinking, as his movements became more noticeably sensual.

Then, one by one, like dominos falling, each person in the group began to move, to writhe, until the group was a weirdly flowing, pulsing mass of humans, twenty strong, eyes staring, mouths open as they began to moan.

"What in the..." Val muttered softly, her voice tinged with equal amounts of fear and awe. She turned to Trevor but he grabbed her arm. Not looking at her, he said, "Watch. You don't want to miss this." It was then she noticed how preoccupied he was, that faraway look people got when working on their internal systems.

Shit. He's doing this. Hacking, or somehow controlling the whole group. She couldn't help but be a little impressed. The red woman must've been the bait and his point of entry to their systems.

Just as Val turned back to the mass of people, the woman pulled back the flaps of her trench coat. Val couldn't see what the woman revealed but from the group's reaction, she had a solid guess. As one, the group climaxed. It looked as though they'd been collectively punched in the gut.

A loud *oof* came out of them, followed by a machine-gun stuttering of *oohs*, *ahhs* and occasional *oh-shits* as their bodies spasmed, doubled-over. Some even collapsed.

Val sat there on the bench, eyes wide and mouth agape.

This has got to be the weirdest shit anyone *has ever seen.* She didn't know whether to laugh or run away.

Regardless, she was enthralled. After a good long moment, the wave of their ecstasy finally crested and crashed to the ground. The cacophony ebbed to a pleased murmur. Many of them dropped to sit on the ground, eyes closed, beatific smiles alight on their faces. Others ran their hands through their hair, heads drooped, chests heaving.

"What—was—*THAT*?!" Val shouted.

"That, my dear, was GIE. Invented by yours truly." He looked very pleased with himself. "And very lucrative."

"Okay, that tells me nothing. What did my eyes just witness?" she asked.

"It stands for Group Involuntary Ejaculation. Just one of a few of my business lines."

Val was a product of Legacy and hardwired not to be surprised, but this caught her off guard. It was brilliant, snarky and sexual all at once. She burst out laughing and found she couldn't stop.

"Wish I'd thought of it," she said in-between breaths.

"Eh," he waved it off. "Pretty soon it'll be copied by everybody and I won't be able to make money off it."

"Yeah, but still. It's brilliant."

"Thanks," he said, then sat up. "Hey, has it been two hours yet? Saul's timer has *got* to be up soon."

Shit. How could she forget? It was nice to distract herself, but she couldn't let herself lose sight of the real goal. Not even for a moment. *Trevor said he could help. Hold him to it.*

Val checked the time on her system. 5:27 AM. "Another half hour," she said.

"It's almost daybreak." He stood and stretched. "Wanna grab a coffee?" he asked, offering her his hand again.

"Actually? Yeah," she said, "that sounds really good right about now." She put her hand in his, the thrill of human contact once again coursing through her.

"Great," he said, winking. "I know just the place."

He took her out from underneath the Arc where, lo and behold, the sun was rising through strata of orange, pink and purple clouds. She had the semi-drunken feeling of being up all night and coming out to see the dawn at first light. Throwing back her head, Val sucked in a lungful of oxygen. It was nice to breathe fresh air. Or at least *fresher* air.

He bought her a coffee. A bona fide, historically accurate cup of coffee brewed from beans. Grown from the earth, no less. Cost three times the printed and diffused stuff.

They sat out on the patio of a café called *TBD—To Be Determined*. A franchise known for their stout defensive perimeters. She sipped at her coffee, the scent and steam curling around her face, studied his features.

He had a long and straight nose that bordered on big but went well with his jaw line and full lips. Wavy, dark hair partially covered his face. His eyes, that piercing green, followed hers.

"You never answered my question." Trevor said. He took a sip of his coffee and closed his eyes. "Damn that's good."

"What question?" Val asked.

"Why in God's name were you working for Saul in the first place? Not to sound like a creep but I've noticed you all week…"

"You mentioned that part already," Val interjected.

"Hey, I couldn't help it—you stick out like a sore thumb. Anyway, I just couldn't put two and two together. Just doesn't seem to fit."

Her mind went to Kat. She promised she'd do whatever it took to find her sister, and the last week had pretty much proven that. How many avenues had she tried? All of them had led to dead ends.

Except, she thought, studying Trevor, *except* maybe *this one.* Maybe he *could* help. She'd witnessed his abilities at the North Market. She could use those skills…

"Hey, I'm sorry," Trevor said, breaking up her line of thought. "You don't have to tell me. I was just curious. Whatever it is, I can see it's eating at you."

Val took a sip of the coffee, looked out over the patio's railing at the throng of people as they passed, refreshed and ready for the day to begin. The physical exhaustion of the long nights coupled with the emotional exhaustion of the even longer weeks since Kat's absence seemed to take its toll then, pushing down on her shoulders, trying to grind her into the ground.

"No, it's okay," she finally responded. "If you're going to help me, you'll need to know. You *are* still willing to help, aren't you?" she asked. She hated the feeling of being vulnerable, it made her feel dirty somehow, but there was no way around it.

"Absolutely," he said, "just tell me how."

"And whatever this is, whatever…attraction…you might feel, is secondary. I don't want you to think I'm leading you on." The issue had to be very clear between them, strictly defined in no uncertain terms. It was all well and good to flirt and be distracted for a moment or two, but Trevor had said he could help and that was the only thing that mattered to Val right now. Everything else was peripheral.

"Is it that obvious?" he joked. But she didn't smile. "Okay, yes. I promise to help you, even if you wrongly decide you're not into me."

Thank God that's over, Val thought, breathing a sigh of relief. "Thank you. Seriously, thank you so much."

"Don't thank me yet. Wait til I do something truly amazing. It'll happen sooner or later, mind you, but you'll have to be patient like a good girl and wait for it. *Then* you can thank me."

"Alright, deal," she said, offering him her hand. He took it and shook it in a strangely professional handshake, which made her giggle. It felt good. She could stand to do more of that.

"Well for starters, you were right," Val started. "Selling myself isn't something I'm accustomed to. I mean, before this week I'd never even set foot inside the Arc after dark." She took a deep breath, looked down at her coffee then back up at Trevor. He waited patiently, holding his mug in both hands.

Here goes...

"A few weeks ago, someone broke into my apartment and kidnapped my sister."

Trevor was stunned.

When he could finally speak, the words poured out of him so quickly they were almost unintelligible. "Oh my god you've got to be kidding me, please say you're kidding..."

She shook her head, regarded him with sad eyes. "I only wish. It happened in the middle of the night. Guy all geared up busted down our door, knocked me out and grabbed her. She was targeted, but I don't know why."

"So that's why you were subbing for Saul," Trevor concluded.

She nodded. "My sister and I, we live in this weird dead zone. None of the ArcSec offices have jurisdiction over it."

"So no one will investigate." Trevor shook his head in disgust. "Truly pitiful."

"Kinda my reaction too," Val said.

"So," Trevor said, palms up on the table. "How can I help? What do you need?"

"Honestly? I'm not sure yet," Val said. "I mean, I saw what you did at the North Market, figured you might be able to decrypt the packet Saul gave, if you hadn't already tried..."

"No, uh-uh. I don't mess with Saul's security anymore. Tried to crack his timers once before, long time ago, learned my lesson the hard way. Suffice it to say, his security's legit. Plus, I was already in the doghouse," he said, striking a somber tone. "Didn't figure you'd appreciate that."

"So you do have boundaries," she said, appraising him

"With pretty girls I'm trying to woo? Sure."

There it was again, that surge of warmth inside of her. It didn't entirely dispel the specter of her missing sister, but it *was* nice to be able to feel something different other than abject terror and despair, if only for a few moments.

"I'm not usually this charming," he said, winking.

"Sure you're not," she laughed.

He put his hands up in mock innocence.

"Swear to God!" He crossed himself in what was supposed to be a catholic ward against evil but ended up looking like a baseball signal. "Honest! Usually I'm quite dull. Like I said, total deadass, remember?"

"Would a dead-ass give a group of total strangers a simultaneous orgasm?" Val asked.

"Well, yeah, that doesn't exactly qualify. Aside from that. Naturally. But hey, that was only the beta."

"The beta?"

"Yeah, as in 'test-phase.'"

"Of course I know what a beta is, dummy," she responded. "Explain yourself."

"So I'm thinking of starting a business of sorts. Based around GIE."

Val snorted, sprayed coffee everywhere. "Like, a legitimate, pay-your-taxes to Legacy Admin business?"

"Nice." He wiped the droplets of coffee from his arms. "What a resounding vote of confidence," Trevor said, dead pan.

"Sorry," she said, barely restraining her laughter. She wiped the table with her sleeve. "Sorry, go on."

"Clearly by your expression you're not...sorry, that is."

"Ok, no, I'm not. But seriously? A company with its sole purpose being getting people off?" She paused, then reconsidered. "Ok, never mind. Once I said it out loud it makes sense. It's called porn."

"I'm not crazy about the porn designation. Nothing against porn, but it seems," he paused, "small-minded. GIE's a group experience, don't forget that."

"How could I? It's burned on the back of my retinas."

"You have to think bigger," he said. "Imagine a networked group session. Like a seminar. Or theater."

"Oh, so is this art? Or is it an educational venture?"

"Everything's educational, Val. It's just in how you perceive it," he lectured.

"No shit?" she said. "Can I call you professor then?"

"No," he said flatly.

"Professor Trevor," she ruminated, hand on a chin. "I like the sound of that."

"Oh my god, please don't. That's terrible."

"Okay...professor." It was her turn to wink at him.

A breeze blew through, rustling the few leaves that remained in the trees that stood watch over the patio. The wind moved over her skin, sending a thrill through her skin.

And for the briefest of moments, Val forgot everything and allowed herself to feel pleasure.

TORI

TORI
AGE: 33
SEGMENT: COMPLEX PROPERTY; NOVAGENICA CPX
POSITION: VP OF HUMAN ASSETS, SUBSCRIPTION
DIVISION
SOCIAL SCORE: NA, CPX-SPONSORED

TORI HOVERED near the doorway of Patient 165D's room. She was young. The screen by her door said thirteen.

The lighting in the room mimicked the beginning of dawn. Blue light filtered down from the ceiling. The girl slept soundly. Out of all the children in Trial Group Delta this one was the most promising.

So far, she remained untouched by the sickness. At least no outward signs, not yet.

"No thinning of the skin, complexion remains hale. Mucus production normal."

Dr. Viswanathan spoke softly at his right shoulder. He suspected she was trying to sneak up on him, but the dry, powdery scent of the floral perfume she liked to wear preceded her presence.

Tori searched her eyes, restrained his surging optimism. They had been here before. Patient 17 in Group Alpha had charted the same response for two weeks—the same amount of time 165D had been admitted. And then, in the span of two days, 17A was dead. Gone.

"No sign of sepsis yet, so that's good."

"Please," he held out a hand as if to stop her, "no more good news. I can't handle it."

She chuckled. "You're funny, Tori."

He cocked his head, not unlike a confused dog. "You know I'm not kidding, right?"

"Mm-hmm."

"I'm serious. Don't get my hopes up."

A wry smile touched the corner of her purple lips. "Be truthful. You just don't want to jinx it. Instead of using that big brain of yours you're letting your spinal cord take control, using your primitive brain."

"If I were using my primitive brain I would be fucking everything in a five foot radius, not standing here with my stomach in knots over whether we'll find the solution in time for the big coming out party. The Exodus will be mandated soon," he said. "Our trusty AI has informed me that our guy on the inside, the one proposing the bill to the Legacy Admins, is extremely confident. They meet soon and he seems to think it's all wrapped up, neat and tidy with a bow. The Admins'll pass the bill, and we need to be ready when they do."

She nodded toward the girl through the glass. "Nothing you say or think can help that girl in there."

"Patient. That *patient* in there. And you're selling me a bit short, don't you think?"

"What I mean is no hocus pocus, no positive thinking, nothing other than sound science can save her life. It's the only thing that will help us find the solution. Math. Statistics. Computing. Nothing more."

Tori just looked at her. Of course she was right, but he didn't want to give her the satisfaction of acknowledgment.

She raised her eyebrows and peered over her spectacles, managing to somehow look down at him despite her smaller stature. "So I will continue to give you updates and her prognosis which, at the moment, is positive. The entanglement is progressing along its normal course. The only thing that remains to be seen is if her immune system will permit it."

"Ok, fine," Tori sighed. "Now please go away." He shooed her with a hand.

With a motherly sort of look that was a mix of both amusement and admonishment, she retreated down the hall, her white rubber shoes squeaking on the vinyl floor.

No matter what Viswanathan said, Tori would continue to reserve his optimism. If 165D made it another week he *might* allow himself a small celebration. A couple of drinks, maybe, and a thoroughly primitive virtual session with Sato.

That got him thinking. He wondered, once Legacy was gone and the CPXs of the world took over, whether the parlors would be outlawed.

Virtual was forever, and totally unmanageable from a legal perspective so that would always be an option. But the real was so much more fun. And Sato was the best, but his services could only be found in a place as dirty and lawless as the Arc. He hoped the parlors would stay, would even lobby his bosses on their behalf if it came to that, but if not he would accept the fact and move on. Everyone had sacrifices to make.

When Tori was twelve, the Arc stood only five levels tall, stretching to just above a mile high. Even then, as it was now, Level 1 had been a shithole.

But it was a shithole his twelve year-old self loved—a lawless place that allowed him to run with a gang, explore the seediest places the Arc had to offer and try new drugs most people twice his age hadn't tried. That and it was easy to hide from Mom. Not that she was ever home enough for it to matter.

The rats were an ever-present component of Level 1 life. They were as big as small cats, but more numerous and much less friendly. Aggressive. Sometimes he'd see them roaming the halls that led to the other apartments, sniffing and scratching at the bottoms of the hollow doors. It was a struggle keeping them out of the tenement building Tori and his mom shared with some forty other tenants.

He remembered that summer of his twelfth year vividly. It was the summer of the first red rain. A storm system laden with infectious bacteria that turned the clouds red had swept across the area, infecting tens of thousands of people. It mostly resulted in persistent eye infections that could be cured with the right mix of antibiotics. But some, he remembered hearing, didn't fare as well as the infection moved on to the brain and caused fatal swelling. The system had rotated through slowly. Like Jupiter's Great Red Spot, the storm had taken weeks to sweep through.

It had also been the summer he'd joined, and left, the gang. He joined mostly because he was bored, but Tori found he had a talent for theft—light fingers and an ear for eavesdropping. And since his mom was hardly present, he kept busy and informed by shadowing marks on the streets, listening in on their conversations before lifting a little something for the gang from their pockets.

He'd heard about climate change from the marks again and again, but the words meant nothing to him. Frankly, he wasn't sure it meant anything to them either. It was just something they seemed to say when the world did something they didn't like. It didn't bother him much either way, because on Level 1 he was mostly shielded from the rainfall. If you didn't count the constant *drip drip drip* of sewage-smelling runoff from above.

After two weeks of thinning crowds, the marks that did venture out into the red rain were the hardened sort. Wary. Their gang, mostly comprised of kids that didn't have a place to sleep at night, stayed out on the streets later and later, looking for sheep and finding only wolves. A dangerous thing for petty thieves.

At the end of the second week, they'd become desperate and had resorted to hitting the North Market at night. A stupid, dangerous game that ended badly. Especially for Merkle. Tori, at least, had escaped with his life. Merkle hadn't been so lucky.

The next day the rain finally stopped. Tori didn't return to the gang. Not that day or the next. It wasn't until the third day after the end of the red rain that Tori realized he hadn't seen his mom for a week.

He had awoken with his head pounding, body aching. Not long after, the diarrhea and vomiting began. He wasn't sure which came first and was too delirious to care.

He slept fitfully that night, on the cool, grimy tile next to the toilet. When he woke up, dazed and feeling worse than the day before, he could barely push himself up. Still no mom. It was then he knew that if he didn't get up, didn't manage to get out the front door, that he would die there alone in the apartment. His mom would eventually come home, exhausted from a major binge on either work or Buena—the week-long high—and discover his body lying next to the toilet. It could be tomorrow, it could be two weeks. Then she'd have to haul his lifeless corpse down to a decomping vat. Just like his buddy from the gang. Little Merkle. He shuddered at the memory, pushed it away.

A sort of rage and panic overtook him. He wasn't going to die there, he didn't want to die there, alone, with his whole life ahead of him.

Dressed in just a stained t-shirt and underwear, Tori had pulled himself up to a hunched position and stumbled from one wall to the other. Pushed the bathroom door open. Just get out the door. Then worry about help.

He could feel the breath rattling in his chest as he opened the main door to the apartment. He shuffled outward, toes catching on something.

He fell straight forward onto his face. It felt like a light went out inside him. He shut his eyes

only to wake up in the same place, same position, his cheek swimming in a thin pool of vomit. He didn't remember how he had gotten there, but he knew he had to move. Now or never.

He grunted as he pushed up, supporting his weight on sore hands and knees. His joints screamed with pain. Something inside told him there was no way he'd be able to get fully upright, so he pushed forward, starting to crawl. His left leg nudged something heavy, coarse fur scratching at his bare skin. He looked down to see a giant rat, lifeless on its side, tongue hanging out of its mouth, bloated and stiff.

In that moment he'd had the weirdest vision of his own dead body lying beside the rat's, its tiny paw nestled in his hand.

Terror overtook him.

He scurried down the hall on all fours, hands, knees and feet prodding him forward. Not quickly, but steadily. He focused on the door at the end of the hall. That was all that mattered. He pushed on, not thinking just moving, until the flat of his palm rested up against the cool metal of the door. Hand on the knob, he pulled himself up to his knees, pulled the door open and crawled through, letting it bang shut against his side.

It was hard to tell whether it was day or night. Living in the heart of the Arc meant living in the shade of a mile's worth of concrete, steel, carbon fiber and a multitude of other novel materials that made the megalith possible. There was no natural sunlight here, just the lamps on the underside of the next level's superstructure that poorly mimicked daylight.

He looked up to the lights and determined it was night. The lamps had dimmed to an antique, mustard-yellow glow.

Tori kept moving, crawling over the steps and down the sidewalk. The concrete scraped his skin raw, tiny pebbles lodging into his palms and bare knees.

The clinic was just a two-minute walk down the street but it took him an eternity to get there. He retched twice as he crawled, not stopping, letting the spittle drip down his chin. He could feel something leaking out of his anus, hoped it wasn't blood. There was no time to check. Move or die.

His feet and hands carried him forward, teetering but still deciding to support his feeble body. Legs and feet flowed around him, like water around a rock in a stream. The people didn't see him, except to avoid him, and he didn't bother looking. He had grown up here, knew what to expect.

But the clinic was different. A tiny little office in a rubik's mall, but it was owned by a Complex. One of the big ones. If he could just make it there, they might be able to help.

Through the glass door he saw the interior, which was meticulous. Clean and bright. Beckoning to him. Tori struggled forward, the fire burning in his shredded hands and knees screaming at him to stop.

He slumped forward against the door, slapped a red hand wetly against its unmarred surface, leaving a bloody imprint. The front of his shirt was stained pink with what was left of the fluid in his stomach and his underwear was stained black.

He had sat there at the edge of a different world, one that could help, one that cared, and stared back at the heartless void that had spit him up here. People not bothering to look, not giving two shits if he expired right there on the sidewalk in front of them.

His thinking hadn't been clear, but this moment had forced a lucidity upon him—he saw it all as though outside himself. It was in this moment, in this fertile soil that the seeds of his resentment were sown.

Legacy must die.

The rest was a blur. He had been taken in, cleaned, given fresh clothing. Cured. These simple acts were more than anyone had done for him in all the twelve years of his life, at least so it seemed to him. And when he was strong enough to walk again he found himself in a brand new world.

His new home, Novagenica Complex.

He never saw his mother again.

The girl lay in bed, still. Another day had passed and still no reaction. Her immune system hadn't yet reacted to the entanglement.

He had borne witness to so many of these illnesses, the children always thrashing and moaning in their beds. It brought back the events from 15 years ago vividly. His own purgatory.

But 165D wasn't there yet. There were still no signs of illness.

"This is the critical moment."

He hadn't heard Viswanathan sidle up next to him. The pale light enhanced the dark circles under her eyes.

"How long?" he asked.

"Two days, maybe three at the outside."

"You sure? Looks healthy to me," Tori responded, allowing himself a small bit of hope.

The doctor chuckled lightly, without humor. "There you go again. I'd like them all to survive, but that isn't realistic. We just need one, you know that as well as I do. The master program was very clear in its result."

"You mean the AI that sings cheerfully in my ear every morning? *That* master program?" Tori replied.

"Don't be silly," Viswanathan said. "You know full well it's a situational being. It manifests itself as needs require."

Tori blew air out compressed lips. "Sure. Okay."

An orderly, short and lean with the dark skin and ropy sinew of a farm worker, approached.

Viswanathan acknowledged him with a nod. "Keep an eye on 165 here. She'll probably worsen soon and I want her kept clean. Understand?"

The man, with careworn lines etched into him over the years, nodded. But despite the aged face, his expression was lit with a kind of vigor, almost joy, like a lamp glowing from within. It struck Tori as strange.

Another low-level subber. What do you have to be so happy about? All the power in the world and Tori couldn't seem to shake the anxiety he felt.

It's just the sick ward. It's getting to you.

Tori shrugged it off as the orderly took note of the door number before heading off down the corridor. He touched Viswanathan's arm briefly, let go. "Notify me of any changes. Immediately," he said. The doctor nodded and Tori turned and left. He noted out of the corner of his eye, through the silvered pane of glass simulacrum, movement from the room. He stopped, looked. The girl turned, almost restlessly, and all of a sudden trepidation began to seep into Tori. Controlling his surging anxiety, Tori managed to peel his eyes off the screen.

He turned and left, for good this time.

RIKU "CHIEF" OGUNWE

RIKU "CHIEF" OGUNWE
AGE: 40
SEGMENT: COMPLEX PROPERTY; ARCSEC CPX
POSITION: ARCSEC CHIEF CONTRACTED BY LEGACY
ADMIN: MIDWEST ARCOLITH, LEVELS 1-3
SOCIAL SCORE: NA, CPX-SPONSORED

GOLDEN LIQUID glimmered in the tiny glass. A lone light bulb swung on a cord above, throwing shards of light this way and that, glimmering dully on the rough, peeling lacquer of the table.

Ogunwe inspected the glass, lifted it between thumb and forefinger. Five other glasses joined his in the air. He glanced at the faces around the table. Sweet, Morales, Carpenter, Sophin and Reichmann.

The ones I can trust, he thought. *I hope.*

Ogunwe cleared his throat. "To the old school, may it die a hard death."

"The old school," the officers intoned.

They clinked glasses and shot back the contents. The whiskey burned. Ogunwe closed his eyes briefly and enjoyed it.

"Ah, damn, that's rough," said wiry, short-haired Detective Sweet, coughing. Her spiky, silver hair shook as she coughed.

"Terrible, you mean," her partner, Kimberling Morales, coughed.

"Ladies can't handle it?" came the booming, condescending voice of Jonas Reichmann. He was built like a brick shithouse, short arms constantly elevated as they stuck out from a barrel chest. Ogunwe shook his head. Reichmann couldn't help himself. It was in his bones to be a Grade A asshole.

"Call me a lady again," Sweet challenged him. There was no humor in her voice.

Ogunwe regarded Sweet. Hard lines, like eroded canyons, showed the habitual mask of defiance she wore, that seemed to define the majority of her long years. Playful was not her thing.

Reichmann put his hands up. "Slip of the tongue."

"Uh-huh."

"Chief?" Sophin said. His shoulder-length black hair was pulled back in a ponytail; the hair on the front of his head hung down around his face, framing a tan brow wrinkled in concern. Dark eyes questing. Only thirty-two but he looked a good ten years older. "What are we doing here?"

"Don't ya mean, what are *you* doing here, Sophin?" Reichmann jokes. "I mean, how'd an ugly immigrant like you get hired on to ArcSec in the first place?"

Sophin showed a mouth full of crooked teeth. "For my brains, tough guy. You wouldn't understand."

"Touché," Reichmann said, tipping an imaginary cap to Sophin.

Sophin turned back to the Chief. "For real, Chief. What's up?"

"Leave it to Sophin, asking the hard questions," Reichmann answered again. Sophin rolled his eyes, sighed, but Reichmann continued. "Isn't it obvious?" he asked.

"Monday morning, level two," he ticked the points off on his fingers, "shitty hole in the wall and shootin' whiskey on the job? I don't claim to be the sharpest knife in this here drawer, but surely one of you brilliant detectives can see it.

The group waited in silence.

"C'mon, guys, it's a gimme!" he said with mock exasperation. "It's a giving up celebration!"

The group grumbled their laughter. The whole scenario had them on edge, and rightfully so. Ogunwe glanced at Carpenter, sitting in shadow in the corner, who'd maintained his silence throughout the exchange.

"Reichmann, I'll never understand how you made detective," Morales said through a half-grin.

"You and me both, sister," Reichmann responded, then shouted over his shoulder. "Yo bartender, another round!"

"Hey Chief," Sweet said, nodded at two men who sat at the bar, puffing on illegal sick sticks. Grey smoke hung in a haze around them. "Want us to bust 'em?"

"Hell, I'd just as soon join 'em." Reichmann said.

"Jesus, do you ever *shut up*?" Sweet spat.

"Do you ever *lighten* up?" he fired back.

"Enough," Ogunwe said gruffly. They both fell silent. "To answer your question," he nodded at Sophin, "*this* is why we're here." Ogunwe pulled a file folder from his lap, tossed it on the table. Sophin probed it with thick, calloused fingers, opened it to reveal a sheaf of copies. In bold, just below the ArcSec letterhead, the copies were labeled *PASS DIRECTIVE*.

"There's a copy for each of you." Ogunwe said. "Read it, then we'll talk."

Sophin passed the copies around. While their heads were buried in the report, the bartender delivered the second round of shots, the old-fashioned glass making a solid *thunk* as his shaky fingers set each one down on the table.

Since his meeting with Briggs, Ogunwe had wavered between bouts of despair and fits of rage. Through all of it, there was one constant that solidified his resolve, pushed him past the fears of what could, what *would* happen to him if Briggs found out.

Olivia.

Without her monthly treatments, she would wither away. Die. He wasn't about to let that happen, ArcSec oath or no. Her safety, her life, was more valuable than his.

Ogunwe watched emotions arise unbidden—concern, anger—across the officers' faces as they read. Carpenter was the first to finish. His already pale Complexion had grown alabaster, bloodless.

"This is a disaster." He glanced up at Ogunwe, expression like a bewildered child looking to his father for answers. A plea for it not to be true. Ogunwe's heart wrenched; he knew exactly what was going through their heads and hearts. They were feeling what he had felt when Briggs had handed down the orders.

Ogunwe pushed it down. Now was the time for decisions. The gnashing of teeth and beating of the breast would have to wait.

All gazes fell on him, expectant. He nodded, and it was though they all deflated, shoulders sagging, worry painted in the lines around their eyes.

"Briggs sent the directive a week ago." He wanted to say more, but instead let the words weigh on them. Reichmann broke the silence.

"Well, shit," he said, lifting his shot glass, "guess I was right. Here's to giving up." He pounded the shot back.

Morales's face flushed red through brown skin. "What do we *do*?" Her words came out quickly.

"What *can* we do?" Sophin replied plaintively. He had an elbow on the table and was rubbing his forehead. "Ninety-five percent? I mean, really? We'll all be out of jobs within a couple weeks."

"Or stuck doing guard duty at a transfer gate," Carpenter said.

"Baby-sitting, you mean," Sweet responded.

Reichmann grunted. "No fuckin' way, not me. I'll be a shit-stirrin' paddle boy before I enforce this piece of garbage." He threw his copy of the report on the table in disgust.

"Most sensible thing you've said all day," Sweet said.

A chorus of agreement went up around the table.

Carpenter was looking shrewdly at Ogunwe, blue eyes unblinking. Some of the color had returned to his face. "What's your angle, Chief?"

"No angle," Ogunwe said. "Seems pretty simple to me. This," he tapped a long finger on Reichmann's copy of the report, "isn't right.

"When I joined ArcSec fourteen years ago it was a way into a new life. I was clueless, but I made a point to learn. Learned to do my job well. More I did it, more I believed in it. Before ArcSec, I couldn't have given two shits about security. But I've always held the belief that a job should be done well. And in our world that means adhering to the principles of right and wrong.

"You can't work security and not believe in right and wrong. It's in our core." Heads nodded around the table. "I put criminals in boxes because I knew what they did was wrong, and that someone had to be on the other side, on the side of right, to protect the innocent ones. So when I moved up from confinement to enforcement, it was the natural extension for me. To protect victims from perps, no matter what their bank statement says.

"PASS changes all that. Everything. Takes my fourteen years of service and-"

"Took a massive dump on it," Reichmann interjected.

"And set it on fire for good measure," Sophin said dejectedly.

Ogunwe nodded. "All the work we've done, all the years we've put in to try to make our home a better place, it'll be undone in a few weeks' time."

"We hear ya, O," Carpenter said. "I myself couldn't agree more, but there's a reason you brought us all here. So what's your angle?"

Ogunwe stared down at the center of the table, felt his body rocking of its own accord. He had to be honest. They couldn't be manipulated into this. "You're here because I trust you, unequivocally."

"Seems like a judgment error on your part, there, Chief," Reichmann said, kicking back to put his feet on the table. Smoke wafted over his head, swirled through the dim light of the bulb. He fanned it toward his face with both hands, nose high in the air, inhaling deeply.

"They've got some paint thinner in the back, Reichmann, if you need something stronger," Sophin said.

Ogunwe cracked a smile. "Don't prove me wrong, Reichmann," he scolded, wagging a long index finger at him.

"Hey, don't say I didn't..."

"Jesus, please shut the fuck up, just for a second and let the Chief speak," Sweet said, her hair bristling like a dog's hackles. She held a palm up in Reichmann's direction. "Go ahead, O. You said you trust us. Why? What for?"

Ogunwe paused. He wanted his team's commitment, and to get it he needed to maintain the appearance of a steady, level hand that was neither excited nor reluctant. No matter that inside he was an agitated ball of barely contained nerves.

"We have to consider our options." He held up a finger. "We can do as we're told. 95% rule goes into effect, most of my staff lose their contracts or opt to take a huge cut in benefits and, as you guessed, most of you get switched to sentry duty."

"Two," he held up another finger. "We can all give Briggs and ArcSec the flying fuck-you, walk out and talk to whichever media feeds will bother to listen."

Carpenter frowned, shaking his head. "I don't see how either of those help anybody."

"Yeah," Morales agreed. "Media's too fragmented. There's no feed out there with any semblance of authority. I mean, maybe a couple that don't deal in strictly editorial, but ArcSec would just tell 'em to kill it."

"Or dump money into a smear campaign," Sophin interjected. "Ruin our lives."

"Seems like that's kinda the vibe they're putting off now," Reichmann said.

"That brings us to the third option," Ogunwe said. "The last one, far as

I can see." He sat up tall in the straight-backed chair, surveyed them one by one.

"Before I get into this, I want it to be perfectly clear that all decisions we make from here forward are unanimous. All or nothing. Now's the time for dissent. And if that's the case, we leave this table without another word, no harm done, have a nice life. Either way, I'm putting in to protect all your jobs."

"Thanks, O," Carpenter said.

"Aw, c'mon Chief, spill it already! The suspense is killing me," Reichmann said, taking his feet off the table and leaning forward.

"Alright. Our last option," he held up a third finger, "is to revolt."

A stunned silence overtook the group.

"Revolt?" Carpenter said, his face looking as if he had eaten something foul.

"Yes!" Reichmann clapped his hands, startling Morales at his side. "Take 'em out! Guns blazin', reboot the system!"

The others ignored him.

"What do you mean, Chief?" Carpenter continued as though no one had spoken. "What does that even look like? I mean, it's not like Legacy has any power over anything anymore, and subscription to join a Complex is voluntary, so I'm not sure what you mean."

"That's why we're having this meeting. To decide what it looks like, together. But first we need to decide whether or not to move ahead in the first place."

Ogunwe dug in his pocket and brought out six green and six red marbles, then pulled a tiny cardboard box from underneath the table. There was a small hole cut out of the top.

"I'm going to leave this box here. We'll all go to the front of the bar and line up. One by one, we'll drop either the red or the green marble in the box. Red for dissent, green to move forward with option 3. You will not speak until after the votes are counted." He gave them a hard glare. "Understand?"

Everyone nodded.

"Alright, good. Let's get started."

The six of them each took a green and red marble each and moved to the front of the bar. The sour stench of old beer clung to the rough and cracked, wooden surface. The two smokers eyed the group, their gazes lingering no longer than a moment. Not a safe part of the Arc for prolonged eye-contact.

Ogunwe's group shuffled forward, one by one, to the table that held the box. The process took no more than a minute.

"Well?" Morales said, taking a deep breath.

Ogunwe went to the table, motioning with a hand for them to follow. The others joined, crowding around his side. A nervous dread boiled up inside him. This was it, the moment of truth. What would happen if he opened it up to see a red marble staring back at him? He hadn't considered it fully. Thought he knew his people, how their minds worked. But what if he was wrong? Like his dad always said, the suspense was worse than the knowing. He flipped the lid off the top of the box.

Six green marbles sat in the bottom.

Ogunwe felt a spike of adrenaline flood his system. A feeling of *rightness* overcame him. That all events had come together and culminated in this moment. That all was as it should be. A sense of unity. He couldn't help but smile.

"Hot damn!" Reichmann shouted, the others flinching. "Bartender!" He whirled his finger in the air.

"Now what?" Carpenter asked, worry suffusing him, from the tight lines around his eyes to his hunched posture. The general mood dampened. Everyone grew quiet as those two simple words brought them back to Earth.

"We celebrate, dick," Reichmann said, jabbing Carpenter with an elbow. "Way to be a downer, dude. I'm moving over here where the party is." Reichmann moved in-between Morales and Sweet, which earned him a hard-eyed glare from Sweet.

"Now," Ogunwe said, addressing Carpenter, "we start to plan. We need lines of passage to circumvent PASS. It'll be critical to be able to bypass the checkpoints, so we can move people and materiel back and forth, or in this case up and down, across the front. But we'll get to that, Matt. For now, let's just enjoy this." Carpenter smiled weakly, but the worry in his eyes lingered.

The bartender brought another round of whiskey.

"Chief O," Sweet said, lifting her shot glass off the table, nodding at Ogunwe. "It seems like we have a cadre," she said, showing a rare, small smile.

"More like a cabal," Sophin said, clinking his glass against Sweet's.

"Cabal needs a name, Chief," Reichmann eyed the liquid in his glass, rolled it this way and that. "I propose The Golden Road."

Sophin made a show of his shock. "Reichmann, I'm...speechless. That's really good."

"Shove your sarcasm up your ass, Sophin," Reichmann said.

"No no, I was being serious. It's reminiscent of the underground railroad," Sophin elaborated, speaking swiftly. "A way for people to secretly move toward freedom."

"Aw shucks, that's sweet," Reichmann said, tucking an ear against a shoulder like a shy child. "I just said 'golden' because of the bourbon."

"It's decided," Ogunwe interrupted, joining his glass with theirs. "To the Golden Road."

The rest of them joined in, leaving only Reichmann, swirling his glass under his nose. There was a mischievous twinkle in his eyes.

"To the Golden Road," he whispered. He clinked his shot glass against theirs. "To the revolution."

"To the revolution," they whispered together.

To you, Olivia, Ogunwe thought, then threw the shot back. The fire crept down his throat. It came to rest in his belly, stoking a warmth and a feeling that was largely foreign to him. Optimism, a wild, free sort of belief in their purpose. For the first time in a long time, he felt good.

Out of the corner of his eye he saw Carpenter down the shot, a forced half-smile plastered somewhat sloppily on his face.

Now what?

Carpenter's open question and worried eyes tried to haunt Ogunwe, bring him down. He shoved it away. There would be time for that. For now, let them celebrate.

Before things got tough. Before they got dangerous.

MANOLO

MANOLO
AGE: 57
SEGMENT: COMPLEX PROPERTY; NOVAGENICA CPX
POSITION: LABORER, LEVEL 6; LIFELONG
SUBSCRIBER
SOCIAL SCORE: 629

THE ANIMATED LIGHTS OF THE MARQUEE pushed against the twilight in an orgy of pink, yellow, green, and violet. The sign was vertical and ran the height of the theater. The building had been stylized to represent the old, stone structures of the early twentieth century. The marquee was the only modern touch, a thing born of necessity. It showcased the multitude of interactive plays that were currently on offer at the theater.

Manolo walked slowly beneath the marquee, bathed in its light. His gait felt aimless as he wandered down the village's main street, hands pushed inside pants pockets, elbows jutting out. A warm glow spilled from doorways and storefront windows. Bells jangled as customers—all property of Novagenica—flowed in and out of the shops, conversing, laughing, seemingly inured to the world's pains.

The air was crisp, verging on cold, and carried the aroma of cooking. The dense, earthen scent of roasting coffee permeated another doorway.

Manolo glanced surreptitiously to his right where Maria, short, sturdy and uncharacteristically icy, walked at his side.

"I was beginning to think you'd died," she said.

Manolo cringed, but kept walking, elbows still awkwardly jutting. He looked like a shy, uncertain boy on a first date.

Just get it over with, he thought. *Doesn't matter that she was Esme's friend. She was yours too, and you don't just cut people out of your life. She deserves an apology and you know it.*

But Esme...Esme had a big heart, and she could've made friends, and did, with even the most asocial of creatures. Manolo was different. If the roles had been reversed, he just knew that Esme would have surrounded herself with the love of friends and family instead of shrinking away into isolation like Manolo had.

He opened his mouth, but before the words could come Maria spoke.

"Now that I know you're alive, I can't decide which is more hurtful," she said, scratching her head. "You being dead, or me knowing you were alive the whole time and choosing to ignore my messages."

The words struck Manolo worse than any physical blow could have. He'd never seen Maria so angry.

Hell, you've never even seen her angry at all. Period.

Which made it all the more shocking. She was right, of course. He'd been so wrapped up in his own emotional state, in his own reasons for misery, that he couldn't even consider how his actions affected his friend.

Manolo pulled his hands from his pockets, let them drop to his sides. "You're right," he said, posture open and etched in compunction. "I'm sorry."

Maria regarded him from behind her circular, horn-rimmed glasses. Lips pursed, brow furrowed.

Suddenly she took him by the wrists and pulled him in, squeezing hard around his middle. Her body was firm, warm against his. Comforting. The inimitable relief of forgiveness flooded through him, flushed out the shame.

They stood there in embrace for a moment, her tightly curled hair tickling at his chin and nose. He turned his head, rubbed at the tip of his nose with the back of a hand. A full head shorter, she leaned away and looked up, locked eyes with him.

"Whatever it is, we can talk about it. But not until you make me a promise," she said.

"Whatever you want," Manolo agreed.

"Don't shut me out again," she said, and punched him in the ribs.

The blow caught him by surprise, and for a moment he found himself trying to catch his breath.

Damn, that hurt, he thought. *But hell if I'm gonna let it show.*

Instead, he released Maria and simply said, "Point taken."

But she wasn't going to let it go so easily. She wagged a finger at him.

"I mean it. Life isn't meant to be lived alone. You're a man, so you're naturally stupid, but this sort of stupidity I can't accept."

"Okay, okay," he said. "I promise. Just don't punch me again."

She stared him hard in the eyes, unblinking, until she was satisfied he was telling the truth, then started off down the sidewalk once more. They strolled, with no destination in mind and no purpose but to reconnect.

But that's not entirely true, Manolo admitted to himself. He *did* have an ulterior motive, but one that he couldn't just spring on her. She'd sniff it out from a mile away if the subject didn't arise organically.

"Have you ever," Manolo started, then stopped to consider his words. It wasn't the thing he wanted to talk about, but instinct told him Maria wouldn't let him off the hook unless he opened up about his struggle. "Have you ever told yourself something about yourself so many times it starts to become true?"

"Sure," Maria said. "I do it all the time. Every morning I wake up and say to myself, 'Maria, you're one tall, sexy, blonde bitch, and damn it if you don't look thirty years younger than your actual age. And voila," she flung both arms out to the side, "works like magic."

Manolo chuckled. "See, this is why I can't talk to you. You can't take anything I say seriously."

"You want serious? I'm all business." She made an effort to scowl. "Lay it on me."

Now that's *the Maria I know. All laughs, and totally unlike any engineer I've ever met.*

Manolo's bald head skimmed the bare lower limbs of a river birch. Its trunk, white with flayed strips of gray bark, grew from a decorative grate in the sidewalk. He ran calloused fingers along the length of a branch, feeling nothing but pressure through the hardened tips of skin.

How do I start? It was almost too hard to admit, but he needed to be honest, to release his burden, and Maria was the only remaining friend, the closest thing to family he had left in the Complex.

"I've been telling myself that life isn't worth living without Esme," he said, the words pouring out. It was like the words were tied to a string, which were tied to a cork that had sealed in all the caustic thoughts. The words came out, pulled on the string and popped the cork, releasing the bitter flood, emptying him of the negative emotions.

"Okay," Maria said. "And...you know that's a lie, right?"

"Right," Manolo said, squinting and rubbing at his forehead. "Of course, I know you're right."

"Naturally," she nodded. "It's the birth-right of the feminine persuasion."

"But on the other hand," he said, "it feels like a desecration of her memory." He sighed, felt the weight of what he'd been carrying around for so long finally lift.

Maria's reaction wasn't what he'd been hoping for. Instead of empathy, she simply snorted. "Men," she said.

"What?" he asked.

She frowned. "I'll tell you what, mister. You're all so egotistical, and for some reason you assume women are fragile. Do you *really* think Esme was so wrapped up in *needing* you and your approval that she'd be offended if you moved on after her death?"

"That's...that's not what I'm saying," Manolo protested.

Maria stopped, crossed her arms over her chest. "Then what are you saying?" She drummed her fingers on one arm as she awaited his response.

Manolo's jaw worked soundlessly. In one fell swoop, she'd challenged his assumptions and laid bare the falsehoods that lurked beneath.

Maybe I don't *understand my reasons as well as I think I do.*

He cocked his head to the side. "Huh."

"Huh what? What's that supposed to mean?" she asked.

"It means, I don't know, maybe you're right, Maria."

"Psh," she waved a hand at him, "what's new?" And just like that, she dismissed his struggle as so much egotistical posturing, as allowing himself to get wrapped up in himself so wholly that he couldn't see the bigger picture. Part of him wanted to be offended, while the other part felt like the sky was finally opening up again.

Manolo chuckled, and they continued their stroll beneath the glow of the village lights.

They stood at the edge of the village, where Main street dead-ended into a hedge. Another street bisected it, ran North toward the giant apartment building that housed the majority of Novagenica's low-tier subscribers. Behind and beyond the bright façade of main street, it hulked in the dark, vaguely threatening at the edge of Manolo's vision.

They'd whiled away a couple hours catching up, Manolo talking and

Maria joking, but it was getting late. He still hadn't figured out how to broach the subject, and in response his body tensed.

"Time for me to head back to the ghetto," Maria said, pointing backwards over her shoulder with a thumb.

"And me to greener pastures," Manolo said.

"You should get out of that house," Maria said. "It's not doing you any good to stay there anymore."

House.

The connection hit Manolo instantly. Housing and access were bound together as one in Novagenica. He had to seize the moment before it passed.

"Speaking of houses," Manolo said, trying to sound casual and rein in his urgency, "how come you stay in that ghetto, all packed together like rats? Engineer like you could live anywhere you want, even up in the northwest." Manolo indicated the management enclave with the angle of his jaw.

"Oh they offered it to me, for sure. Could've brought the whole Sinao clan with me too. But you think I'd enjoy living there? My whole family, out of place, no community?"

"Huh," he pretended to ruminate, looking down at the sidewalk. "Guess not. So how does that work, then? I mean, do you still have access to pretty much anywhere in the Complex, even though you turned down the house in the hills?"

"Oh sure," Maria smiled, turning her hand in a broad arc. "I just wave a hand and all the doors open for me."

"I'm being serious," Manolo said, letting his disappointment show through. It was hard to know when she was telling the truth or being facetious.

"So am I," she said bluntly. "I literally wave a hand and the doors open. All of 'em."

"Oh," he said, taken aback. *Wasn't expecting that, but don't waste the opportunity.* "You know that tall, monument-looking thing out in the fields? I pass by it every day and I wonder what the heck it is."

"You mean the obelisk?" she asked.

"White, few hundred feet high, narrow and pointy?"

"Yep, that's it. That's comms and surveillance," Maria said.

I knew it, he thought, but said, "Surveillance?"

"Not surveillance in the old-fashioned sense. Really just the place where all the sensor relays meet."

"Ever been in there?" he asked, tone striking the balance between curious and conspiratorial.

"Sure, tons," she said. "The main construct for Novagenica's AI is housed there. Doesn't need much tinkering, it tends to right its own wrongs. I just show up to look pretty."

Now for the real ask.

"Do you think you could show me sometime?"

Maria slashed a hand through the air. "Out of the question. Can't do it. Brass would skin me alive."

The air went out of him. He kicked at a rock on the ground, sending it clattering down the street. *There goes that,* he thought, masking his disappointment with a well-timed yawn and stretch. "Oh wow, really?" he said, covering his mouth with the back of his hand.

"No, not really," she slapped him on the shoulder. "Have you forgotten who you're talking to? I'm the wizard-bitch with a magic wand. I can unlock *all* the portals, honey. Stick with me and you can go anywhere, see anything."

Manolo slapped his forehead. "Oh Maria," he said.

"Yes," she affirmed. "Yes."

"I'm gonna hold you to it now," he said before turning to the hedge.

"I'll do you one better," she responded. "You're off tomorrow, right? Meet me there at 10:30 and I'll give you the tour."

"AM?" Manolo asked, suddenly uncertain. He hadn't planned on anyone seeing him enter the tower.

Maria wrinkled her brow. "Of course AM. When else? You're not running some secret ops, are you?"

Manolo rolled his eyes. "You know me," he said. Unsure of what else to say, he headed across the street toward the hedge. When he got to the gap, he stopped briefly and looked back. The edge of the village was where the lights stopped. Maria hadn't moved. She stood on the corner, her body painted half-dark, half-light. He couldn't make out the expression on her face, but her body language spoke of doubt, of a question she wanted to ask.

Let's circumvent that for now.

He walked back across the street, put his arms around her and pulled her in close.

"See ya tomorrow," Manolo whispered. "And thanks," he said.

She breathed into his chest, clutched the sides of his shirt. "10:30 sharp," she said, looking up. "And you're welcome. I'm glad you called."

Her expression contained something wistful Manolo couldn't parse, but whatever the initial question was, it remained trapped behind her lips. She turned and left, headed northbound into the night. Manolo rushed back across the street, ducked through a gap in the hedge. Onto the lush, wet grass of the greens that surrounded the two Research buildings and the silent, stone Barracks that watched over them.

His scuffed white sneakers collected dew as he crossed the lawn. The wetness seeped through and gradually soaked his socks. He looked up, saw the faded memory of stars, mostly drowned out by the light emanating from the Arc far to the north. It was a moonless night, which allowed the stars to shine a little brighter, and would've been perfect for what Manolo had in mind. Now he was forced to enact his plan in the plain of day.

Secondary plan, he corrected himself. *In case the girls' idea of playing pretend doesn't work.* The thought settled his nerves. There *was* a way out of the Complex. He just had to know the right people.

"I'm the wizard-bitch with a magic wand."

Manolo grinned. It seemed he *did* know the right people.

The tower was like the sword of a god, bursting through the earth's crust, impaling the blue line of sky. It thrust out of the center of cornfield, impossibly tall and imposing for how narrow it was. A path cut through the swaying green stalks, led to the foundation, which was sunken in a square around the tower's base.

Manolo waited on the eastern side, in the full sun, as Maria walked up.

"Good morning, sunshine!" she said. The golden light glinted off the tight curls on her head, made them shine. "You're nice and early," Maria said. "I appreciate that. Not like my worthless son-in-law," she grumbled, "I don't know how my Marisa ended up with a dolt like him."

"Punctual to a fault," Manolo said, "especially when I don't have the AI breathing down my neck, telling me where to stick my feet."

"That can be annoying," Maria agreed. "Lucky for you, I can just turn it off," she said, walking right up to the vault-like door set in the tower's base with no indication of stopping. Manolo put a hand out to stop her, but the door *whooshed* to the side just before she hit it.

She turned around, a shadow in the dark confines just beyond the doorway, beckoned with a hand. "You coming?"

Manolo hurried inside, his sneakers clanging on a metal grate. The door slid shut behind him with a sound like *doom*, its closing reverberating through the small space, and he couldn't help but feel a touch of claustrophobia, like the door was the lid of a coffin, shutting him in.

Just your nerves talkin', old boy. Get a grip, before the machine sees what you're feeling.

Instead of looking inward, Manolo focused outward, on his surroundings. The inside of the tower was a hollow cavity, and Manolo and Maria stood on a square metal platform supported by four horizontal shafts, one per side, that jutted into the walls. Through the platform's gridded floor, Manolo saw empty space. A dark nothing.

Is that just the bottom of the platform or is that...

He gripped the cool metal of the waist-high railing, leaned out over the edge for a better view of what lay below.

Yeah no, that's nothing. There's nothing below us.

"Careful," Maria said, as if to respond to his thought, "that's a hundred-foot drop. And I am *not* planning on picking up Manolo soup today."

Manolo hastily pushed back from the rail. "Holy..." he muttered. "Why's it so deep?"

"Geothermal. It's how we cool the unit," Maria said.

"The unit?"

"It's what we engineers call the machine intelligence running this place. It's the unit's world, we're just livin' in it. So," she flung her arms out. "What you wanna see first, the bottom or the top?"

"I don't care," Manolo said. "You decide."

"Hold on to your shorts then, this baby moves fast," Maria said.

Manolo gripped the rail in sudden fear, *Manolo soup* repeating in his head. The platform shuddered and whined, then began to crawl upward.

Manolo rolled his eyes at Maria. "Hilarious."

She shrugged in that *what can I say?* manner.

Manolo studied the walls as the platform crept upward. The surface on the inside was crafted of the same stone as on the outside, the only difference being that the inside was covered in lights. They were every-where, blanketing every square inch of the interior. They consumed his vision, in one color—white—blinking away. At least that's what it looked like at first. But when he stepped back, angled his neck to look upward, then down, trying to take it all in at once, the blinking turned into some-thing that had rhythm, sweeping up and down the walls. Like language, or music.

"You see the lights? Don't answer that, stupid question. Each one is a processor. The walls themselves are made of ultra-light, ultra-strong mate-rial laced with conductive patterns."

Manolo shook his head. "I have no idea what that means."

"It means that the building itself is the machine. You can't point to one thing in here and say, 'this is the heart of its intelligence.'"

"The tower's alive?" Manolo asked, feeling the downward pull of dread and sudden doubt weighing on him.

"Uh," Maria hemmed, "that might go a bit far. The AI *is* a super-intel-ligence, but to say it's alive? I'm not comfortable with that."

Well that's a small relief, he thought. *Still, the whole building? And they were inside it? Creepy.*

"What about the lights?" Manolo asked. They continued to twinkle across the white surface, like whitecaps across the surface of an ocean of light.

"Just a way for it to communicate issues. It can tell me where the prob-lem is with maps in my head and precise locations, but the lights are a low-tech failsafe. Redundant. Plus," she added, head craning back, mouth agape. "I think they're pretty."

On cue, the lights swept up and down the sides of the walls, then exploded in a display of fireworks. Maria leaned out over the railing and patted the smooth white stone. "Aw shucks," she said. "I love you too."

"Is it...listening to us? Right now?" Manolo asked.

"It's listening at all times, everywhere. Definitely trained on anything that might approach the outer walls, but also on the day to day of every-thing that happens inside the walls too."

"Don't you find that a little, I don't know, invasive?" Manolo asked.

They reached the top. The platform stopped and she looked at him oddly, head cocked to the side, eyes concerned behind her glasses.

"Only if I had something to hide, Manolo."

It was an invitation for him to confess and he knew it. If, on the one hand, he took it, then the jig was up. The AI would know everything about his motivations and he'd be locked in the Complex forever. On the other hand, if he lied and said nothing then he wouldn't get what he needed.

Somewhere in the middle will work best. Tell the truth, but not all of it.

His head was within a couple feet of the ceiling. He reached a hand out, felt its surface. Smooth and cool. Here, the stone wasn't studded with lights, but had a heavy wire the thickness of a soda can running up the side and through the top, where the sides of the prism terminated.

"Ah," Manolo hesitated. He pressed his palms together, fingertips rubbing at his lips. "I have a problem, and it's a bit embarrassing. I just don't want it recorded for eternity, to be judged later. Is there a way I can, you know, turn the ears off? So it's not listening?"

"You? No. But me?" Maria's expression softened. "Of course." She stared into the distance for a few moments, off to the side, pupils dilating, then came back, turning her head back to him so he knew she was back. "There. Done. What's on your mind, friend?"

"You're sure it's okay?" Manolo asked.

"Yes." Maria rolled her hands in tight circles, urging him to get on with it.

"I've been denied a day-pass, so many times I'm losing count. Marco says it's because the machine thinks I'm a flight risk."

Maria tapped at a tooth with a crimson-painted nail. "I knew I should've hooked Marisa up with that boy of yours. He's a smart one."

"So he's right?" Manolo asked.

"Of course he's right. The unit controls everything in here. Those applications you fill out don't even get seen by human eyes. They get parsed by the unit and overlaid on top of your existing psychological model. But you should've just come to me. I could've told you you'd be rejected. Don't waste your time."

"So, what, that's it? I'm stuck here for life?" Manolo asked. The prospect of being stuck inside 's white walls was like a noose cinching slowly around his neck, a centimeter per year. "I don't get to see Marco or the girls where they live? Not for their performances, holidays, nothing?"

"Slow your roll, bucko. Did I say any of that?" Maria said, reaching a hand up to his shoulder, giving it a squeeze.

Manolo looked down. "No, but it sure as hell sounds like it."

"Hey, what did I say last night? When you asked me if I could get you in here?"

"I dunno," Manolo said.

"I'm the wizard..." Maria said slowly, making a show of enunciating every syllable dramatically. "C'mon, you can do it. Say it with me."

Manolo found Maria's eyes, twinkling with their usual mirth and mischief.

"The wizard-bitch with a magic wand," Manolo repeated.

"Good!" Maria clapped. Then, pretending as though she held a wand in one hand, she whipped her arm left, right, up, down, then directly at Manolo, making *poosh* sound as she did so. "There you go, princess. Good for a day pass, courtesy of your fairy god-engineer-mother-type."

"It can't be that easy," Manolo stated.

Maria's face went flat. "Did you just miss what happened here?"

"Well no, but..."

"Look, if you need details," Maria said, taking his hand and patting it, "I can't just issue you anything. That's still the job of the unit. What I did was basically make it deaf to you. It can see that you're there, but won't register anything the algorithm defines as negative, or flighty. Feel how you want to feel, say what you want to say."

Abruptly, the platform began to descend again and Manolo moved his hands back to the rail, putting his rear-end up against it for support. He tried to maintain his composure, to look thoughtful. To hide the pure joy he felt. Being invisible to *the unit*, as she called it, was better than just being given a day pass and being shut of the whole ordeal.

This granted him an anonymity, an ability to do more than just escape. *But what are you gonna do, Manolo? Old-timer gonna take a cane to the system, beat it to death? Be realistic. Escape is enough.*

The platform reached ground level again, and the lights swept around the interior of the building in a cyclone.

"Happy with your tour?" Maria asked, moving to the door. It *whooshed* open for her.

Manolo reached out and patted the walls again. "It really is amazing."

"I know. Enough processing power in these walls to maintain a constant census of the world. Given enough sensors, that is."

Something about the comment struck Manolo as off, made him feel sick. He remembered the way it felt to have the footprint meter in his head, telling him where to go and how quickly to get there. And the warning siren, when he'd spoken to the patient, dropping him to his knees. The sheer invasiveness of it made him feel nauseous all over again.

Imagine that all over the world. Human automatons in lockstep, doing the bidding of a machine that cared nothing of their emotions, their lives, their aspirations. The thought gave rise to a sense of horror. *Could that be possible?*

He felt a tug on his sleeve, pulled back.

"Hey, you okay?" His eyes focused. It was Maria. She'd caught him woolgathering, and by her concerned expression his face was an open window to his thoughts.

Manolo rubbed a palm down his forehead, slid it down his face. "Yeah, sorry, was just thinking of what you said." *Change the subject, now.* "Is it really going to work?" he asked as she pulled him in tow out of the dim interior to the bright sun of mid-morning. They stepped out amid the corn stalks, rustling in the breeze.

"Give it a week and apply again. That doesn't work, then I don't know what to say," her eyes went wide, googly as she shook her head, "you must be cursed or something."

Manolo felt the sun on his face, closed his eyes, felt it on his eyelids. The breeze tickled his skin. He ran both hands along the side of his bald head, breathed in the clean, fecund scent of dried earth.

Not cursed, he thought. *Finally free.*

RAY

RAY
AGE: 36
SEGMENT: COMPLEX PROPERTY; PERSENSE CPX
POSITION: REDACTED
SOCIAL SCORE: NA, CPX-SPONSORED

"HOW DOES THIS LOOK?" Ray asked. The sotto voce he spoke in was little more than a subvocal hum. He sat on the patio of a café on Level 4 called *To Be Determined*. A chain that advertised the stoutest of defensive perimeters.

He peered through the narrow, vertical stripes of the security fencing. It was eight feet tall, made to look like wrought iron, with bougainvillea ivy wending its way through the bars. He leaned forward on the edge of his chair, sniffed at a purple flower. It smelled vaguely sweet and light, like honey and rose petals. He flicked the security bar with a fingernail, listened to it chime. A light chuckle bubbled up from within. The bars were an affectation, a feel-good security blanket. Nothing more. Nothing was safe, especially against the likes of Ray.

The early rays of morning light pierced the outer shell of the arc, angled through the buildings in narrow shafts. He sat back in his chair and observed the early morning crowds, a human torrent along the pedestrian ways, as they headed to work. Heads nodded in time to private concerts, while others remained cocked in that peculiar way they did when watching a feed no one else could see.

Mouth-breathing sheep, the lot of them, he thought with an inward sneer. He was surrounded by the soft, by a population that only wanted a distraction out of life. Everywhere he went, whether it was the upper echelons of Legacy territory—like where he was now—or in society's ditches, like the Arc's basement where his coffin apartment was located, people were the same.

Leading aimless, pointless lives. It was even worse, now that ArcSec was enforcing their PASS directive. Without work, some on 1 and 2 were being evicted out into the streets, others beginning to starve. The social unrest was beginning to flare, like a wildfire, and the only thing that could cure it was to allow itself to burn the whole thing to the ground. Start anew.

The birth pangs are here, Ray thought. *It's gotta get worse before it gets better. A whole lot worse.* Oddly, the thought gave him joy.

"Is that all?" he asked, again in what sounded more like heavy breathing than language. He'd compiled the list in a simple text program and left it open on his system so it was viewable.

There may be an issue with our partners at Novagenica. Predictive data hints at probability of a defector, the AI responded. *But this is still inconclusive. For now, it's a good place to begin.*

To begin. The words stirred a sort of excitement in Ray's loins. He wasn't like some subbers, who found themselves distracted by prurient desires. But finding himself at the cusp of a new beginning, of something long-awaited *did* arouse a visceral sort of fire within him. The energy wasn't of a sexual nature, but it was close, seated somewhere above his balls and just below his gut.

His eyes roved the list, were drawn to the top. A good place to start, considering he was already on 4.

"Show me the address," he said quietly.

Done, the AI responded. The map showed transit time as a fifteen-minute walk. Excellent. He could be in and out quickly, before breakfast.

Our sensors have been paired with all records to build a dossier. Would you like to read it yourself? the AI asked. *Or we can summarize it for you on the way, if it pleases you.*

"Do that," he said. He was eager to get moving.

The first, and most notable detail, is the target's fondness for martial arts. He trains five times per week. Once every two months, he fights in the pits on 1. Unbeknownst to his wife, whom we surmise would not approve.

"Hmm," he grumbled. He was betting on an easy mark, not a challenge.

Perhaps it would be wise to uneven the odds. You are obviously far superior to any target, but perhaps in the name of excluding chance, you would consider an enhancement?

It didn't take much to convince him. The AI was right. He had a lot of work to do and couldn't risk being slowed down or injured.

He bumped *Speed* from his internal reservoirs, the drugs hitting his bloodstream instantaneously. Light and life blossomed from within. It was like he hadn't existed prior to bumping.

And then there was light.

He felt ready. He felt *alive.*

"You're right," he said out loud to the AI. "Of course you're right." Enough with the pretense. Enough of the hiding. The action was starting, and it felt good to discard that particular layer of subterfuge.

He balled his fingers up, squeezing his fists together, felt his forearms bunch up. He was ready to smash something.

The man at the table to his left, young with thick, black hair and a perfectly manicured beard stared in Ray's direction. His hands were cupped around his coffee mug, which steamed in the early morning light. Ray grinned, showed the man his teeth. The man looked hastily back down at his mug, spilling some in the process.

"*Shit,*" he cursed, flicking the coffee from his hands on to the patio pavers.

Sheep, Ray thought again, looking at the man, daring him to make eye contact. No such luck. The man was thoroughly cowed, visibly shrinking into himself, huddled around his coffee cup as though it were his lifeforce and needed protection.

Ray stood, rolled his neck around to loosen the stiffness that had gathered there. The drug cocktail infused his every cell. He needed to move. Now.

Through the wrought iron and bougainvillea, he spied the junction where 18TH street headed east, away from the transfer gate. His target was just a few blocks down, in a joint-venture Condo Complex, operated by Corporate Realty, WWD, but leased by Blue Sky Ad Ventures, a marketing corporation with aspirations to Complex-hood.

Martial arts aficionado or no, the target was still a sheep. Soft, compliant, like all the other Legacy citizens out there.

Gird your loins, Jerome old buddy, 'cause here comes the wolf.

JEROME

JEROME
AGE: 28
SEGMENT: LEGACY CITIZEN
POSITION: VP MARKETING; BLUE SKY AD VENTURES,
LLC
SOCIAL SCORE: 901

JEROME AWOKE TO A BLESSED LIFE. *Every day,* he thought. He stretched inside the warm covers, ivory and baby blue, felt the satisfying pull of muscles loosening, opened his eyes to Elia's face.

Normally she was up before him, sipping at her coffee, legs curled up underneath her in the plush chair in the corner, reading from one of her favorite fashion feeds. Since the pregnancy she'd been sleeping ten, sometimes twelve hours a night.

He took in the supple curves of her body, enhanced by the filmy sheer nightgown that clung tight to her skin. She lay on her side, back turned to him. He ran hand along the hollow of her side, up the curve of her hip.

"Mmm," she purred.

"Morning, love," he said, curling his fingers around the front of her hip, massaging gently but firmly.

"Mmm-hmm," she repeated.

Beautiful wife? Check. Well-paying job? Check. Nice place high up the Arc? Check. The foundation for a happy family? Check.

He was reminded of their courtship. The early days. She'd been young and beautiful—not that she wasn't now—but with a certain wild streak Jerome had found both enticing and troubling. On only their second date she'd taken him on a daytrip to 1 and he'd been terrified out of his mind that he'd end up bleeding on the ground with a knife plunged in his side. He hadn't even known, it was a surprise, she'd said. If he'd only known he could've worn something more appropriate, something that could stop a knife or projectile. Then again, if he'd known where they were going he probably would've faked an illness, made some lame excuse not to go. So he'd proceeded, living with the terror, unwilling to cut bait and run. Something had held him there, something he hadn't fully understood at the time.

After that date, his assistant, an early Sanjay, had taken every opportunity, night and day, to chirp in his ear that Elia wasn't right for him. And eventually, as the relationship turned serious, Sanjay had been so bold as to tell Jerome not to marry her. Stupid algorithm didn't understand a damn thing about people. What it didn't get, and what Jerome himself didn't understand initially, is that maybe he *needed* a little of her wild side, a little of that fire to make him a better person. But then, Jerome couldn't blame Sanjay for its inability to fully fathom the myriad nuances of human relationships—the assistant was an early model and hadn't yet accrued all the necessary data inputs globally to be able to piece together a better picture of humanity.

When it came to business, Sanjay had been a wiz. It'd been instrumental in helping Jerome maneuver the byzantine management structure of Blue Sky, helping him make the right political moves at just the correct moments. But love? Not so much. Ever since he'd been a kid, these were the only things he'd ever wanted—loving wife, great job, nice home. So he put himself in a frame of mind, every day, to achieve them. If he analyzed it deeply enough, he'd probably find it stemmed from all the things he'd been missing in his own upbringing. But he'd never been one to dwell on the past, and he didn't intend to start now. His sight was set on the present, the blessings it held, and on their future together.

He continued to massage, knead the soft tissue, feeling a familiar arousal that, sadly, had been stymied since the pregnancy.

"Hey," Jerome leaned in, spoke softly in his wife's ear. "Remember when we used to have sex? Those were good times."

"I know," Elia said, voice muffled from the pillow. "I wish I felt better. This kid is kicking my tail already."

Jerome moved his hand upward, under the gown, found a breast and gently squeezed. She giggled, shied away. "They're sensitive."

Jerome mock sighed. "Such a distant memory," he said, rolling over and off the bed.

The morning went as most mornings did. Get up, shower, cup of coffee, read the news while seeing a man about a horse before departing for work.

He inspected himself in the mirror, a softness to his body that hadn't been there six months ago. Sympathy eating, they called it.

Not like you were getting any exercise then, either, he thought. Jerome knew his strengths, and physical activity didn't top the list. His skills regarding a credit hold, however—unparalleled.

Well, once the kid was born he'd have to do something about it. He turned from the mirror, dropped his boxers and sat on the toilet. Cup of coffee in one hand, he flicked through news feeds on his HUD. Everything was about the deteriorating situation on 1 and 2. Video feeds of crowds huddled in the cold. People rioting. The occasional body lying in the street. He flipped past it.

"Seen enough of that crap to last a lifetime," he muttered, his voice echoing in the narrow, tall room. *Maybe some cartoons,* he thought. That seemed like a nice contrast to all the negativity in the world.

He pulled up an episode of *Valkyries of The Holy Mountains,* knowing exactly where it would lead. Anime was really made for one demographic, people past puberty that were attracted to women. And this one was no exception.

This particular episode focused on the Valkyrie of Tateyama. A fire haired, scantily-clad warrior of a woman. Per the usual, her armor—silver-plated with red ornaments—was not made to protect, but to reveal. It barely contained her ample breasts, which the artists would make you think were her primary weapon they were so exposed. As though she needed to keep them close to the air and free from restraint to brandish them at a moment's notice. Their blazing glory brilliant and blinding in the reflection of the light off Tateyama's metaphorical peaks.

"Who knows," Jerome said to himself, shrugging. "Seems to be working. Let's just dim the lights a touch for mood." He used his system to dim the lights in the bathroom to a warm candlelit glow.

"There," he said, nodding with satisfaction. He sat on the toilet, got a firm grip on himself, and went to work.

He was barely a minute in when the door creaked open.

"Oh, hey babe, gimme a minute," he said, the words coming in a rush. He hastily threw the side of his foot up against the door to stop it.

But the door continued to open, banging his foot aside. It happened so quickly he didn't have time to pull his pants up. There, at the entrance, stood a man. Middle-aged, bald with a reddish face, clad in all black. The man looked at him curiously. Jerome's manhood wilted in his hand.

"Who the f –" he began to say, pushing himself up from the toilet. Before he could complete the thought, quick as a viper the man reached out with an arm, swept it across Jerome's body. Jerome looked up in shock at the stranger, trying to process what had happened.

The man looked at him almost as if he was disappointed. Like he was a father expecting more from his son. Then, quick as he'd come, the man turned and left, not looking back. Jerome heard his soft footfalls on the hard tile floor, then very faintly the front door clicked open and closed.

Jerome came back to his senses, pulled his underwear up. Inspected his body as he walked toward the bedroom.

"Babe?" he said aloud. No response. Halfway along the long, darkened hallway that led to their bright, airy bedroom Jerome felt a burning sensation in his left arm. He looked down to find a small cut there, maybe a couple inches long.

Huh. Looks like he got me. Jerome took a couple more steps before his legs failed him, body collapsing to the floor. He grunted as his chest collided with the tile. He lay there a moment, bewildered as his heart pounded wildly. His right hand went to it reflexively, as though he could calm it with a touch. Damn thing felt like it was trying to break through his ribcage.

Need help, he thought feebly. He tried to cry out but nothing came out but a dry croak, a weak gasp. *Where's Elia?*

His breath was getting shorter. With effort, he managed to get his chin on the cold tile of the floor, get it under his head so that he could see through to the bedroom. Atop the bed he could see his wife on her back on the bed, baby bump silhouetted by the light emanating from the far wall. Her arms were outstretched, splayed to the sides on the sheets like an angel accepting her flock. Unmoving.

Huh, he thought, then his breath gave out and his head lolled to the side on the hard floor, eyes staring but no longer seeing.

TREVOR

TREVOR
AGE: 18
SEGMENT: LEGACY CITIZEN
POSITION: UNKNOWN
SOCIAL SCORE: 821; FLAGGED FOR VERACITY

"YOU'RE NOT THE ONLY PERSON *in this Goddamn world, so stop acting like it.*"

Trevor lay in bed, in the complete black, unable to get Estelle's voice out of his head. The words rolled over on an incessant loop, paired against the scrim of his friend's angry face. He sighed, the guilt renewing itself afresh.

Looks like you finally got something through, he thought. *Bravo, Estelle.*

Up to this point in time, the direction of his life had largely been governed by one signpost.

Nicolette.

His sister had been adopted years ago by a couple on 3. Trevor had been left behind in the Talksmall-run orphanage, holding only a faded picture of Nicolette and a copy of the "No Contact" contract, a condition of her adoption. Even then, he'd understood. The adoption pretty much guaranteed her a better life than what they currently had going for them. Trevor didn't hold it against the adoptive parents. His feelings didn't matter—Nicolette was what mattered.

But ever since he'd reached his years-long goal of acquiring enough cred to pay for Nicolette's enrollment at MAIT—a high-tech prep school that served as a feeder for the collegiate-level university—he'd been wayward, rudderless.

And then he saw Val.

When he first laid eyes on her, his mind hadn't consciously processed the vulnerability she exuded. He hadn't overtly thought, *man, this girl needs help, I better lend a* hand, but he'd still sensed it all the same. She'd seemed...

In need. And face it, dude, you're a sucker for a girl in need. Also doesn't hurt that she's pretty. His lips parted in the dark, the image of Val's face from yesterday morning swimming up in his mind.

Yet, Estelle's words had gnawed at him. Eroded at his assumptions starting that night she'd pulled him by the ear out of the railyard, and every night since.

Living in the Arc had largely succeeded in hardening the part of him that recognized the pain in those around him. Suffering was so prevalent, such a way of life that he had to put up walls or get crushed by 300 million tons of misery.

So he compartmentalized, made rooms for different emotions. Rarely opened the doors to the ones that threatened to overtake and immobilize him.

But now, it seemed maybe that was all wrong.

After all, when we ignore the pain of others, we're just giving everyone else permission to ignore our own pain and suffering.

He'd move mountains for Nicolette, no questions asked. And he saw that same desperate need in Val.

But maybe, a small voice whispered in his mind, *maybe it's time to recognize the fact that* everyone *needs help.*

"Stupid brain, that's beside the point," he said, sitting up in his cot. He could lay there all day weighing philosophical pros and cons, but he had other things to think about right now, more pressing concerns.

Like how to make use of the information Saul had given.

He remembered looking at the data when they'd finally opened it, sitting outside at the café. He'd almost laughed out loud, but knew that if he'd done so it would've demeaned everything Val had done to earn that information. And she'd never forgive him.

So he'd remained silent, looking concerned. At first glance it looked total garbage, exactly the sort of thing to expect of that shit-sucking leech Saul. Just a small snippet of text:

BLUE SKY AD VENTURES, LLC.

Not a link, nary a tag, zero embedded information. Just the name of some unknown company. One molecule in the ocean that was comprised of current and historical companies, living and dead alike.

Trevor had returned to his coffin and worked for a solid day now, his and Val's little arcside adventure feeling like it occurred weeks ago as opposed to twenty-four hours.

Countless searches and still nothing. At least, nothing out of the ordinary that could connect Blue Sky Ad Ventures—a small marketing company based on Level 4—to a kidnapping. Trevor loved puzzles, but he was starting to get to that edge where the puzzle became less fun and more frustrating.

Trevor fumbled for the light switch. He could've just had his system enhance the dark, but he liked the idea of finding things based on touch alone. He shut off his system, groped at the wall. The kind of darkness you could find in a coffin complex was a rare sort of currency in the arc. No windows—not in the coffins themselves or even in the entire building. Unfortunately, with his system off he was able to focus more clearly on his other senses. The sour smell was enough to spur him to action.

Trevor's fingers found the smooth plastic of the switch and flipped it. He swung his legs over the edge of the cot, rubbed his eyes and opened them.

Home sweet home, he thought. "What a shitbox."

Mustard-colored light rained down from a sealed industrial light in the ceiling, illuminating the four-by-nine confines of his coffin apartment. The tiny room had a toilet/sink combo unit that Trevor never cleaned and a small table and bench that folded down to form the bed he sat on.

"My very own...rectangle. I always wanted one," he sighed.

Generally he spent as little time as possible in his coffin apartment, but his commitment to helping Val find her sister was serious. He'd known it would take time, which would require him to have access to food, water and a toilet for extended periods of time, none of which were guaranteed outside the confines of these walls.

Most days he awoke around noon, brushed his teeth, pissed and got the hell out of there until it was early morning and time to crash. It was times like these, interminable stints cooped up inside the coffin, that reminded him why. His back ached and he felt the surreal sensation of the walls beginning to shrink around him.

He stood and stretched, pulled on black pants. Throwing his leather jacket over his shoulder, Trevor fled the tightening vice of claustrophobia and his own stale smell.

The Little Death Coffin Complex—not a true CPX, just a holdover term from the early decades of the century—was sandwiched between The Moist Dream—designer drug and fetish outfitter—and a thrift pawn shop that went by The Rattlesnake. Both storefronts outshone The Little Death, which boasted no more than a simple lighted sign depicting a coffin dripping with a milky fluid—most likely ejaculate. The narrow lobby was typically home to several multi-colored and multi-shaped ladies and boys of the night.

Trevor stepped over one such boy, a teenager by the look of him, who lay half upright, head leaning up against a wall, legs extended. His face was filthy and careworn, even in sleep Trevor could see the hardship set there in a permanent mask.

Poor kid, he thought, frowning. *There was a time I thought I had it bad. Now I know I was just feeling sorry for myself.*

Out into the dark and fetid, sticky air. The atmosphere glowed with all the wavelengths of light bouncing off the fog that seemed to perpetuate the fully encased lower levels of the Arc, none so severe as Level 1. He was originally from 2, but something about 1 sang to him. Everything that made it an abomination also made it interesting, like a curiosity he could touch, sample, taste. Where others found it distasteful, he found it invigorating.

Trevor stopped at the edge of the street, took in the scenery. Pimps, hookers, resident dealers, the occasional Talksmall party enforcer and daytrippers alike thronged the street. Clogged with people, scooters and bikes, it was like entering some rapids. He inserted himself into the flow, managing to not get run over.

Where to?

He needed help but wasn't exactly sure where to get it. People like Saul, sole proprietors, didn't join gangs or parties, didn't pick sides. They were the Swiss of the criminal world. Saul wouldn't be of any more help, no matter what Trevor offered him in return. He needed resources and that typically meant a party.

He had an in to Talksmall. He'd start there.

Hopefully, he thought, *Estelle isn't still pissed at me.*

The headquarters of Talksmall was based in a subterranean warren that had been carved out from beneath a block of tenements. Individual basements connected into an enormous network used for meetings, housing, planning. It also served as the base for their educational apparatus, First Liturgy of the Arcolith. But nobody bothered with all those complicated words. Around here, it was just called FLOTA.

None of the entrances were marked, and for good reason. Modern gangs were like political parties, angling, like the CPXs of the world, for membership. But Talksmall didn't advertise. Their mission wasn't to win hearts and minds through campaigns and hollow shells of promises, instead opting to put all their efforts into installations that dispersed control into the hands of the many. Whether, installing and supporting zero mark-up printers or disseminating information it all amounted to the same thing—they won membership by winning hearts through (mostly) benevolent acts. It was only just recently that they'd gotten into the schooling game. Easy recruits.

Trevor descended a small flight of steps and rapped on a rusted metal door. It looked low-tech but he knew better. An entire suite of sensors was checking him out, his biomarkers, built-ins, known associations within the party, the whole kit and kaboodle.

He waited patiently for a minute before the door swung open effortlessly. Estelle stood before him. She had one eyebrow arched, which he took to say *Your move.*

"Let me first just say I'm sorry, okay?" Trevor said. "I know I screwed up. You were right."

Estelle squinted, seeming to analyze every bit of his face. "It's not me you should apologize to. A whole lot of people got their skulls cracked because you thought of something fun to do."

Trevor fiddled with his fingers. "Yeah, I know."

Estelle's features remained cut in stone. "So, this just a social visit or..."

"Apology tour?" Trevor blurted, then shook his head. "Sorry, stupid mouth. Little bit of both, I guess," he said. "I wanted you to know I really do care."

"About something other than yourself?"

"Mostly what you think of me," he said, head down.

She snorted air out of flared nostrils, waved him inside. "C'mon, let's talk."

She took him down narrow and straight, branching corridors, past rooms full of children.

"So the permit came through then?" Trevor asked, motioning to a room full of children, some sitting, others jostling for a position on a circular rug on the floor. Necks craned back, enveloped in a story being read aloud by a female teacher.

"Yeah," Estelle said. "You're looking at the newest church in town. Wasn't ever really in question, just a matter of maneuvering. Got the permit *and* got Legacy to pay us for the protection of school-routes," she boasted.

Trevor whistled, long and low. "Sweet outcome. You sure you guys aren't partnering up with The Statisticians?"

She looked over her shoulder, shot him a glare.

"So that's a no then."

Trevor let it drop. Any talk of political partnering for Talksmall was sure to receive an icy reception. He was never quite sure why but didn't press the issue.

She led him to a commons room, maybe twenty by forty, offered him a chair at a circular lunch table. The lighting was low, conspiratorial. Groups of people sat in twos and threes at tables scattered around the room.

She pulled up a chair, turned it around backward, and sat, arms folded on the backrest.

"So what's up?"

Trevor took a deep breath, spilled the whole story. Val, her little sister, the info from Saul, the dead end. He left out the field trip to show Val his new invention—GIE—on the grounds it was extraneous. That and the fact he was here to convince her he was reformed. GIE wasn't exactly proof positive of that fact.

Estelle nodded, her face clouding over.

"So, I guess I'm asking if you have any resources, or know anybody who can help." He shrugged. "I've looked at this from every angle I know."

Her eyes, brown with touches of green around the edges, searched his for a long moment. Then, responding to some internal cue or resolution, she leaned back, looked up at the ceiling for a moment.

"From the sound of it, this was professional. Not the sort of thing you'd see from a solo-jobber or junky gang. They'd just snag her in the middle of the day, too fucked in the head to worry about Legacy enforcement or ArcSec. Which means she's valuable, for some reason or another."

"Mm-hmm," Trevor agreed.

"By process of elimination that means you're dealing with, most likely," she made a look of apology. "A Complex."

"Shit," Trevor said, throwing his head back to stare absently at the yellowed ceiling. He put his hands together started rubbing his knuckles.

He felt Estelle's cool fingers on his hands before he saw them, taking them into her own, comforting him.

"Hey, look at me," she said. Trevor obeyed. "I have a friend who might be able to help."

Something inside him leaped. "For real? Like, actually *can* help?"

"Smartest person I know," she said, nodding.

"Aside from myself of course." Trevor said, putting all the cocky youthfulness into it he could muster.

"Naturally," she agreed.

"So who..." Trevor trailed off, distracted by someone approaching in his peripheral vision. Trevor turned to see a man with long, straight black hair walking straight toward them.

"Hey E, who's your friend?" The intruder put his hand possessively on Estelle's back, leaned down for a quick kiss. Estelle probably didn't see it, but this act was as old as the distinction between male and female. Marking his territory.

Pissing on your boundaries, eh buddy? Trevor thought. *Metaphorically speaking, of course.*

"Hey babe," Estelle said to the man, starting to stand. "This is Trevor."

"No, no, sit please. I don't mean to interrupt, just let my curiosity get the better of me." He held a hand out. "Jayne Moon."

Trevor took his hand and shook. "Trevor."

Jayne dragged a seat over from a nearby table and sat in it backwards. "So how do you two know each other?" Trevor felt a small stab of irritation in his chest. For a guy who didn't mean to interrupt he was doing a lot of it.

"Old friends," Estelle started, "I was doing a voluntary vaccination program, ages ago, for Legacy's Health and Human Services, right before it got phased out. I was knocking on apartment doors in the Vine district, Level 2? You know..."

Jayne nodded. "Yep, sino-jazz club district, right? Legal psychedelics, blind tigers, classic opium dens."

Trevor and Estelle both nodded. Despite being annoyed, Trevor couldn't help but note a certain sharpness to the guy. A quick intellect for sure, but more than that he had a way of watching and listening that let you know he wasn't missing any cues or details.

"So I'm knocking on doors," she continued, "and I come upon this malnourished boy and his little sister. I mean, I can see his ribs through his t-shirt, but his little sister, she looks fat and happy. After about thirty seconds I figure out these kids' parents aren't just at work, they're gone. Hadn't been around for three weeks. Just disappeared."

Trevor nodded, the small smile showing on his face just a mask for the pain that gripped his whole chest. Love and trust broken just like that.

"So I brought them in to Talksmall's adoption program."

Jayne nodded, turned to Trevor. "Sounds like you've already got Talksmall blood in you. Any interest in joining *TS* for real? Looks like you're the right age to join."

"Nah, thanks but no thanks," Trevor said. "I'm more of a solo kinda guy. Like to do my own thing."

"Your choice, of course," Jayne nodded. "Just consider it. We're doing good things here. Just like you and your sister benefited from that adoption program, Talksmall has its hands in a ton of similar programs designed to help kids like you. You could help a lot of people that were in the same..."

"Enough of the hard sell," Estelle interrupted. "He's just here to catch up."

Jayne blinked, then offered a sheepish grin and shrugged. "Sorry. E says I push too hard sometimes. Just believe in what I do," he said, standing and extending his hand. "Nice to meet you, Trevor. See ya round."

"Likewise," Trevor said

Jayne looked to Estelle. "See you in five for the next report," he said, somewhat coldly, then walked out of the commons area down a white-painted hall.

Trevor raised his eyebrows. "You guys a thing, huh?"

Estelle nodded brightly.

"Methinks he's not too used to the rejection," Trevor offered.

Estelle wrinkled her brow. "Maybe, but when you're in his position you need to be persistent. You'll understand when you get older, when you have more responsibility. His life exists in a pressure cooker." She paused, getting a far-off look. "But he's a good man."

"If you say so. A little too pale for my taste, and kinda pushy," Trevor said.

"Good thing you're not dating him then," she said.

"Damn good thing," Trevor agreed.

"You check your inbox yet?" Estelle asked.

"For?"

"Name and address of my contact. The guy that's gonna help you in your search?" Estelle tapped him gently on the back of his head. "Any of this ringing a bell?"

Trevor put a finger-gun to his own head, shot it. "Ohh yeah."

Estelle rolled her eyes. "Dude."

Trevor checked his system. The message receipt icon blinked slowly. "Yep," he said. "Sure as shit, there it is. Name and address." He opened the message, reading the contact details she'd promised.

IAN BAKSEN
3Y-609 JARBOE, #86
SUNSHINE, L3—MA 64030-3Y-776186

"Superb stuff. Thank you, m'lady." Trevor stood, making a sweeping bow.

She slapped him on the shoulder. "Don't say I never did anything for ya."

"Seriously, though," he said, dropping the pretense and drawing her into a bear hug, squeezing with all he had. "Thank you. You're a lifesaver."

"And seriously," she whispered in his ear, "that's what I do. Just try to make sure it's worthwhile."

He held her out at arm's length, looked at her. "I will."

He took off running the way he came in. "Thanks again! You're awesome!" he yelled as he left the room.

"Just let me know how it turns out!" she shouted back. "And don't be a stranger!"

And then he was gone, running down the maze of corridors. He had to get a hold of this guy, this Ian. A tiny bloom of hope, after so much darkness, began to unfold in his mind.

He ran harder. There was work to do.

Hundreds of heads bobbed and swayed before him, all moving forward a shuffling half-step at a time. The stairs to Level 2 were the worst.

There were only eight official transfer stations in the Arc that allowed you to move from one Level to the next. And since Levels 1 and 2 were indistinguishable from one another with regards to class structure, there was a lot of mingling and movement of population that occurred between the two.

At two in the morning, it only took a few minutes to traverse the stairs. No traffic.

At mid-morning it grew to a thirty-minute process.

As his foot hit the first step, his system chimed internally. Val. He had snuck a picture of her using his built-ins and so it was this image—her taking a sip from a white coffee mug at TBD—that came up as her call rang inside his eardrums.

"Hey sugar." The instant it came out he knew it was wrong, even felt wrong. But what can you do. Sometimes teenage boys were known to say stupid things.

Silence awaited him. He was waiting for her to admonish him when he realized she'd done it without saying a word. She'd hung up. *Shit.*

He rang her back.

"Sorry sorry sorry," he blurted out. "Crap just comes out of my mouth sometimes. Verbal diarrhea. Can't help it."

"Try harder," she said, and he felt relief cascade through his body.

"Yes ma'am," he agreed.

More silence. He thought he'd screwed up again when she spoke.

"What are you doing?" she asked.

"Right now?"

"No, in seventeen seconds."

"Sorry, stupid question. Um, nothing at the moment, except rubbing up against a few hundred of my closest friends."

"Sounds like something you'd be doing."

"Yeah, ya got me. Actually, I'm trudging up the transfer gate to 2. Was about to call you," he added, which was, of course, a slight stretching of the truth. He was going to call, sooner than later, but he wanted to talk with Estelle's contact first.

"Oh yeah?" Her voice perked up. "You find anything?" He could hear the hope in her voice, didn't want to be the one to crush it.

No choice, stud. You can't lie to her.

"Not yet," he said, "but I think we should meet up, put our heads together."

"Ok. Where to?" she asked.

He toggled his system's status to share and dropped a pin on the address where he was headed.

"Meet me here."

"Yeah. Be there in thirty."

"Over and out."

The diner on 2, *Insert Name Here*, consisted of orange and green freight containers stacked in random order around an atrium that stretched forty feet up. No view to sunlight here on 2, but the enterprising owners of the diner had managed to scale the underside of the Arc's superstructure and wire up focused daylights.

Around ground level, potted plants and grow lights dotted the area. The air was humid but smelled surprisingly clean, earthy.

Trevor waited for her at a small table on the second tier of containers, steaming tea held between his hands. No matter the season, the lower levels were always on the chilly side. Like living underground.

"Whoa, this place is amazing!"

Val peaked above the roof of the container that served as the floor of the second tier. She reached an arm up, grabbed a handle that was bolted into the container and pulled herself up over the edge.

She wore a tight, faded blue t-shirt that showed off her athletic physique. Her hair was wild, blond locks pushed by a gentle breeze to play back and forth across her face, over her lips. She was as beautiful as ever, beaming with wonder. Something small leaped in Trevor's gut.

He gazed at her as she sat.

Finally, after a small eternity of taking in her surroundings, she looked at him.

"What, do I have something on my face?"

It took him a moment to respond.

"No, sorry. It's just..."

"What?" she gave him a quizzical smile.

Trevor didn't realize he had been holding his breath until he exhaled. There was this strange yearning, mixed with a certain queasiness in his gut that was totally foreign to him. He knew it was tied to the way he felt about Val and that if he just blurted out how he felt about her then it would abate, but he had the sneaking suspicion that it would make things weird. That was the last thing he wanted. Besides, there was work to do. Her sister's life was on the line. Romance could wait.

Don't forget she pretty much told you to your face that it wasn't going to happen, he reminded himself. *So stick to the priorities. Find the sister first. Then, just maybe, you can woo the girl.*

"Um..." Trevor searched for something to say, but his thoughts were scrambled. "Do you, uh, want some tea?"

"Sure," she said brightly. "That sounds good."

He toggled his system, ordered her a jasmine tea. "Done," he said.

"Thanks," she responded, rubbing her arms against the cold.

"My pleasure, m'lady."

She shook her head. "You're a dork, y'know that?"

"I'll chance to take that as a compliment." Trevor grinned.

Seeing her smile back at him made him yearn to kiss her. He managed to hold the grin, but on the inside found himself groaning.

What have you gotten yourself into, man? he admonished himself. *She's gonna chew your heart up and spit it out.*

Trevor knew what he'd willingly committed to—to helping Val find her sister Kat. The only condition Val had stated, as if she'd been in a position to state any condition at all, was that there couldn't be any expectation of romantic involvement.

It's the right thing to do, but still...

It was going to be hard to keep their relationship one hundred percent professional. Mostly because he felt pulled toward Val, found himself delighting in her every expression, hanging on the nuance of her every word.

"Your choice," she retorted.

"Not much of..."

"Ok, well," Val interrupted, clapping her hands. She rubbed them together vigorously, leaned forward. "Let's get to work."

"Ok, yes," Trevor agreed. "So here's what we have." He shared his system's view with Val so they were looking at the same thing:

BLUE SKY AD VENTURES, LLC

"It's a small company based here in the Arc, Level 4."

"Ok. That's good, right?" she asked.

"I don't know. From what I've been able to look up—which isn't much—they specialize in a kind of economic bartering system. They place ads that in return subsidize other services."

Val's forehead furrowed.

"That's confusing as shit. Can you dumb it down some?"

"Umm...you know what a subsidy is?" She shook her head.

The serving platform arrived at that moment, stopping their conversation. The platform appeared at the side of the table, rising gently on a rubber chain that pulled it up alongside a chrome pole. It held a steaming cup of tea, which looked like it was made of plastic but was probably designed to break down within a year. That way the company could process it through a printing exchange, get a credit on another comping job.

Val reached out before the cup could pass and grabbed it with a deft swipe. She blew the steam off the top of the green-tinted liquid, sipped at it cautiously.

"Hot," she sucked in air through her teeth. "Ok, subsidy. Explain."

"I guess the best way I can describe it is it's the same concept of everyone getting their Standard Universal Credit," Trevor said.

"The dreaded monthly SUC." Val said in a haunted voice, fingers wiggling above her head like she was the boogeyman.

"Uh huh, just like that. Whatever that's supposed to be," he said, teasing. "So anyway, after paying for rent and food, most people can't afford the little extra things they want, or even need. Not enough jobs to go around and, in any case, not enough people that want to do the jobs that are available. So Legacy subsidizes, in effect, your existence, by giving you the monthly SUC. Beyond that, they even subsidize your housing by making the rent cheaper, giving credits back to housing owners for offering cheaper rent. This starting to make sense?"

"Kind of?" Trevor was mesmerized as she pushed back her hair behind her long neck, leaned forward and sipped at her tea.

Oh hell, you're gonna make this as hard as possible on me, aren't you? he wondered, then did his best to ignore the baser urges. *Stuff it, dude, stick to the priorities.*

"So they used to do this with farmers all the time," he said, putting both hands behind his head and leaning back. "Legacy, that is. Before we comped our food from printers, farmers were like the backbone of the country. So Legacy made sure that farmers had a very good incentive to stay in business."

She looked over the rim of her mug at him, nodded for him to go on.

"For example, if the market price of, say, milk, is a dollar lower than what it takes for the dairy farmer to stay in business, Legacy steps in and pays the farmer the difference, sometimes more. They would even pay these guys not to farm their land."

"So just like the dole, then."

"Similar, but no, not *exactly.*"

"We're getting paid not to work."

"Well, yeah, but," Trevor sputtered.

"It's exactly the same," Val asserted.

"Well, no. They had resources, like land, and they got paid not to utilize them."

Val held up her hands, turned them to and fro. "See these beautiful work-makers?"

"Clearly," Trevor responded.

Val made a show of lifting each leg and inserting a hand under each thigh. "Resources, not utilized."

Trevor cocked his head, his mouth forming an absent *O.*

"Have you ever lost an argument?" he asked.

"No," she said casually. "Have you?"

"Just twice now."

"It's okay, sweetheart," she said in a patronizing way, patting his hand. "You can stop counting. There'll be lots more to come."

Trevor's stomach did a little fluttering thing at her touch. His eyes lit up. *Lots more to come?*

"Does this mean you're keeping me?" he asked, taking the conversation in a direction he'd promised not to go. But he couldn't help it, he *wanted* to joke, and have fun, and flirt with her. Besides, she was the one leading him there.

Just roll with it, dude. Let her control how this goes.

"As long you behave," she said.

"I'll be a good boy. Promise."

"Good. You can start by helping me figure this thing out."

The thought of her missing sister was a sobering one.

Trevor looked more closely, past the points of admiration, past everything he found lovely about her, and found an underlying tension in Val. It was there in her eyes, the set of her jaw and shoulders. She looked under a lot of pressure. There was still a strength to her, but tiny stress fractures were beginning to show in the façade.

I can't even imagine what she's feeling right now. Just the thought pulled at his heart. *She did everything by the book, to provide for and protect Kat, and then someone comes in and decides to rip their lives apart. It's not fucking fair.*

He took a deep breath, tried to dispel the sudden anger he felt at everything, everyone. The system in general. *Don't forget, she's got a lot more to be pissed off at than you. Just do what you can to help.*

"Right," he said, drumming his fingers on the table. He stared off at the jumbled freight containers, stacked two and three stories high to serve as platforms for other tables. Ivy grew over the tops and sides, lending the place an overgrown and forgotten industrial look.

"Hey, there was something you said that got me thinking of something," she said, snapping him out of his trance. "You were talking about the farmers and the subsidy, before we comped our food from printers."

"Yeah?"

"So where are the farmers now?" she asked.

"Some are still out there, but these days it's just cheaper to grow the stock. Why do you ask?"

"Well that's what I was getting at. Who's making the stock? It has to come from somewhere," she said.

"Lab grown these days. But you're right, the labbers of today are like the farmers of old, because vat-grown food gets subsidized also. Sometimes by Legacy, sometimes by non-profits or some other third party that wants something in exchange."

Val held her mug in both hands, stared into it, nodding thoughtfully. "Alright, now let's apply that to what we know of Blue Sky Ad Ventures. You said they subsidize comped food from printers. So what if they acted as a sort of subsidy marketplace, offering discounts on the product to people like us in exchange for something else?"

Trevor clapped his hands together loudly. "Yes! Just like that! And what we're giving them in exchange could be information. It used to be attention, like watch this thirty second ad or whatever, but anymore all anybody wants is our information."

"Like what kind of information?" Val asked.

"I'm not sure," Trevor said. "But I have a lead on someone who might be able to help."

"Who?"

Trevor looked around. Even if someone was listening in, no amount of whispering would prevent them from hearing what he had to say. "Don't know him personally, but he was referred to me from a trusted source."

"Ooh, the mystery is too much. I have to know who."

"His name is Ian. And from what I'm told, if anyone can help find your sister, this is the guy."

VAL

VAL
AGE: 19
SEGMENT: LEGACY CITIZEN
POSITION: UNEMPLOYED; GOVERNMENT DOLE
RECIPIENT (STANDARD UNIVERSAL CREDIT)
SOCIAL SCORE: 478

I T WAS DARK. Like Level 1, but not quite so close. Not so suffocating. Mid-level buildings lined the streets, lighted brightly. From what Val could tell, they appeared to serve either one of two purposes: overcrowded Legacy housing projects or overcrowded Legacy housing projects repurposed for street food vendors.

She wrinkled her nose. The air smelled tangy and sweet, with a hint of putrefaction. Like cooked ethnic foods sprinkled with dried bits of effluence. The streets were a carnival of color and activity, people bustling in every which way and the air was rich with the screams and squeals of children.

"So this is Vine district. Mostly residential along here. Grew up right around the corner," Trevor said, pointing to a break in the housing that branched off to the right, down a narrow split. "Not bad if you know what streets to avoid."

"Yeah? Would this be one of 'em?" Val joked.

"Nah, we're on Main. Equivalent of Main Street below us on 1. Safe on whatever level, for the most part. Take a wrong turn into a free drug zone, though," he laughed without humor, "and you'd think it was either the apocalypse or Mardi Gras, depending on your tolerance and moral constitution."

She rubbed her hands together. "Ooh, sounds fun."

Trevor cocked an eyebrow at her, his expression sobering. "Believe me, sister, one pass through a free drug zone is worth enough anxiety to take five years off your life. FDZ's no joke." True to his words, his face was flat, unreadable, serious. Unlike him. Weird.

No, what's really weird is you thinking you know this guy well enough after a couple days to really judge him. She shrugged as if in response to her inner monologue. *What other options do you have? Take the help where you can get it. Besides,* she eyed him sideways, *you could pick worse company.*

The transfer gates sat far off in the distance, barely visible through the haze of fog and food-smoke. She and Trevor walked on in silence, steadily approaching the gates. They were close enough now to see the crowd gathered around the base of the stairs.

"Shit," she muttered under her breath. "Another dead end."

She felt the frustration begin to build in her chest. After weeks of nothing, they were so close to getting answers yet every transfer gate they tried was blocked.

"Shit," she said again, more forcefully.

Trevor looked to her, concern on his face, but said nothing.

Val felt as though she'd walked ninety percent of the arc. What should've been a quick, straightforward journey from Level 2 to 3 had turned into a two-hour, rambling exploration of the variegated districts of Level 2. They were trying to get to 3 where Trevor's contact—the so-called wizard—lived, but were met with roadblocks.

Tall, clear walls barred the way at every transfer gate.

Val snorted what felt like hot steam out her nose. "What now?"

"Let's go check it out anyway," Trevor said. "Maybe there's a way around."

Up close, the transfer gates weren't much more than staircases. Immense structures of carbon and manufactured stone that switchbacked several times, leading from 2 to 3, or vice versa. They started fifty feet wide at the base on 2 and tapered as they neared the junction with 3.

Tricky bastards, Val thought. The narrowing effect was clearly essential to the design, a sort of aesthetic deterrent. It probably took a whole lot longer to go up than it did to go down. The gates were clearly designed by developers on 3 or above, which from what Trevor was saying acted as a middle world, holding on to some traces of its Level 2 lineage but moreover aspiring to the heights and wealth of its bigger brothers. Middle class, but with its nose in the air.

Each transfer gate in the Arc came as a set of two, one for upward traffic and the other for downward. They were built around a park plaza that was populated with low-light grass and copper trees, oxidized to green, with a fountain spewing water in the center. Her gaze followed the staircase, back and forth, all the way up to Level 3, where another park was situated. From there, a hollow shaft in the Arc extended another ten stories up. Sunlight glanced off the upper reaches of the space, threw shards of light across the buildings that flanked the park.

It was awe-inspiring, and the contrast between *up there* and *down here* had never been starker. Val shook her head, looked at the sign posted to the inside of the clear wall.

THIS TRANSFER GATE UNDER CONSTRUCTION. NO ADMISSION. ALL CROSS-LEVEL TRAFFIC SUSPENDED UNTIL FURTHER NOTICE. THE OFFICE OF LEGACY ADMINISTRATION JAIM DEL POTRO; ADMIN, LEVELS 1-3, DISTRICTS 67-130

Behind the clear walls, a contingent of five ArcSec soldiers stood in full riot gear, expressionless and clad all in black. Their presence left Val with a creeping sensation of dread.

"What's going on here?" she whispered. "Now guards?"

It felt like construction was an excuse and that these walls might just be a permanent addition. The crowd that gathered there around the base of the stairs began to grow louder, more agitated. Val moved closer to the edge of the group, listened in on what people were saying.

"This is the final straw, they can't treat us like dogs anymore," one man whispered, hunched over conspiratorially.

"It's time to organize. Here we work for scraps and then they just up and decide we ain't good enough to scrub the ground they walk on," another man said, then spit a thick wad of phlegm on the ground. "Makes me *sick*."

The words aroused a familiar sentiment in Val. Something she'd been pushing down since Kat had been kidnapped, that floated up to the surface with surprising speed: *Why?*

Why Kat? Why was this specific twelve-year-old girl kidnapped, among the millions in the arc?

And the answer was, there *was* no answer. It was nothing more than dumb, awful luck. They'd been born into the wrong circumstances, lived in the wrong area, a couple of girls with no prospects, no family and few skills. The answer was, she'd been kidnapped because it was easy, and life was expendable when you didn't have the benefits and protections afforded by wealth.

The thought stoked Val's rage.

She worked her way around the back of the group, spotting faces she recognized from the other transfer gates they'd been to earlier. The mass of people were a bubbling stew, throwing hands up in the air in exasperation, shouting, growing louder and more vociferous in their protests when, from the edge of the bystanders, Val saw a group of rough-looking men approach from the rear. As one, the crowd hushed and watched with curiosity as the men made their way silently to the clear barrier. They converged on different places along the wall simultaneously, faces set in stone.

In unison, they put one hand on the wall and reached inside jackets to pull out hammers in all colors and sizes. Without uttering a word, they cocked the hammers back.

"*ONE.*" One of the men shouted. "*TWO. THREE!*"

On "three", they swung in time, hammers crashing down into the wall.

The impact boomed, and Val could feel the ground vibrating under her toes.

The ArcSec guards drew their batons and the men stopped to inspect their work. The crowd cheered their support.

"*Tear that fuckin' thing down!*"

"*Yeah!*"

"*Let us through!*"

"*Open up, ya' bastards!*"

Val started to press forward, into the crush of people shouting and urging, when she felt a tugging at her shirt. "Hey." It was Trevor, pulling her back, away from the crowd.

Val spun. "Don't," she commanded, swatting his hand away. "Don't do that."

"Hey," he said. "I'm trying to keep you safe. This whole scenario smells bad."

Val looked over her shoulder to the men who were once again striking the walls. The booming continued in rhythm. "Bad?" she asked, looking at him like he was a complete stranger. "How is this bad? These people are right," she said, pointing at the crowd. "The pigs who put this wall up should be punished. *They're* keeping me from getting my sister back. This wall needs to come down, and I want to be here when they break through."

Trevor shook his head. "They probably won't, and no you don't." He regarded her with sad eyes. "I'm sorry about your sister, I really am. But that wall is the only thing protecting us from those ArcSec guards. Trust me, we're betting off finding another way around."

She studied Trevor, growing increasingly frustrated. This was no time to play it safe. Couldn't he see? They were *so close.*

"You can go if you want," she spat. "I'm staying right here." She turned her back on him and slipped into the crowd. She heard him sigh but didn't bother to look back.

If he follows, he follows, she thought. *If he doesn't, well, I'll figure it out on my own. Like I always do.*

Immediately, she was met with glares as she shoved her way through the press of bodies. A woman in her mid-thirties with a dirt-smeared face, grabbed at Val's hair, pulled her head backward.

"Hey, stop!" Val yelled, spinning and grasping at her own hair, then pulling it away and out of the woman's grasp.

A man, at the woman's side, snickered. "Hey look, a boojie baby trying to get back home." Val was bewildered. What did she say to that?

These people think I'm *bourgeouis?* she thought. *I'm just trying to get through like them.*

Suddenly Trevor was there, pushing back, creating space without actually touching anyone.

"She's with me, and she's no boojie baby. Arcside, Level 1."

They eyed Trevor with disbelief, Val with distaste. "Yeah, sure, whatever you say."

Val was grateful to Trevor, but it did little to abate her anger. He wanted to help, that was plain as day, but there was no way he could truly understand what she was going through.

When you lose a sister, she thought, *then we can talk.*

She spun and wended her way through the crush, contorting her body around elbows and shoulders, under arms and fists pounding the air. The booming of hammers continued, along with the shouting of the crowd. She could barely hear herself think.

Just as she made it the front of the crowd, the army of hammers crashed into the barrier as one. This time, the sound of impact was different, a kind of ripping and cracking, like the sky was tearing open.

The crowd sensed the difference and surged forward, screaming.

At the fore, Val was swept up. Propelled up against the barrier, she had barely enough time to extend her arms to break the impact. The side of her head struck the clear barrier, bounced off. She was lifted off her feet as the mass of people continued to beat in waves up against the barrier, pushing, squeezing.

She heard her name being shouted but she couldn't respond. Her ribs were held in a vice with no room to expand. She was being crushed.

Her vision began to go dark, and the world tilted.

She found herself on her back, looking up. The crowd of rioters surged forward, over Val and the fallen barrier, rushing the staircase. Halfway up the stairs, the ArcSec guards had retreated but stood at the ready, fanned out across the staircase's width. Batons in one hand, guns in the other.

Ready for a fight.

Wait, what? What Val saw confused her. She focused on the weapons. *Those are real guns.*

The first wave of rioters reached the ArcSec soldiers and were dealt with swiftly, the batons *swishing* through the air, back and forth, cracking heads, arms, ribs.

Val pushed up to her knees as the mob roared unintelligibly, flowed around her and up the stairs, undeterred. She felt a hand on her shoulder, one under her armpit, lifting up.

"You okay?" It was Trevor.

Thank god, she thought. She was about to thank him when she looked farther up the stairs, caught a glimpse of the first landing where the stairs switched back. A flood of black-clad soldiers pounded down, heavy boots stomping, shaking the ground below their feet.

"Ah hell," Trevor cursed. He pulled Val to her feet violently. "Hurry!" he shouted. "Let's go!"

They ran and didn't look back, not even when the sharp, clear thunder of gunfire echoed across the space, the reports banging back and forth between the buildings.

After zigging and zagging, taking turns at random, for what felt like a half a mile, they stopped. Val put her back up against the eroded brick of an alleyway, hunched over, hands on knees, chest heaving.

"Oh my god, oh my god, Trevor," her voice built to a crescendo as the gravity of what had just happened set in. "What was that?"

"I think we both know what that was," he said.

"But *why?*" she practically screamed. She'd never experienced anything like that in all her life, and a kind of hysteria overcame her. "Why would they do that? Just for construction? Those people just wanted to get to their jobs on 3." She shook her head, "I don't believe it, it doesn't make sense. None of it."

"I agree," Trevor said. "Something's up, and I'm gonna see if I can find out what the hell is going on."

"How are..."

Trevor put a finger in the air. "Gimme a minute," he said. "I'm calling someone."

Val tried to even out her breathing as she listened in on Trevor's conversation.

Who could he possibly be calling right now? And how can he be so calm?

She paid attention to his body language. He was nothing like the agitated ball of nerves she'd become. He stood upright, confident, in charge despite the chaos. She was glad he was with her. If he hadn't followed her into the crowd, she'd be dead.

That's twice now, she thought, shuddering. *You saved my life without a thought for yourself.*

Overwhelmed with sudden gratitude, Val wrapped her arms around Trevor's arm, pulled herself in tight to his side. He looked down at her, face softening for a moment before continuing his conversation.

"Yeah, that's right," he said. "Whole contingent of ArcSec guards." A pause. "What the—I *promise* you I didn't start it. I was trying to get us the hell out of there." Frustration in his voice. "I don't know, Estelle. I was too busy running."

Estelle? Who's Estelle, and why would he call her right now? Maybe she's his contact.

"I can't imagine many survived. I don't mean to sound like the biggest asshole you've ever seen, but right now that doesn't matter to me. I mean, it does matter, but what I need is to find out why. Something bigger's going down, and I think you and your friends at Talksmall are gonna have your hands full, like yesterday." Another pause. "Yeah. I'm still trying to get there. Any leads on how to get to three that doesn't involve a transfer gate? Alright. Call me back."

"Does that exist?" Val asked.

Trevor looked down at her, and Val suddenly became aware of his body next to hers. Its warmth, its strength. It was comforting, in more ways than one, but it was also a distraction.

She released her hold on the leather sleeve, took a step back.

"Does what exist?" he asked.

"Passages between the levels that don't require use of the transfer gates," Val said.

"Eh, might just be an urban myth. But hell," he threw his hands up in the air, "we're kinda running out of options. I'll try anything at this point."

"What are our..."

Trevor put his finger up in the air, mouthed the words *Sorry,* and turned to the side.

"Hey," he said to the air, "what ya got?" A pause. Val wished she could listen in on the conversation; not being able to hear didn't do anything for her nerves. Trevor's expression grew dark.

What's she saying?

"Oh, that's bad," he said. "That's real bad news, Estelle."

It felt like a ten-pound stone had been dropped in Val's gut. "What's bad?" she tried to ask, whispering, but he just angled his body away again.

Tell me! she wanted to shout, but instead turned away, paced up and down the dim alley. She trailed her fingers along the brick. The mortar was so deeply eroded that she could stick her fingers in-between the brick all the way up to the first knuckle.

"I don't care if they call it 'pass', or 'no fucking pass for you'," Trevor said, "just...ugh, I know. Can you tell me how to get to 3?"

Val urged her body to stillness, focused intently on Trevor's next words. More than that, it would be the way he said them that would tip her off as to whether or not this was a hopeless venture.

She was disappointed when she was met with silence. She pivoted to find Trevor standing within inches of her, eyebrows raised.

"Ready?" he asked.

"For what?" she asked. "Is there a way?"

"We're in luck," he grinned. "It seems Talksmall has a new 'partner'. At least, that's how Estelle put it. And they must be pretty high up."

"Why?"

"We're wastin' time. Let's walk, and I'll tell ya."

"Where are we headed?" Val asked.

"The Golden Road starts at the free drug zone," he said, not bothering to explain.

"Okay," she said, scratching her head. He walked out of the alley and into a wider thoroughfare packed with people. Val followed.

The Arc was known for its size, and from a distance it was a monolith, a massive man-made structure that erupted out of the earth, singular and striving toward the heavens. But as she explored its intricacies, she understood it was more a jumbled assortment of domiciles and buildings, storefronts, drug-houses, abandoned buildings, apartments, business offices. These all came in an array of shapes and sizes, competing for space and haphazardly arranged according to timing and necessity, nestled inside a massive superstructure.

All of which was knitted together by black interstitia—the dark alleys and nefarious spaces in-between.

She saw it more in places like this, on the lower levels where the Arc was more of a formless idea than a plan, where the growth had been rapid and reckless.

"Completely fucking lawless," she said, thinking aloud.

"Hmm?" Trevor asked absently.

They had gone up, down, around and through tight-walled corridors that stank of fresh waste, shoulders brushing up against the walls on either side of them, slicked with black slime. Through ever-narrowing thoroughfares where the competing smells of fried chicken and ethnic cooking dueled.

Just like on 1, there was no natural light in these back-corridors. The light came from lamps, sometimes working, sometimes not.

"You said something about Talksmall's partner being high up," Val prompted as they walked along a narrow street lined with crumbling tenements.

"It's the timing that gives it away," Trevor said. "Seconds after I get off the call with Estelle, she gets a mysterious notification about an underground network, called the yellow brick road."

"Oh," Estelle said, "so that's what yellow brick road means."

"I guess. Not exactly sure, to be honest. But whoever, or whatever they are, they knew about the barriers going up at the transfer gates. Said it was connected to a program nicknamed 'PASS'."

"That's appropriate," Val said. "And unimaginative."

"It's actually an acronym," Trevor said. "Stands for practical application of social scoring. The partner told Estelle the purpose of PASS is to keep people with low social scores from traveling back and forth to the wealthier districts."

It all clicked into place for Val. "Motherfuckers." She pounded the base of her fist against a building. Dark green slime rubbed off.

"Huh?" Trevor asked.

"Those motherfuckers. They're creating a walled society. Keeping us scum out, leaving us to kill ourselves, to fight over the scraps," she hissed. "To get kidnapped, tortured. That's what our lives are worth to them," she pointed up, indicating the upper levels.

Trevor rubbed at an ear but remained strangely silent.

What is up with that? Do you not agree? She wondered. *This is going to bother me. If he's going to continue helping me, I have to know.*

Before she could ask, Trevor came to a halt, touched Val's arm in an indication to stop.

His eyes probed the landscape in front of them, unblinking. "FDZ," he said. "Free drug zone."

The first thing that stood out was the guards. Around twenty in total. Dressed in navy blue uniforms, a white 'TS' printed in block letters on the front and back. They wore mean looking helmets, with downward angled slits for the eyes. About half of them held what looked like wooden bats. Probably printed but just as effective. Weighted on the inside. The other half, likely officers, had handheld pharmaguns swinging on cables at their sides. The screens on the guns were lighted up. Punch in the drug on the screen, then fire away. A whole pharmacopeia at their fingertips.

Beyond the guards stood the fences. There were two of them, both ten feet tall and made of the same clear material used at the transfer gates. They stretched across the road, from one dilapidated building to another, completely barring passage. Each fence had only one entry point, which was placed at a different extreme than the other, ensuring that the guards stationed in-between the fences would have time to slow the flow of people down.

Trevor pointed at the walls. "In case of a stampede. They try to weed out the Kopa users but nothing's a hundred percent."

Val had heard of Kopa, and what she did know didn't make her feel any better. One of the few infectious drugs, you could "catch" a high from someone else with the exchange of bodily fluids. The stampedes happened as a result of the drug's tendency to erode personal boundaries coupled with a component that drove an almost insane need for more of the drug. She'd never seen one and didn't plan on it.

Don't have much choice now, do ya? she thought.

They approached the entrance to the FDZ. Val chewed her lip, felt for the telescoping baton in her pants pocket, its cool surface comforting. She didn't consider herself brave, just ready to do whatever necessity required. An instinct of self-preservation. In Legacy, it was just how people grew up. The face of danger presented itself in many ways and many lights, every day, and it didn't discriminate based on age, gender or race. But when Val looked through the clear walls to the streets beyond, her breath caught in her chest.

The streets were packed with junkies, mostly in some state of sitting or lying down. Soot-stained faces and fingers, tattered rags for clothing. Some walked in circles, limbs jittering. Others talked to no one at all, jaws slamming open and shut, looking like a ventriloquist's doll with wide and wild eyes. It looked like a carnival of crazy people. Or, to steal a trope from decades past, a street full of zombies, all with a slightly different, and yet the same, affliction.

"We *have* to go in there?" she asked.

Trevor shrugged, gave her a look that showed he wasn't thrilled about it either. "Not much of a choice," he said. "They said, 'look for the red light at the end of the Golden Road'. Instructions were pretty clear. The Golden Road starts in there."

"I'm liking this adventure less and less," she said curtly.

"Just imagine it's a party," Trevor said, grinning. "C'mon, let's make this quick. And keep your eye out for anything that looks gold or yellow, probably painted."

He pulled her along behind him, passing through the open slots in the fences. They were inside now, not stopping to gawk. They cut a path along the street, wending through clots of people chattering to one another and themselves. One man fixed Val with an insane grin as they passed. His gaze was mesmerizing, red-rimmed green eyes dilated beyond anything natural.

"Uh, don't do that," Trevor said, voice low.

"What? What'd I do?"

"Make eye contact. That's a no-no. It's like encouragement to enter their crazy space."

"Ohhkay," she said, then forced herself to keep her head down, eyes on the pavement. She did her best to ignore the stench of those they walked between. A sour smell pervaded throughout, punctuated with the dull smell of dried fecal matter and the oppressive pungency of urine-soaked clothing.

Almost there, she thought, glancing up from the cracked and stained pavement. A flash of something bright caught her eye. A rectangle, miniscule, no larger than a couple inches, was hand-painted in a slightly darker shade of yellow, on the windowsill of a building.

"Hey! Come here," she gestured at Trevor. "Look at this," she pointed to the barely visible icon.

Trevor got his face up close, first looking at the broken-out window then below it. He put one hand to the sill, rubbed the yellow rectangle painted there.

"Well holy hell," he said. "Looks like you've found the Golden Road."

She smiled and was about to say *thanks* when something grabbed at her shoulder. She didn't think. Her reaction was so natural, so recurrent as to be born of instinct, that she didn't have time to think, simply act. Val spun, whipped the baton out in one fluid motion. Her eyes locked on the target. It was the man with the crazy grin and green eyes. His jagged, grime-caked nails pawed at her sleeve, teeth chattering. A bead of saliva dripped down his chin. She heard Trevor scream *NO!* just as she snapped the baton down on the man's forearm.

The man howled in pain, pulled his arm back and cradled it. As one, the junkies around them seemed to finally take notice of Val and Trevor.

The message in their collective gaze was striking in its combined silence and clarity.

Interlopers, it said.

She felt another tug on her arm, and she spun and raised the baton. This time, Trevor caught her hand mid-swing.

"Oh," she sighed. "It's you." Adrenaline flowed through her muscles, making her jumpy but ready to fight. Trevor forced her hand down to her side, lowering the baton.

"Look," Trevor put his arms on her shoulders, physically turning her around, forcing her to see. Like palsied predators creeping through the grass, the crowd of on-lookers began to close in on the two of them, one jittering step at a time.

Oh god, what did I do?

"Put that thing away and *run!*" he whispered fiercely. "And for god's sake don't let anyone bite you!"

Trevor clutched at her arm and charged out of the tightening circle. He put his head down, rammed his shoulder into a woman's chest, bowling her over. He stumbled and nearly fell, but Val pulled forcefully up on his arm, righting him. She followed as he pinballed from one druggie to the next, shoving, punching, knocking people over left and right. She followed in his wake, in the cleared path lined with human carnage.

Behind her, men and women howled with rage and frustration. She didn't look back.

Keep running, keep running. Her legs pumped as she followed Trevor's brown-jacketed form, twisting through tightening passages.

As they ran, she caught the occasional glimpse of a yellow-painted symbol. The bricks, painted on the sides of buildings, sidewalks and even on store signs, led them through the unseen warren of the Arc. If the transfer gates where the staircases were situated were the known world, these tiny, maze-like circuits of streets and alleyways were the uncharted territory.

They went by at a blur, with little time for Val to take stock of her surroundings.

Finally, Trevor slowed, then stopped, head beneath one of the yellow markers. He put a hand out against the wall. "Hey," he said, breath coming in gasps, "can we agree to let me lead the way from here on out?"

She had nothing to say, except, "I'm sorry."

"Yeah I know, it's just that some fights can't be won win with brute force. I know you want to believe in the injustice of it all, in righting the wrongs and all that, but the world is more subtle than that. It's," he paused, searching for the right word, "ambiguous. You can't just trample over everything in your path."

There were so many things she wanted to say, like how easy it was for him to say that when he didn't have anything at stake. Not like her. But she let it go. The fact was, he was right and she'd been wrong. She'd put their lives in danger. No arguing that.

So they continued on, following the odd rectangle: this one etched on a concrete wall, that one painted with a glittering indelible on the asphalt below their feet.

Soon they found themselves at the end of a long alleyway. Just beyond its mouth, a dark entrance was sunken into a building that formed one of the alley's walls. A red light glowed dully from the top of a short flight of long, low steps, barely pushing back against the dark.

The red light at the end of the yellow brick road, she thought. *That's it.*

Val peered into the gloom and saw a hulking mass prone on the landing, outstretched arm flopped limply across the top steps.

"I think this is it," Trevor said, ignoring the entrance and the person that could have been either passed out or dead. He walked past it, the red light sliding against the side of his face before he plunged into the near blackness of the alley.

Val followed. They didn't have to walk far before they found themselves at a dead end, walled in on either side by brick structures that ran straight up to the underbelly of Level 3. Val pressed her hand to its rough surface, looked up.

Halfway up the wall a three-foot square section had been cut out. Black bars secured the opening. From there, her eyes traced three-inch-wide cubes that jutted out at odd places all throughout the wall. It was as though the wall had been pressed through a mesh, then solidified.

Hand and footholds, she thought. *That's how we get up.*

"Hey, come look at this." Val waved Trevor over. He jogged up to her.

"What's up?" he asked.

"Look at these." She touched one of the extrusions, then pointed up to the opening. "And that."

"What am I looking at?" he asked.

"The way up to three. We have to climb," she said.

"Oh," Trevor said weakly. "Oh no," he repeated, a hint of revulsion entering his voice.

"That's the only way out I can see," she said.

"*Shit*," he whispered with genuine fear.

For Val, fear had exited the equation the moment Kat had been kidnapped. She wasn't stopping until she had her sister back. If that meant she had to scale a vertical wall to get there, then that was what she would do.

"It's okay. I'll go first," she said, trying to soothe him. She turned to the wall, put a foot on one of the blocks, and started to hoist herself up.

She felt a hand on her shoulder.

"Wait," Trevor said when she turned around. His eyes looked worried, full of concern.

"Aww, you're sweet. But don't worry, I'll be ok."

"Ok, well, how do you know you can get through that gate?"

"I don't."

"Hmm," he murmured. "This is sounding kinda half-assed."

Val shrugged, then smiled. "Wouldn't have it any other way."

Her muscles tensed as she readied them for the climb.

"Wait!" he reached out again.

"C'mon, Trevor. The first time was cute, but now I'm getting annoyed."

"I know, I know, sorry. It's just –"

She cocked an eyebrow as if to say *I'm waiting.*

"I—I like you. Please don't die."

She chuckled lightly, leaned across his body, pecked him on the cheek.

"I like you too. And I don't plan on it. See ya topside!"

The climb was about thirty feet, which didn't seem that far until she got a little over half-way and decided to look down.

"Bad move, Val," she said under her breath. Her stomach fluttered involuntarily. "Just keep moving."

The hand and footholds made the climb relatively easy. It was the mounting terror, like a vice tightening around her chest, of not being able to get through the gate that threatened her more than anything.

"You okay?" Trevor hollered from below.

"Fine."

She continued upward. Within another minute her hand gripped one of the black iron bars. She planted her other hand on the water-slicked ledge that formed the lower wall of the opening.

She pushed and the gate gave way. Not expecting it to move, her upper half pitched forward and her toes slipped from their holds, flying out from underneath her body. Her torso slid backward and the gate slammed shut as her weight pulled it back down.

Time slowed as her entire body slid backward. Her feet kicked at the air.

Her ears could hear the wordless scream that slipped from her throat, but her mind was in a hyper-panic mode. She flailed with her other hand, grasped at the bars, found purchase.

Allowing herself to look down, she prodded the wall with her toes until she found a hold.

"Holy shit! Are you okay?!" Trevor called.

"Umm," she said. She wasn't about to waste energy on conversation at the moment. *Let's try this again.*

Making sure she had good footing, she climbed up onto the ledge again, putting a knee on it, one white-knuckled hand gripping the bars for dear life the whole time.

Slowly, carefully, she pushed on the gate and crawled through.

It wasn't until she was all the way through, collapsed in a heap on the ground, that she allowed herself to breathe.

"Val!"

She put her head to the gate. "I'm okay! Come on up!"

"Alright!" he shouted back. "Give me a minute!"

It seemed to take forever, but finally Trevor poked his bushy head of hair above the opening, wearing his trademark stupid grin. Still, a part of her found it cute.

"Hey," he said, winking. "Come here often?"

"Yeah, wiseass. Save your jokes for this side of the gate."

His grin faltered. "Right. Help me with this one?"

"Yeah. Do it like I did," she said.

"The first or the second time?"

She shook her head. "You want help or not?"

He climbed up farther, put a knee up on the ledge then ducked his head and pushed through. Grasping his hands, Val readied herself to catch him, her feet firmly planted against the wall. She wasn't sure whether it would do any good if he actually slipped. He had to outweigh her by sixty or seventy pounds and despite what the movies depicted, most men wouldn't be able to catch and haul up a two-hundred-pound person.

Luckily, she didn't have to. Trevor crawled safely through, the gate clanging shut behind him. He stood, breathed out a huge sigh of relief, and offered Val his hand.

"Wow," he said, once she was on her feet. "I was *not* expecting to have to do that today."

"I suppose we could've waited for construction to finish, but you wanted an adventure," she said.

"Yeah well I have a feeling the adventure isn't over."

Level 3 was roughly twice the height of 2, with three distinct decks, all about fifty feet high. Ian's apartment was situated on the second deck in the Sunlight neighborhood.

Val heard the click of the magnetic lock, watched the door swing slowly open. She glanced at Trevor. "No hands. Fancy."

Inside, a hallway filled with light and a glimpse of a glass door that led to a balcony.

"Welcome to the levels of the rich and famous," Trevor said in a sharply articulated voice. He led the way down the hall, which terminated in open living space that was bright, airy. There was a couch and a chair, both white. The balcony they had seen from the doorway was situated on the far side of the room. Sunlight poured in.

A tall man wearing a dark-green shirt and khakis stood in a nook off to the left, hunched over a square table, placing items in a crate. There were four crates stacked on the table, each brimming with odds and ends. Plates, cups, what looked like books.

"Welcome." The man, Ian she presumed, spoke. His voice was soft, clipped. He didn't look up. "Estelle said you'd be coming. Surprised it was so soon, with the construction."

"Yeah, don't get us started on that," Trevor joked.

Ian looked up, gaze unblinking. He did this for what felt like a minute, then went back to his business of sorting and packing.

Trevor glanced at Val, eyes wide, questioning. She shrugged.

"Um, hi," Val tried. "I'm Val, this is Trevor."

"Ian. Pleasure." Pack. Sort.

"I pinged you earlier," Trevor said, "about the girl?"

"Yes. I received it."

"Okaaaay," he said, waiting for a moment. "So I guess I'll jump right in here. Val's sister went missing a few weeks back…"

"Check the morgues?" Pack. Sort.

"What? No," Val spat.

"Most logical place to start," Ian responded.

"Well maybe, but no…" Trevor said.

"She was kidnapped," Val interjected forcefully.

Ian's head snapped up so quickly Val took a half step back. "Kidnapped? By whom?"

"We don't know." Exasperation crept into her voice. "That's why we're here, asking you." She was starting to think this was a mistake. On top of that, who the hell was this chick Trevor knew that recommended talking to this weirdo?

Ian seemed unperturbed. He levelled that cool gaze on Val.

"Were you there when it happened?" he asked.

"Yeah." She held her palm out, showed him the angry scar that was only now starting to fade.

He nodded. "What do you remember? Where did it occur? On the street? At home?"

"At home, middle of the night. Some guy busted down our door, knocked me into next week," she said, trying to sound casual, almost flippant about it. But the memory was anything but casual, or funny.

Val took a deep breath, the images of that night materializing her mind, unbeckoned and unwanted. The man, the dash to the kitchen. The knife. Its sharp bite into her palm. The wild, desperate attack and the explosion of pain and light. The morning light on her face, throbbing skull.

Kat's empty bed.

She felt her throat start to swell up, tears flooding her eyes. "He knocked me out. Hit me hard. I woke up in the morning and… she was gone," she managed to choke out.

Tears rolled silently from her eyes. She wiped them away. Suddenly, Trevor's arm was around her shoulder. He pulled her in tight and she relented, folded into him, let the tears wash freely down her cheeks.

Ian was, apparently, oblivious to the sudden show of emotion.

"Do you live in a Legacy building? Subsidized rent?" he pressed. "Why do you think they took your sister?"

She shook her head against Trevor's chest, smearing tears and whatever else was streaming from her face into his green sweatshirt. It had that days-old, stale boy smell.

"Not they. Just one guy," she said, her voice muffled.

"No," Ian said. "You're mistaken. Maybe just one person did the taking, but from what you're telling me there's more than one person responsible."

"How do you know that?" Trevor asked.

"Think," Ian said. "One person can pick a kid up off the street. It's easy, even common in the Arc. The effort involved in getting to your sister means there was a plan and a reason she, specifically, was taken. Haven't you bothered to ask yourselves why?"

Val pulled back from Trevor, sniffed, wiped her face with her sleeve then made a futile attempt at cleaning the front of Trevor's sweatshirt. Ian was staring at the two of them again, looking as though he was waiting for them figure something out.

"Forgive me," he said, averting his eyes and bowing his head. A small circle of thinning hair showed on the crown of his head. "My manners aren't always what they should be." He disappeared into the kitchen, made some rummaging noises, glasses clinking together, and returned a minute later with two clear glasses of water.

"Please," he said, pushing them forward as an offering.

Val accepted the glass and took a long drink, swallowing half the glass in one pull. The water was cold, cleansing. "Thanks," she said, wiping her chin.

Ian nodded almost imperceptibly, made a small contrite smile. "Did your sister recently have her genome sampled?" he asked.

"No," she drew the word out, pronouncing it slowly as she thought over the past several weeks. "I don't think so. I mean, I did, when we went to comp some food, but…"

Then it struck her, a punch to the gut that sucked all her wind out. The glass of water nearly slipped from her grasp.

"Oh shit," she whispered, placing the glass on the table. Her hands flew upward, went to her mouth, her fingers splayed there like a protective grill.

"What?" Trevor and Ian said, simultaneously.

"The church," she said.

"What church?" Trevor asked.

"Deus Ex."

"Day-oos what?" Trevor blurted.

"Deus ex Machina. The hand of god," Ian said quietly. "What about it."

"It's where we print our food. Kat went to class there sometimes, so they let us use their feed. Cheapest stock anywhere."

"And?" Trevor prodded.

"I tried my account first, but I was short credits –"

"So your sister used her ID to pay." Ian finished her thought. There was a huge lump in Val's throat. She swallowed, nodded in assent.

"So that's how they got it. Machine must be compromised," Ian said.

"Motherfucker," Trevor said. He grabbed Val's hands with both of his, locked eyes with her.

"Remember the subsidies?"

Val just stared.

"What are the odds that Deus-whatever-it-is uses Blue Sky to subsidize their feed? You yourself said they had the cheapest stock anywhere."

"Oh Christ," was all Val could manage. A sinking feeling overtook her. She was responsible. If she'd only had enough credits whoever it was wouldn't have taken Kat. On top of that...

"You know how many orphans the church lets print from their machine?" Val asked, a rising horror at the thought of all the kids that were in danger.

"You think the church knows about it?"

"I can't imagine they do, but... I don't know." She shook her head. "I don't know."

"Question." The word came like an arrow shot between them, interrupting in a clipped, curt manner. Val turned to find Ian looming at her shoulder. Up close, his height was much more imposing than when he had remained on the other side of the table.

"Who," he continued, oblivious to his intrusion of her personal space, "or what, is blue sky?"

"Blue Sky Ad Ventures," Trevor interjected. "They sell ad space in return for cheaper product to the end user."

"The subsidy you mentioned." Ian said. He stepped back, pulled a leather-backed chair out from the table and sat. He motioned for Val and

Trevor to take a seat as well. He then put a finger to the side of his mouth and began chewing assiduously at a nail, looking thoughtful.

"So, ever hear of 'em?" Trevor pressed.

"Who? Blue Sky?" Ian asked. Trevor nodded in confirmation. "No. Not yet," Ian said. And then the corners of his lips turned up ever so slightly. "But give me a couple minutes, and we'll know all we need to know. I'm going to share my output," he said. "Maybe you can help spot something."

Val didn't know how much of a help she'd be as computers and programming weren't her thing. And by that, she meant she knew hardly anything about them, except how to use them. Just because everyone grew up with a system didn't mean everyone knew how they worked.

Ten seconds of observation was all she needed to understand why Trevor's friend, Estelle, had recommended Ian. Val had seen what Trevor could do, but Ian was in a class by himself.

After thirty seconds of watching the data flow across her field of vision, Val gave up. She kept the shared window open, but went to sit on the couch and wait.

Hopefully they find something. *Anything.*

Ian moved through the data like a wide net through a rushing river. He consumed it, parsed it, moved on. As far as Val could tell, there was no discreet movement from one process to the next, just a flood of data, relentlessly streaming.

"There. See that?" Ian asked. He and Trevor sat at the table, heads tilted up at odd angles, looking at images projected onto their internal field of vision.

"Um," Trevor said.

"That." Ian marked the data so it flashed in their view. It was a name.

"Jerome Theodore Aguilar?" Val said tentatively. "Who's that?"

"Project manager for Blue Sky's 3d printing feed platform," Ian said.

"Oh-kay, and why is this significant?" Val asked. She didn't mind looking like the stupid one. The only thing she cared about right now was getting answers.

Ian cocked his head, looked at her like a bird studying a person. "Were you not paying attention?" he asked, rather seriously.

"To what, exactly?" Val asked, getting irritated. "The billions of alphanumeric characters and links we just sifted through, or the idiotic..."

"Uh, let's just say no, we didn't see it," Trevor interrupted. "You move a little too fast for us."

Ian waved a hand as though shooing a fly. "First I determined whether Blue Sky itself had been infiltrated and had their code altered. I eliminated the possibility of a hack, then started looking at their update system. The code gets updated on a regular schedule—once a week, every Wednesday. By Jerome Aguilar. This schedule holds all the way back to several years. However," he held up a long index finger, "there was a double-update two months back. One at 3 AM precisely and another also at 3 AM and 39 seconds."

Ian sat back to supposedly let the significance of this statement wash over them. Trevor seemed to get it, which was good enough for Val.

"So this second update was initiated at the same time to mask it, so the logs would show an update at 3 AM and another at 3 AM. If anyone noticed the second update it would most likely be dismissed as an aberration. A fail and retry," Trevor said.

"Yes," Ian confirmed. "and the first did in fact fail, so Mr. Aguilar's tracks are *mostly* covered."

"Mostly?" Val was getting curious now. Despite being kind of an aloof dick, she couldn't help but admire Ian's prowess. That and his willingness to help at all in the first place. She looked at Trevor. Not a common thing, but it seemed she was finding the right company.

Ian brought up a picture of a man. He had caramel brown skin, short curly hair, green eyes. Happy eyes, Val noticed. A cheerful twinkle to them.

"This is Jerome Theodore Aguilar?" Val said.

"This is Jerome Aguilar," Ian confirmed. "On the Wednesday of the second update—September 23RD—two things happened. First, Jerome Aguilar received a sizable one-time payment that has yet to be repeated. Second, that machine, and probably many more in the Blue Sky network has been siphoning off DNA data and sending it here."

Their view flashed, displayed *UNIVERSAL SYSTEMS, LLC.*

"This is the end of the line. The name isn't important, it's just the very bottom in a tangle of shell companies that..."

A maze of connected script flowed left to right across her view. There had to have been thousands of companies, all connected by lines that, together, made a complex webbing that outstripped anything a spider could accomplish. If she looked at it the right way, it almost looked like an organic substrate, like viewing something at the microscopic level.

The script stopped flowing. One entity remained at the top, bolded and flashing in a huge font. Val's stomach clenched then sank, as though weighted with stones.

"Christ on a motherfucking kebab," Trevor whispered.

Ian nodded, lips pursed.

The words continued to flash but failed to coalesce in her mind. She felt like the victim of a horrific accident, the impact throwing her into a state of shock, capable of sight yet incapable of comprehension.

Gradually, the flashing, like a warning, triggered some instinctual awareness inside. Snapped her back to reality.

They were just letters, but unfortunately they spelled NOVAGENICA, CPX—the world's largest and best-defended pharma Complex.

Val reeled, stumbled backward before Trevor caught her arm.

It wasn't *random after all,* she thought. *It's worse. Novagenica kidnapped Kat for a reason.*

It was the perfect culmination of the day's trials. Further evidence that those in power took what they wanted and left the rest to rot.

But why would Novagenica take my sister? What could they possibly want with a teenage girl?

But Val knew. Deep down, she'd always known. They'd found something they needed inside of her, and instead of asking, or even paying, they simply used their power and *took* it. Took Kat.

For what? For tests? To try out a new drug?

The thought of Kat stuck in a cage, a lab rate, made Val sick, filled her with a renewed sense of urgency. She *had* to find a way to get to Kat, now more than ever.

Val steeled herself inside. It felt like a victory and a defeat all at once. A pyrrhic victory, maybe. Sure, they'd discovered the people behind Kat's kidnapping. At the same time they'd been dealt a crushing blow. She'd need a small army just to crack the Complex's walls.

Val retreated to the balcony, slid the door open, stepped outside. Clutched the railing, elbows locked, muscles standing out on her forearms. Looked out over the staggered array of buildings, homes and shops. Felt the cool air slip across her skin, drank in the warm sunshine. Turned her face upward, letting the warmth sink in, closed her eyes. Some sort of meat sizzled on a grill nearby, the unmistakable scent of it carried past her on a languorous current of air.

The incongruity of it all left her with a strange sense of vertigo.

How can these people just go about their day, knowing what happens below them every day of every week?

They departed Ian's apartment in silence. Ian had waved away their offer at payment, sidled back to his crates and the task of packing them. Trevor and Val ambled slowly back down that sunlight-filled hallway. It seemed dimmer now.

They stopped at the stoop of the building.

People walking all around. Pleasant, smiling faces of happy people with happy lives. Good jobs, plenty of food, nice homes. Safe homes.

Just a couple hours ago you were running for your life from a massacre. These people don't even care that it happened.

The thought struck Val like a slap to the face. Today more than ever, the realization that more money didn't just mean more or better *things*, had been rubbed in her face, like a dog's snout in its own feces.

Wealth meant safety, security. It was a funny thing for a person to buy, safety, for their family, their children. Here on 3, people didn't worry about their children being kidnapped. Go even farther up the Arc and Val was certain the concept was totally foreign to its residents.

Val felt a heat start to rise in her, a slow burn that began in her toes with a sort of tingling and moved upward, from dull anger to seething rage.

"I'm sorry," Trevor said, put a hand on her arm.

She shrugged it off, the anger boiling over. Thunder raged in her ears. "Don't be sorry," she spat. "Be *fucking pissed!*" Her chest heaved. "You think any of these people have to worry about this shit?" She swept her hand outward. "No fucking way, dude, no way. It's just us poor folk that get bombed, shot, kidnapped, tortured and experimented on. Stolen, like Kat." Spittle flew from her lips and she counted the atrocities one by one, slapping the back of one hand into the palm of the other.

Trevor didn't say anything, but his face had hardened, a serious look that, if her whole body hadn't been afire with fury, might have been frightening.

"It's so obvious. I'm just an idiot. It's like...like I've been sleeping for eighteen years and all of a sudden, boom!" she slapped her forehead, "I'm awake. All these people understand it, that's why they live here. Down below, level 1, level 2, Legacy territory outside the Arc, basically you have two choices," she said, fuming.

"Side with Legacy and you get your monthly credits, but other than that you're treated with indifference. You can die or disappear and nobody cares. Sign with a CPX and that's a whole different story. Not better, just different. There you're a serf. You're just manpower, energy, a symbol of work. In a Complex, they just think of you in terms of *joules*."

Val found herself panting. She was flushed and winded. She needed to get a hold of herself. Freaking out wouldn't accomplish anything, much less help them get her sister back. She focused on her breathing. Inhale, exhale. Repeat. Gradually, the heat drained from her face and her body began to calm, her pulse and breathing returning to normal.

She looked down, surprised to find Trevor's fingers intertwined with hers. She pulled his body close to hers, took some comfort in the warmth. "I just don't get it. Why doesn't somebody say something, do something?" she asked plaintively, looking up at him.

The hard look had not entirely gone, but was overtaken with something more like determination. "I don't know. But a better question is what are we gonna do about it?"

He's right, she thought. *Look to the future, don't whine about the past.*

"I don't know," she said. "But whatever we do, it'll be difficult." She searched his eyes, so bright and green and full of life.

And compassion.

"You sure you're up for this?" she asked.

"Hundred percent. Just gotta change into my boots first." He made a show of looking around. "Now where'd I put those damn things? Must've left 'em in the trusty coffin."

Shit. Your coffin's on 2. The thought reminded her of their harrowing journey just to get to 3 where Ian lived.

"Trevor?" she asked. "How do we get back down? To two? I do *not* want to climb back down that wall."

Inexplicably, Trevor started to laugh.

"What?" she looked at him warily, halfway mirthful and the other half confusion. "What's so funny?"

"While you were out on the balcony I asked Ian where he was packing up and moving off to. Got *way* more info than I wanted. Started rambling on about PASS—which, of course, we already know about—but then he went on about some conspiracy between ArcSec and PerSense. Considering what we saw today, it's probably all true. He said he was moving to 2."

"But all the gates are blocked," Val said.

"Exactly what I told him. Asked him how he was gonna get down there. 'Easy', he says, 'through The Blind Tiger.'"

"What's the Blind Tiger?"

"Opium den in the free drug zone. You know that red light? That's the exit."

She slapped herself on the forehead. "You've got to be kidding me."

"Nope. Anyway...*then* he proceeds to message me a map of the Arc, and not your average garden variety. This baby details all the connections and correlations between buildings and levels, including the paths carved out by our mystery society, the Golden Road."

"Well that's just swell," Val said. "Would've been nice to have a couple hours ago."

"It'll come in handy, for sure. But that's not even the best part. For shits and giggles I ask him if he has any tips on getting into Novagenica's super mega walled fortress and he just looked at me like I was an idiot. 'Join a tour group,' he says."

"Whoa, that's actually a really good idea," Val said.

"What I thought," Trevor said. "So what do ya think? You in? Wanna take a tour?"

"Am *I* in? I should be asking you the same thing. Kat's *my* sister. You don't have any skin in the game."

"I've got more than you know," Trevor said, looking at her for a moment, admiring. The message was apparent: *you, Val, you're what I have at stake.*

She looked down, unsure of what to say, suddenly and acutely aware of the feeling of his fingers enmeshed with hers. She'd never had a boyfriend in the traditional sense of the word and wasn't used to this sort of interaction.

But I could *get used to it*, she thought. She placed her free hand over his. "You're sweet," she said and watched his expression soften, deepen into something she couldn't quite describe but felt it connecting them.

She knew she couldn't do this alone. And now she knew she didn't have to.

KAT

KAT
AGE: 13
SEGMENT: LEGACY CITIZEN
POSITION: UNDERAGE; GOVERNMENT DOLE
RECIPIENT (REDUCED BENEFIT)
SOCIAL SCORE: 300

THE LIGHT inside Kat's perpetually shuttered room crept gradually from black to light blue, the hue of impending dawn.

"How do you feel, Kat?"

The doctor was Indian. Through the blue light Kat could see her image on the screen set in the door. Pale light from the corridor made her short, jet-black hair gleam. Her voice sounded tinny when filtered through the speaker, but Kat could still hear a hint of an accent. The doc was decent, as far as kidnappers went, but Kat wasn't about to give her an ounce of satisfaction.

"Fucking awesome," Kat spat, rolling over in bed, away from the door. "Like a million bucks and a ham sandwich. That's," she rolled back over, glared at the woman in the white lab coat, "how goddamn terrific I feel."

To her credit, the doc handled the verbal assault with aplomb, her face betraying no hint of tension. Sure, the door helped to deaden the vitriol. Wasn't the same cursing someone out through an intercom. "Oh good," the doc said, then paused, undoubtedly scrolling through a live feed of Kat's bios. "Because I'm showing a slight uptick in temperature. Not enough to be called a fever. Just wondering if you felt any different."

Kat pursed her lips, shook her head. Truth was she was starting to feel shitty. An ache that had begun in her back was wrapping its pulsating fingers around her middle and pushing upward, seeking out new and unsullied territory. It was getting difficult to breathe.

But she wasn't about to relay any of that to the doctor because, well, because fuck her, that's why. Kat still didn't know the woman's name—no post-kidnapping introductions, go figure—so instead she just called her doc. She'd think of something derisive but didn't really have it in her at the moment.

The doc pressed her lips together in a thin line, nodded, then turned to go.

Kat swung her legs off the edge of the bed, noticing how winded she felt with just this small action. She looked up to notice the doc had stopped in the middle of the corridor, eyeing Kat skeptically over one shoulder. The slightest hint of concern in her furrowed brow.

"Do let me know if anything changes, dear."

Kat gave the woman her best hate-smile. "You deserve this," she said, then flipped the woman off.

The old stand-by, Kat thought, feeling the warmth of satisfaction. *Gets 'em every time.*

Sliding her feet into the thin blue slippers they provided, she stood and tied her robe, shuffled to the door opposite the doctor. She pulled it open and stepped into the hallway that ran along the tens of other rooms situated alongside hers.

She hung a right, headed toward the commons.

She didn't particular want to go to the commons, but if she had to choose between that or sitting beneath a microscope, she supposed she'd choose the former.

But really it was a toss-up. The commons, and by extension all the other kids, weren't much more attractive of an option.

In the weeks since she'd been there, Kat had only managed to make one friend. Zara. It hadn't been hard, especially since Zara was a natural talker, one of those types that feel uncomfortable when people *aren't* around, as opposed to Kat, who was pretty much the exact opposite. Zara had found Kat sitting alone at one of the circular tables, plopped down beside her and just started jabbering.

In Kat's experience, Zara was definitely the outlier. The rest of the kids, a mix of teenage boys and girls, were more likely to glare at her than introduce themselves. She'd never felt such comprehensive, naked hostility.

And that's sayin' something when you come from Legacy, Kat thought, heading down the hallway.

The next door down was Zara's. Kat stopped, rapped her knuckles on the door.

"Who is it?" Zara called from behind the door.

"The President of Zimbabwe, here to escort you to breakfast," Kat shouted through the door.

The door swung open to Zara's smiling face. "Knew it was you," she said.

"*No*," Kat said in mock surprise, eyes wide, hands going to her cheeks. "You mean my President of Zimbabwe ruse didn't work?"

Zara elbowed Kat as she stepped into the hallway, closed the door behind her. "Don't be an asshole."

"Can't help it. I come from a long line of assholes. It's my nationality. It's even on my crest. A giant, puckered asshole on a field of pink," Kat said as they started down the hall.

Zara made a disgusted face. "Nice," she said.

Zara was two years older than Kat but stood a half a head shorter. She had the dark, rounded features of a Polynesian, but her long black hair was perfectly straight. To Kat, she looked like an islander version of an elf. Everything about her was tiny, cute and maddeningly well-proportioned.

Kat would've been jealous, but the girl had befriended her almost immediately.

That and jealousy seems just a little petty when you're trapped in a hospital with psychotic mad scientists, she thought.

At the end of the hall, a large open room—gray, like everything else—held tables, chairs and, unfortunately, a host of kids her age.

Too early, Kat thought.

As if she could hear Kat's thoughts, Zara said, "You don't usually eat this early."

Kat paused there at the edge of the room. "You're on to me," she said, looking over the tables. All of them were occupied.

Damn, she thought. *Can't hide now.*

"What's up?" Zara asked.

"I wanted to get away from my room," Kat said. "Creeps me out knowing those psychos can look in on us whenever they want."

Zara nodded. "I know what you mean. Where you want to sit?"

"I don't care, you pick," Kat said, walking toward the bar that held the food.

"Hey, grab me one too," Zara called out from behind her.

Kat nodded her assent, picked out a couple bars and pouches of juice, weaved her way through the tables to where Zara sat, next to a girl and two boys.

Somehow, Zara had managed to coax them into conversation.

Is it just me? Kat thought. *Why won't anybody talk to me? Am I diseased or something?*

"Any luck, Antonio?" Zara asked one of the boys. Rotund didn't do him justice. He looked like a boulder with an afro.

He shoved a whole bar into his mouth, spoke around the pieces. "I tried to work my way around the firewall," he said, bits and pieces sticking to his lips, "but no luck."

"At least they still let us use our systems," Zara said, "so that's a positive, right?"

The boy next to Antonio scowled. His name was Robbie and was scarecrow thin, with a pugilistic lower jaw that jutted out when he spoke. "Is it?" he asked. The skin on his face matched his tone—angry welts erupted across his cheeks. "It's more like a tease. 'Go ahead and use your systems all you want, little kiddies, just keep on believing you can find a way out.'"

"So what are you saying, Sam, that they're taunting us? Daring us to try to break their security?" Zara asked.

"No," Sam explained. "What I'm saying is they're giving us hope. Leading us down a path that has no exit, just further into the maze."

"Delaying the inevitable," the other girl at the table said. Her slouched posture spoke of despondency, her eyes fixed on a point somewhere in the distance. "They don't need to prevent, just stall."

"God that's morbid," Kat joked. She raised her hand as though calling for help. "Can we get some uppers over here on table 9?"

Nobody laughed. All eyes were on her, and they definitely weren't smiling. Even Zara looked troubled.

Shit. Foot in mouth syndrome strikes again. But what did I say that's so bad? she wondered. *You'd think I killed their mothers.*

"Why would you say that?" the girl asked, looking profoundly hurt.

"Hey, I'm sorry, I must be missing something." Kat said, putting her hands up in defense. "I just made a joke. It wasn't meant to be mean or anything."

Sam scowled, then shook his head in dismissal and leaned in toward Antonio, continuing their conversation in hushed tones. The girl, Kat still hadn't gotten her name, stood and left.

Kat looked to Zara. "What?" she asked. "Do I have the plague or something? Why does no one want to talk to me?"

Zara gave Kat a sad smile. "Don't worry about them, it's not you. As for the plague part, I can't exactly say."

Kat dismissed Zara's attempt at a joke. "Seriously, Zara, there's something I'm missing here. Just because we're a bunch of lab rats doesn't mean we have to treat each other like animals."

Zara looked away. "They're just scared," she said. "And in denial."

Well that partially explains the problem, Kat thought. "In denial of what?" she asked. This conversation was taking a turn for the bizarre, but something told her to keep digging, that whatever lay at the end was of vital importance.

Zara sighed. "They think it's a mistake to make friends."

"Well that's just weird." Kat said.

"Not really. Not if you understand this place," Zara said.

"What's to understand? Seems pretty straightforward to me. We're trapped in a hospital with a bunch of strange, anti-social kids." Val took a bite of her bar, waved it in the air. "Must be some kind of social experiment. Lord of the flies for the modern world."

Zara coughed into her elbow. "It makes more sense the longer you've been here," she said, gesturing to the other kids in the commons. "We've all been here long enough to know how this ends."

Zara's ominous tone kick-started a fear deep down in Kat. The saliva in her mouth dried up and she had a hard time swallowing the food in her mouth.

Up to this point, she'd simply assumed the avoidance was personal, that either her attitude or possibly even her looks were somehow an affront to the little closed society that lived there in the hospital. She hadn't considered the fact that it could be systemic.

"Ends?" Kat asked. "How *what* ends, Zara?"

Zara was seized by another coughing fit, causing her to turn red as she hacked repeatedly into her arm. When she was finally able to catch her breath, she levelled somber eyes on Kat.

"Life, Kat." Zara said. "How life ends. Everyone here, without exception, dies here."

Kat cried the rest of the afternoon. Sleep came as it does for the emotionally distraught, out of exhaustion, a need to check out. Reality fuzzed and the next thing she was (somehow) lying beneath her sheets and it was morning.

She skipped breakfast. Her mind had hatched the inklings of a malformed plan, and breakfast wasn't in it. *Seek help*, it told her. *From any quarter.*

There was only one person she'd encountered, with whom she'd physically shared the same space, so that was where she'd start. She knew she was grasping at straws. *But that's all I have.*

It was mid-morning when the man entered the room. Tentatively, as though the floor was made of lava.

Kat sat on the ledge, watched him with feline eyes. Curious, but reserved. In control. She held in check the excitement she felt.

Part one of the plan, check.

This was the same man she'd flipped off. Who had inexplicably collapsed to his knees then crawled out of the room. Like before, he stepped into the room freighted with an armload of white sheets.

"Good morning again," he said to Kat, setting the sheets down on the dresser. Then he did the oddest thing. He cocked his head to the side as though listening for something. After a few moments posed like this, his body straightened and he seemed taller, lighter, as he went to strip the sheets off the bed.

"I wanted to apologize," he said, pulling a pillow out of its case. "For the way I left last time."

Kat squinted. "That *was* super weird of you," she said. She picked at a chipped toenail.

"Trust me," he said, "it was weird for me too."

"What happened?" she asked.

Bent over the bed, reaching for the opposite corner of sheet, he looked at her briefly. "I was punished."

Whoa, the plot thickens. After her conversation with Zara yesterday, she'd confined herself to her room, thought over and over about how she could possibly *get the fuck out*. Anxiety warred with determination. She was not going to die in this place.

She'd hoped, but it was not the same as seeing the opportunity come to life before her eyes. *Use it!* her mind shouted at her. *It's an opportunity, use it!*

"Oh now I'm hooked," she said. "Punished for what?"

"For talking to you," he said.

Somebody threw the electricity on in her head, the lights exploded behind her eyes. But she played it cool. Somewhat.

"Yeah, I can see how that's a crime. I'm wanted in three countries and even more Complexes. I've done *evil* shit. Why they got me locked up in here."

Hunched over the bed, the man pushed upright. He regarded her with sad eyes. "Worst thing you kids've done is have the wrong gene sequence."

And with that, he snapped the sheets taut and left, the old ones bundled up in his arms.

She almost reached out a hand for him to stop, but instead she sat there, watched as the door eased shut behind him, sealing against the wall with a final click.

The hook's been set, she thought.

It wasn't much, but it was something. For the first time that day, shit, since she'd been locked up in that horrifying place, she felt a sense of accomplishment, of worth, of hope.

INTERLUDE

JAIM

JAIM
AGE: 35
SEGMENT: LEGACY CITIZEN
POSITION: LEGACY ADMINISTRATOR: MIDWEST
ARCOLITH; DISTRICTS 35-134
SOCIAL SCORE: 880

THE GREAT HALL of The Sherlock was a study in ornate wood-work. Jaim stepped off the streets on Level 4 where the theme bar was situated, through the massive front doors and back in time to the late 1800s.

The ceiling vaulted some twenty feet up, yet with heavy draperies covering the walls and thick-hewn furniture, the room still maintained a staid, oppressive atmosphere. Maybe it was the smoke. It hung in the air, a dense fog that thickened into increasingly dense striations of white and gray on up to the ceiling.

The great hall was made for carousing, a place where people could congregate, throw back pints and theatrically puff on thick cigars. The bar, a richly lacquered span of wood, took up the right side of the room.

Jaim had come in his heather gray topcoat, black and white herringbone deerstalker perched atop his head. A fire roared in the giant brick hearth at the back of the hall. He ducked down briefly as he wove through the masses of drinking men wreathed in smoke, peering through the flames to see through to the great chamber, which was a slightly smaller version of the great hall that was more intimate yet still open to the public. Like the great hall, the great chamber was occluded with smoke. Sherlock and Watson lookalikes, alongside the occasional Lestrade and Gregson doppelgangers, milled about. Men congregated about the fire in the hearth that connected to the great hall, brandy glimmering in cut crystal glasses they held in one hand; in the other, pipes were nestled in their palms, wafts curling gently upward.

Jaim snorted derisively. While smoking was true to the theme of The Sherlock, Jaim and the other admins spurned the idea of taking their nicotine by setting fire to it. The preferred snuff, like the aristocracy of centuries prior.

Speaking of. He pulled out a gilded snuff box. It had a large amethyst set in its lid, its violet facets glittering brightly. An approximation of the gift Sherlock Holmes had received from the King of Bohemia. Taking a small pinch between thumb and forefinger, he lifted it to a nostril and sniffed, inhaling sharply. The nicotine hit his system immediately, lifting his spirits and heart rate. He felt good, powerful. Confident. The orange scent lingered in his nose.

Putting the box back in his coat pocket, he passed through a hallway that connected the two great rooms and followed it to where it terminated. To the left, the great chamber; to the right, the rented quarters. He went right, down a long wood-paneled hallway which was lit with kerosene lamps. The fumes the lamps gave off were thick and acrid. They joined the layer of cigar and pipe-smoke that permeated every cubic inch of the place.

The din of the rabble died away as the hallway turned left, leading to a set of stairs. At the top, two rooms were situated on either side. Both doors were closed. The only light came from the antiquated lamps on the walls. The flames danced subtly in the controlled climate, throwing irregular shadows down the hall. Jaim knew from his previous visits that the door on the left held the garden; the right, the drawing room.

Contemplating the brass doorknob of the drawing room, Jaim set his face into what he considered a hardened, serious countenance. Adjusted his cap, straightened his back. It was best to look like a shark when swimming amongst them. That or get eaten.

The wooden door creaked as he pushed it open, and warm light spilled through the doorway. Unlike the other gathering rooms, this one was smaller, intimate, interspersed with the odd couch dressed in green velvet. Not one, but two smaller fireplaces stood opposite one another, poking irons mounted at their sides.

"Jaim!"

John McAlister shouted at him from the fire on the far side of the room. He grabbed two glasses of brandy, one nearly empty, from the stone mantle above the fireplace and waddled over to Jaim. McAlister's round, chubby face glowed inside out with the ruddy eagerness of liquor. His blue eyes twinkled from beneath the shadow of the black felt bowler he wore.

"Got ya a glass, my friend," he said, handing the full glass to Jaim.

Jaim took it, shook McAlister's outstretched hand. "John," he acknowledged, maintaining an austere bearing as he surveyed the crowd. For some reason he couldn't quite understand, wearing the 19TH century costume made it easier for him to assume an aura of power.

"Always glad to see ya, Jaim." Despite the fact that McAlister was an equal, no one treated him as such. People-pleaser, suck up, welcome mat. These were the names McAlister's colleagues flung at him when he wasn't listening.

Jaim peered over McAlister's head, taking in the rest of the men he'd invited. Moon, of course, was there, dressed smartly as Watson. He stood at the opposite fireplace in quiet conversation with Greitens. He'd hand-picked McAlister and Greitens specifically as they were the other two Admins who had districts on Levels 1 and 2 and would be the most amenable to Moon's proposition.

Old Greitens wore that maddening monocle he favored. That and the combination of his red, white and brown beard made him look more like a half-blind grizzled Viking than a 19TH century Londoner detective.

Jaim crossed to the windows opposite the drawing room's entrance, pulling McAlister with him. Out the window was a digital representation of the view from 221B Baker Street, the fictitious residence of Sherlock and Holmes. Outside, horse-drawn carriages and cabs sped along a cobbled road beneath a low-hanging, gray sky.

Jaim bent forward, leaned into McAlister. "When did everyone get here?" he whispered in the shorter man's his ear.

"Ah, well," McAlister hemmed, fingers nervously circling the rim of his glass, "kind of drummed up the excitement to a fever pitch when you said you'd be bringin' a mystery guest. You should've seen the way Greitens's face fell when he saw a man push open that door. You know how he likes his women. Myself I can't quite understand why the hush-hush over the whole thing, but after five minutes, Jayne, ah, Mr. Moon had him won over. Certainly is a charmer. Made himself quite at home."

Jaim nodded absently. He was annoyed but managed not to show it. This was supposed to be his prize, his idea, his victory.

But if the plan comes from Moon and the thing flops, you still have plausible deniability.

Greitens poked morosely at the fire, listening with an upturned ear at whatever Moon was saying to him. Jaim approached them.

Let's get this show started.

"Mr. Moon," Jaim said, putting a hand on Moon's jacketed arm. "So glad to see you."

Moon's face lit up with recognition. He took Jaim's hand in both of his and shook vigorously. "Mr. Del Potro!"

Greitens nudged Moon with an elbow in his side. "We call him DP around here." He nodded at Jaim. "DP."

"Greitens," Jaim acknowledged, then turned back to Moon. "Call me whatever you like," he said. "You don't have to listen to this old bat."

Jaim squared up, addressed all three men as McAlister scurried over. "Gentlemen," Jaim said. "Shall we begin?"

"Mm-hmm," Greitens murmured, pushed at a log and took a small sip of his brandy. "Enough with the mystery."

At that moment, the governess, a pretty young thing dressed in a frilly white top and blossoming, lacy skirts, entered the room and asked them for their orders. Any other day and Jaim would've found her beauty distracting—perhaps would've chanced to use his powers of persuasion to get her to sit on his lap and enjoy a virtual session together. Not today. The prospect of power, both the potential increase and its inverse, was too great. It turned out that risk, for him, was an enormous turn-off. He couldn't even think of getting an erection.

"If I may?" Moon said, looking at Jaim.

"Make it quick," Greitens said, looking over his shoulder at the exiting girl. "I want to have time to dandle the governess on my lap." His speech was showing the first hints at slowness. As if he could read Jaim's thoughts, Greitens removed a gilded snuff box from the inside of his heavy coat, took a small pinch of brownish white powder and sniffed it up one nostril. The monocle fell from his eye, made a tinkling sound as it hit the ancient wooden floorboards. "Blast," he cursed quietly as he pitched forward to pick up the eyepiece.

"Jesus Greitens," McAlister whined. "Your mind is all cock. We have a guest here..."

"Aw shut the fuck up, McAlister, ya twit," Greitens grunted, heaving his bulk back into an upright position. His face was florid with the effort. "I'm just playin' around some."

"You'll get your chance to sweet talk a pretty girl, or maybe even two, later," Jaim said. "But for now both of you shut the hell up and give your full attention to Mr. Moon." Jaim looked expectantly at Moon. "Go on," he urged.

Whether it was his own eagerness or the sense of peace radiating from Moon, Jaim couldn't tell, but his extremities tingled with electricity. He felt calm and ready. Energetic, prepared for battle. He took a sip of brandy. Yes, that was the way to visualize it. Riding into battle, charged up and looking forward to what may come.

"If I may be blunt," Moon said, "Legacy is dying."

Greitens coughed on his brandy, great wracking hacks that doubled him over, his hands gripping his thighs.

McAlister bent in obsequiously, hands pattering at the man's shoulders and back. "You okay? Greitens? Greitens, you okay?" he asked in worried tones.

Greitens shoved McAlister to the side. "Get off me!" he roared. "I'm fine, ya sniveling imp, I'm fine." He bent back, eyes wide, hands on his hips, and sucked in a lungful of air. "*Hoo!* Helluva way to start, Moon." He looked Moon dead in the eye. "But you have my attention now."

"Sorry for that," Moon said. "This is not an attempt to be dramatic, rather to simply state the facts. Every day the Complexes reel in more and more subscribers, and every day Legacy, your government, loses citizens. You can see where this headed. It could be two years, five, ten...the *when* isn't exactly known, but the inevitability of it is right before your eyes."

"So?" Greitens objected. "This is the course of history you're talking about. Semantics," he said with a wave of the hand. "It's already written. Nothing you can say or do will change that."

"I disagree, and here's why. It's a simple mechanic. As Admins of the Legacy government, you still retain an enormous amount of power. You can legislate without interference. *You* can draft a bill, however you like, and make it law. Laws alter behaviors, change societies, and modify the course of history."

"Well, within reason," McAlister interjected, his jowls bouncing. "We have to have a majority vote of the Admins to pass legislation."

Jaim made a show of looking at McAlister, then Greitens. "There are five Admins, McAlister. I see a majority right here in this room," he said.

"To cut to the chase," Moon said, "I've prepared a document I'd have you consider. Let me just send it, then you all can read it and we can discuss."

Jaim's system chimed, and a document showed up on his field of view labeled *DS-75—EXODUS*.

To Jaim, it wasn't new. He'd already reviewed the bill, co-written if he was being honest. It was a simple budgetary bill that would re-structure how Legacy paid out the Standard Universal Credit to its citizens. To be short, it forced those outside the floodgates to move into the Arc proper if they were to receive their full credits. This would then condense an unwieldy, scattered population into something controllable. Subsequent legislation would focus on mobilizing that workforce into social projects, also tied to receipt of their SUC. As they'd written it, Jaim had been possessed by wild fantasies of the new society they could build, both literally and figuratively.

But neither McAlister nor Greitens had seen it, so the next few moments would be crucial if they were to succeed.

The two men concentrated on their systems and Jaim watched them with avid interest. McAlister continued to glance up periodically at Greitens, looking for cues, while Greitens's face darkened, drawn down somewhere between a scowl and outright rage.

"What the fuck is this?" Greitens asked. He turned toward Jaim. "What the fuck are you trying here, DP?" He shoved a fat finger into Jaim's lapel. "Tell me. You think this is a *good* idea?"

Jaim put his hands up. "Just..."

Moon, the countenance of calm, placed his body between them.

"It was me," he said. "I wrote this, not Jaim. Will you allow me to explain it?"

Silence, then finally Greitens shifted his attention to Moon. "What's to explain? We Admins pass this and there'll be revolution on the wind."

"You make a point," Moon conceded. "There's always that danger. But ask yourself this—do people actually give a shit anymore? I mean, let's face it, Legacy has become synonymous for all the ways that people don't care for one another. The rising stars in the world are the Complexes. They hold the new social model. A Complex can take care of its subscribers. It's really quite simple—food, shelter, health..."

"We already give 'em those!" shouted Greitens, spittle dribbling down his beard. "What do you think the Standard Universal Credit is for? Fuckers don't even have to work if they don't want to, just live off their monthly SUC. I'd call that a better than fair deal."

There, Jaim thought, *there it is. The hook.* Now Moon just had to pounce on it.

Moon nodded in concession. "You make a point, Mr. Greitens. You make a point. But I hadn't quite finished. At the end of that list of human needs is *purpose.* Purpose is the final piece, and arguably the most important. A Complex gives people work—something that's difficult to find in Legacy—and work gives people purpose. Worth."

The men quietly considered his words.

"Now let's look at Legacy. You've suffered any number of insults, but more than anything it's been a death by a thousand cuts over the years that's led to a long decline. You've been defunded and broken up into a system that is structurally incapable of taking care of its constituents."

"You come here to convince us or insult us?" Greitens said, puffing his chest out.

"Please don't take offense," Moon said calmly. "I'm just putting the evidence before you. But honestly, you want to know what I see?" He let the question hang in the air before them. "Opportunity. Maybe the last one we have before the Complexes take over everything, including your jobs."

While his voice remained level, the words served like a slap to the face. The two other Admins regarded Moon with the greatest sobriety they could muster.

"The way I see it, we don't act now and the system we call Legacy will be gone, dust, in five years, seven at the outside. *But,*" he paused, an index finger in the air, "not if you pass this bill. DS-75 isn't just a one off, a last-ditch attempt to consolidate a population and increase revenue. No, it's one link in a chain of legislation that will re-shape what Legacy means to people. The next link will create projects—Work-for-SUC projects that force people to work to receive their Standard Universal Credit. In each one, you can expect some pushback, but in each one what you're actually doing is making slight alterations to the psychological make-up of Legacy's citizens. You're changing..."

"The future," Greitens said. There was a gleam of understanding, maybe even of optimism, in his eyes.

He can see it now, Jaim thought. *Same as I did when Moon first proposed the plan in my office.*

"That's *right,* Mr. Greitens," Moon said leaning in, face alive, body eager. "That's exactly right. Now let me ask again because I think it bears repeating. Who has the power to change the future?"

Jaim marveled at Moon's mastery. There was a confidence to him. He looked like he'd been doing this since he was a child. Practiced and at ease.

"We do," Greitens said, awestruck. "By fuck, we do."

"*You* do," Moon repeated. "You're Admins, for God's sake. Pass this bill. Bring the populace in to the arc, inside the floodgates, and consolidate power into your hands. Then, bill after bill, you can change the future. I won't sugarcoat it, there may be riots. But you've got Talksmall on your side. That counts for something, especially on 1 and 2."

"Okay, I can see where this heading," Greitens said. The calico whiskers of his mustache quivered as he spoke. "But where will all these people go? You're talking about uprooting hundreds of thousands. Creating a city of refugees."

"This is where Talksmall comes in. Within the DS-75 bill are the budgetary provisions for Talksmall to build permanent camps. We have the space and this bill gives us the resources," Moon answered confidently, as though it was as simple as saying the words and *poof* it was done. "A couple places come to mind. The Riverside Park is a good start. We can use Signal Hill as a model."

"The homeless ghetto?" McAlister said, horror disfiguring his elfin features.

"It's a good model," Moon said. "Sometimes the solution is as simple as stacking a few freight containers together. But I don't want to get distracted on specifics right now, there'll be plenty of time to focus on the details. The real question at heart is whether or not you three are willing to fight for what's right and what is yours."

Jaim felt a stirring in his gut at Moon's speech. He had to hand it to the man. Clearly in his element, Moon was powerful, poised. Even managed to be provocative without pissing everyone off completely.

Scanning their faces, Jaim could almost see the thoughts turning over in their skulls, weighing risk against reward. Moon had effectively painted them in a corner. The looming threat of the Complexes, the inevitability of that burgeoning black mass on the horizon. It would consume their way of life if left unchecked, allow them no means of escape. But there was a way out of that corner. All they had to do was act. Do or die. It was that simple.

The silence wore on for a minute. Moon continued to hold their attention, not backing down, allowing the silence to do its work.

When the quiet was finally broken, its source surprised everyone, including Jaim. "We'll need multiple layers of planning," McAlister said gravely. His flushed cheeks hung like sacs at the sides of his face. "And dead silence. No one talks about this to anyone outside of this room."

Greitens settled his gaze on Jaim, then on Moon. "Can we trust you, Mr. Moon? Will this play out as you say?"

Moon met his stare. "Of course, Mr. Greitens. You have my solemn pledge of the utmost discretion, and the promise of all the resources available to me at Talksmall to make this work."

And just like that it was decided.

"Greitens," Jaim turned to the man at his side. "You're close with Arc-Sec's Commandant, that old maniac Briggs, can you ask him when the PASS directive goes live?"

"Ah shit," Greitens said, "senile bastard loves me. He'll tell me anything."

"I'm thinking we shouldn't publicize the vote until after the checkpoints are in place," McAlister said.

"Lookin' after your own ass, again, eh?" Greitens chortled.

"Yours too," McAlister retorted.

"My ass is off-limits, as you're well aware," Greitens said in jest.

"Don't flatter yourself, old man. 'Bear' isn't my type."

Jaim stepped in before they got too far off-track. "Obviously there are a lot of details to figure out, but tonight isn't the night. I propose that we take the next week to mull over as many of the problems and details we can think of individually before we put the bill out there."

Jaim grabbed his glass from the mantle and raised it. One by one, the other men in the room raised their glasses.

"Our three votes represent the majority," Jaim said. "Do we have 'ayes' for the passage of bill DS-75?"

"Aye," McAlister said, eyes darting to Jaim then Greitens.

"Aye," Greitens said gruffly, then tossed back half the glass of brandy in an audible gulp.

"Aye," Jaim said, and took a small sip from his glass. "It's decided then."

"Now where's that governess!" Greitens shouted, violently raising his glass in the air again, the amber liquid spilling over the edge and onto his jacket. "Dammit!" he cursed, brushing at the stain. He took McAlister by the elbow. "You're coming with me. We're gonna go find that wench if it's the last thing we do."

Greitens pulled McAlister, who smiled sheepishly at Moon then shrugged, away from the fireplace and out of the room.

The door slammed shut behind the two Admins. Jaim and Moon had the room to themselves.

"Well," Jaim clinked his glass against Moon's. "That was well done."

Moon looked thoughtfully at the fire, its flames jumping and dancing in his black eyes, then back up at Jaim.

"Quid pro quo, Jaim. I'll be expecting you to return the favor, of course."

There was a brooding quality in Moon that Jaim hadn't yet witnessed. It suggested at hidden depths and veiled meanings. A tiny seed of doubt cracked open inside Jaim, fed by Moon's reticence.

I don't like the sound of that. But...he's right. I wouldn't have been able to convince Greitens and McAlister so readily. Jaim thought well of himself, but persuasive speaking wasn't a forte of his.

"Of course," Jaim replied coolly. For now, Jaim had gotten what he wanted. If, in time, Moon turned out to be more adversary than ally, well he would deal with that...disturbance...when it arose.

"Don't worry," Moon said, face suddenly brightening, the cloud gone, dispersing Jaim's doubts as well. "I promise you—if we come up against something that proves too difficult to handle, you'll be the first person I contact."

Assuaged, Jaim rolled his shoulders back, loosened his body up. He put a hand on Moon's shoulder.

"I know, but I'm trusting you, so don't fuck me okay?" Jaim said.

Moon offered a sly grin. A wolf in sheep's clothing.

"I'd never fuck you, Jaim," he said, then reconsidered. "Well, not until you ask politely."

Jaim laughed. "That'll never happen my friend."

Moon simply held his gaze with a not-quite grin that said nothing and yet volumes at once.

BOOK THREE
DS-75 — THE EXODUS

DS-75

"THIS IS A MESSAGE FROM THE EMERGENCY BROADCAST SYSTEM. THIS IS NOT A TEST. ON NOVEMBER 16 ALL LEGACY SUBSIDIES FOR HOUSING BEYOND THE MIDWEST ARCOLITH'S OUTER FLOODGATES WILL BE CUT BY SEVENTY-FIVE PERCENT. THIS NOTICE DOES NOT APPLY TO COMPLEX PROPERTY. IF YOU BELONG TO A COMPLEX, IGNORE THIS NOTICE. THIS NOTICE APPLIES ONLY TO LEGACY SYSTEMS AND PROPERTIES. THIS IS NOT A TEST. ON NOVEMBER 16 ALL LEGACY SUBSIDIES FOR HOUSING BEYOND THE MIDWEST ARCOLITH'S OUTER FLOODGATES WILL BE CUT BY SEVENTY-FIVE PERCENT..."

SAUL

SAUL
AGE: 24; AGE IS AN ESTIMATE, NAME AND ID
STOLEN/ASSUMED
SEGMENT: LEGACY RESIDENT
POSITION: UNKNOWN
SOCIAL SCORE: UNKNOWN

SAUL'S SHOP was hidden. Secreted away from the north-south feeders that spilled their traffic into the Arc, tucked back on a side street that rarely saw traffic—daytripper or otherwise. A two-story brick building with boarded-over windows, it sat like a hunched beast bowing in the dark shadows of its taller masters.

Saul exited the side door of the shop into an unlighted alley. He pulled the filthy brown robe he wore tight, close around his body. The night was a bitter one. Level 1 typically held its warmth, but tonight the frigid air crept in at the Arc's edges.

He yanked the robe's coarse, scratchy hood over his head. It wasn't enough to keep him warm, but that wasn't the point tonight. More than anything, he had to look anonymous. And what was more forgettable, more ubiquitous than a beggar?

Shuffling forward on gray rubber slippers, Saul slipped out of the mouth of the alley, head down, eyes up. Not breaking character, he made his way slowly toward the building that was his target.

It was funny to think of a building almost in the same way he thought of a person. But that was the way it worked now. Buildings weren't just a heap of materials put together in a way that made them stand up. They were intelligent, almost alive, being fed by sensory input and able to make decisions based on their health and threat level.

And this building, Saul knew deep down, was one sweet treat.

He continued his slow amble, eventually making his way out to Walnut where the traffic heading deeper into the Arc flowed more heavily. He turned north, joined the flow.

Twenty minutes of laborious, slow progress brought him to the shadow building, as he called it. One half of what used to be a pair of condominium towers, its height vaulted up beyond the neighboring buildings, projecting directly into the underbelly of Level 2. Its sister building, now a vacant pile of rubble across the street, had collapsed before Saul had been born.

Saul scanned the height of the building, peered into the gloom of the alley at its side. No activity yet. That was fine, good even. He could wait.

Hands cupped together before him, Saul moved up and down the stretch of sidewalk that fronted the shadow building in a swaying, almost drunken gait. An hour passed. Two. He busied himself as he walked mindlessly to and fro by watching feeds of doctored children's videos. He always got a kick out of that.

Another forty minutes dragged on before the mark finally showed, looking like he was trying to blend in. Head bent beneath the cowl of his robe, Saul sneered. He shuffled down the length of the building, the covert camera threaded into the hood's fabric focused on the man, who was dressed head-to-toe in black. Despite attempts at camouflage, any trained eye could see the youthful vigor in the man's step, the kind of energy that was a marker of good health and, by extension, of wealth.

That and the fact that he wasn't advertising. No names, no wares, no nothing. A man without a side-hustle in the Arc wasn't *from* the Arc. In fact, he wasn't just not advertising, he was totally dark to other systems. Nothing came up when Saul focused his recognition software on the man.

Saul whistled to himself, low enough that nobody would hear him. *Must be running some pretty pricey stealth-ware.* A surge of greed, of the purest need, like water to a man dying of thirst, flowed through Saul's every cell. Whatever this man had, Saul wanted it. Had to have it. *Would* have it.

He was close now, able to hear the heavy footfalls of the military-grade boots the man wore. Saul turned out toward the street at the black-clad figure's approach. He slowed, no doubt seeing Saul the beggar and waiting for him to make a move. As a feint, Saul started, in small shuffling steps, out into the street.

The mark took the bait. As soon as Saul stepped down into the street, the man picked up his step, approaching the shadow building freely.

There was no time to lose. Saul spun quickly, approached the mark from behind. He was a good head taller than Saul himself and, despite an obvious desire to not draw any unwanted attention, he cut an imposing figure as the stood at the foot of the building.

Close now, Saul jammed his left hand into the robe's pocket, felt the cool coating of the slimy smart goo. He took a handful of the entangled gel, rubbed it between his fingers. Hunched quietly behind the man, Saul reached out with his left hand, tugged at his sleeve. He opened his fingers and slid his hand back, making sure to deposit as much of the gel as he could.

"Help the homeless?" Saul asked. The man spun around, flailing his left arm as he did so, recoiling at Saul's touch. The disgust in his face was almost comical to Saul. He inched closer to the man, hands cupped to receive an offering, until he got a curt, "Fuck off" for his efforts.

"Yes, boss," Saul said, bobbing his head like a feeble old wretch barely in control of his limbs. He turned to the north, stumbled away. Out of caution, he continued to beg his way up and down Walnut for another half hour before cautiously making his way back to the shop.

The two-story building looked ready to fall apart, like it was ready at any moment to crumble beneath its own weight into a pile of rubble. He disappeared into the cave of the alley at the building's side, entering once more through the side door. Saul shut and barred the heavy door behind him with a six-inch thick bar of composite. When it came to security, lo-tech was simpler and more secure than anything else.

His shop was mostly open space. Workbenches lined the walls with a thin, square table in the center of the room. Papers and trash littered its surface.

Throwing off his robe onto the back of a stool, he stood at a tall work-bench. The moment of truth. His plan had been three years in the making and while wasn't the smartest criminal in the Arc, he *was* shrewd. Oppor-tunistic. Probably the best way he could describe himself. Opportunities abounded, but most weren't shrewd enough to see them when they arose.

Saul had smelled opportunity from the moment bought the high-rise. He'd paid a few junkies in bags of Buena to try to find a way in. When that hadn't worked, Saul waited. Sat on it. Someday, he knew. Patience was a virtue that wasn't the sole province of the virtuous.

Three years later the building came to life, startling even Saul. Business had been good lately, and his grand project had gone forgotten. He thought he'd missed his opportunity, the crew in and out so quickly he didn't have time to prepare. Which was why he'd spent half the night in his beggar get-up, drifting lazily up and down the stretch of sidewalk that fronted the no-longer abandoned, mute building.

He connected his personal system to the shop. Time to find the goo. It was supposed to give him access to the mark's clothes, which would in turn grant him access to the man's system. In turn, that system would have built-in permissions for the shadow building.

And voila. Saul would have his prize.

Not to take by force, nothing as mundane and short-sighted as that. Access. That was everything. And when nobody *knew* he had access, the value of that access skyrocketed. He'd be a fly on the wall in one of the most powerful CPXs in the world, soaking in every bit of information with nobody the wiser.

Saul wiped what remained of the gel onto a thin pad on the workbench, then entered his shop's system, directed it to search for the viral goo. It was a singular, entangled mass that didn't require networks to communicate. As the state of the gel that Saul had deposited on the man's sleeve changed, so would the remainder of the gel. His system could use that information to infer the state of the other portion. If the plan didn't work, the system simply wouldn't find the gel's better half. The shadow building had a host of defenses that, despite its power, Saul's shop had no shot of penetrating. He needed the goo to grant him access from the inside.

For fun, he'd used an icon of a treasure chest to represent the gel. Instantly it popped open. A warm, yellow glow emanated from within. Saul peered inside.

"Sweet mother's milk," he said to himself softly. He reached inside the chest and picked up the avatar of the man that had, so politely, told him to fuck off.

TORI FANNING; NOVAGENICA, CPX

This was the mark, for sure, but barely recognizable. His avatar was the cleaned-up version of himself, well-manicured and clothed in a light blue suit. Dashing. "Quite the trophy, my friend. And you're all mine."

Saul directed the shop to absorb the Tori's permissions. As it did so, the list scrolled in text on his field of vision. Saul watched, did a quick inventory as the text flowed.

At first the words didn't make sense. Like he was reading another language that couldn't be translated. He slowed the pace of the display down, started to unravel the meaning of the information he was reading.

When it hit him, he caught his breath. Eyes widening, his pulse raced. He'd finally done it.

All these years he'd been angling for a fish that could get him access to the shadow building. People, entities, *parties* – they'd all pay a shiny nickel for covert access to multiple levels of the Arc. What he hadn't bargained for was getting access to *all of Novagenica*. *That* was the kind of prize that could move mountains, level nations and bankrupt CPXs.

And it was all his.

TORI

TORI
AGE: 33
SEGMENT: COMPLEX PROPERTY; NOVAGENICA CPX
POSITION: VP OF HUMAN ASSETS, SUBSCRIPTION
DIVISION
SOCIAL SCORE: NA, CPX-SPONSORED

THE BUILDING was a rarity, a behemoth that stretched upward into the shadowy upper reaches of Level 1. Surrounded by newer construction, two and three story garish shop fronts that screamed with light. Except for the space across the street, which was a black hole of rubble. And amidst all this, the building, the Novagenica safe house, a seemingly ancient edifice devoid of windows and light.

Tori wove through the foot traffic, a myriad array of scavengers, hookers, hawkers, onlookers and criminals. You could tell it was night in the Arc just from the people. These were the sort that didn't give a fuck if they looked you dead in the eye. The tourists—daytrippers—kept to themselves and seemed to have a singular purpose. Point A to point B, avert the gaze types. By eleven at night, the daytrippers had long disappeared.

Tori wore a black hooded sweatshirt and black jeans. No one would recognize him, but that wasn't the point. The point was to blend in, not seem out of place. He wasn't worried about getting jumped—he had self-defense systems in place—no, it was the prospect of calling attention to the building that he worried about.

Earlier, when he'd first arrived, Tori had stood at the foot of the building for a brief moment, ran his gaze up its length. It was then he'd felt a tug at his elbow.

"Help the homeless?" came a gravelly voice from behind him.

Tori had spun around to find himself face to face with a man in a filthy

275

brown robe. He was missing most his teeth and his mouth smelled like a dumpster. The man had tugged at Tori's sleeve again. Tori pulled away.

"Fuck off," he'd said, shooing the man away, "before you get dead."

"Yes, boss," the man had said, stumbling, with a hideous smile, then shrank back down the sidewalk and around the corner, cackling as he went.

The man was long gone now, but he'd left an indelible stain on Tori's memory. The breath, the broken teeth, the pure filth of the man. Disgusted, Tori ran his tongue over his teeth in an attempt to clean the soiled feeling. He straightened his sleeve and turned around to face the alley to the side of the building. The prospect of going in wasn't much better than facing the homeless man's breath, but there was no avoiding it. He pulled his hood tight so that it covered his mouth and nose.

He plunged forward into the alley, its darkness enveloping him. The stench was overpowering. Foul, fetid, dripping with waste. It seemed every surface was coated in slime. He felt like he'd entered a cave into another realm. Watching where he put his feet, Tori felt more than smelled the heavy yet almost sweet stink of something dead and rotting. It couldn't be a human corpse. The building had defenses and it included this alley as a part of the perimeter. No entry permitted. But rats, cats and other small animals could come and go unhindered.

Probably a dead dog. It smelled big, if that made sense.

The building had been in NG's possession for a few years now. It was unique in that it was one of a few whose height allowed access to the next level of the Arc. Before the superstructure of the Arc had been piggybacked onto the downtown, the city had been one of middling size. Just a small handful of edifices spanned two levels and even fewer still could call themselves skyscrapers.

In those few years its perimeter had been tested once. A push from some loosely-banded addicts to take the fifty story building as their own. The building's defense system rebuffed the attempt swiftly and effectively, leaving the bodies to rot around its perimeter until one of the parties, maybe Talksmall, sent a clean-up crew to dispose of the bodies. Probably tossed them in a decomping vat, reused the water and carbon. The bodies were long gone but the local memory remained. Anymore, if the building felt threatened, it simply replayed footage of the first attack and that seemed to send the potential invaders running.

He continued to shuffle his way down the alley. He prodded at a small hulking object with the toe of his boot. Wrong move. A part of the mass detached itself from the other mass, hissed at him. He punted at it, but missed and slipped, falling backward. Instinctively he threw his hands out to catch his fall. Wrong move again. He fell hard on his ass, but not before planting his right hand in whatever the rat had been chewing on. Felt the thing *squish*. The smell of death exploded, consuming him.

He let out a string of curses, wiped his hands on his sweatshirt then stuck them in his pockets. Turned on their sanitizing function. While he was at it, he converted the sweatshirt to filter mode. The material made a seal with his skin on the bridge of his nose. The air immediately went stale, but it was an infinite improvement over inhaling some dead creature's off-gassing scent of rot.

Tori pushed to his feet and went swiftly down the alley, using an IR overlay on his vision to avoid stepping in anything else that might further ruin his evening. Half-way down the alley, a heavy steel door materialized in the dark. Tori approached and the door clicked open. He kicked it open with a heavy boot, moving inside.

There were few things he despised more than being in the Arc, especially Levels 1 or 2. The sights, sounds and smells compelled a revulsion, a physical reaction to the sensorium. The grittiness of it all made him feel soiled, and he swiped his tongue over the fronts of his teeth again, cleaning the imaginary film there. Like it or not, though, he had to be here. It was the pinnacle of Tori's work and he couldn't afford to fuck it up.

Stepping in to the dark and dusty main floor, he kicked the door shut with the heel of his boot. He was surrounded by floor to ceiling pallets. The structure of the first three floors had been peeled back so that he could look straight up to the underside of the fourth story. What filled that space was stacks upon stacks of medicine. Stockpile for the impending event.

In the center of the floor, three printers were situated side by side, brothers in arms, gleaming black and ready to be put to work. Tori had connected with the building before entering the alley, so he checked now to verify the building could talk to the printers, set them to work when the time came. Theoretically they'd pre-printed enough medicine to deal with an epidemic. The printers were for what came next. You never knew how an individual would respond to crisis, but it was a dead certainty that the multitudes would respond with hysteria. Tori needed to be able to react to the needs of the moment. More than react, he needed to be able to capitalize on it. And that was what these machines were made for.

A set of switchbacking stairs stood in the far corner. He jogged up the steps to the fourth story where his prep team had set up command. Soon enough, the building would be teeming with workers and operatives from Novagenica.

Like the rest of the building, the fourth floor had been hollowed out. Along the perimeter, interconnecting screens had been painted on the walls. He brought them all to life, tapping into the feeds the PerSense team was currently putting up on buildings and hovering platforms all around the outskirts. The Arc proper was already laced with cams. Since ArcSec held the security contract with all Legacy territory in the Arc, and PerSense now owned ArcSec, they'd already given Tori the keys to the kingdom. He could pull up any feed he wanted within a three-mile radius.

He couldn't pull them all up at once. Despite the fact of the building's robust processing power, the universe of data offered up by ArcSec's multifarious sensor network would overwhelm the system. So the engineers had built in a handful of the more useful permutations and called it good. Tori pulled up the "Outer Rim" pre-set and found himself surrounded by an aerial view of the segment of land flanked by the outer and inner floodgates.

Dully glowing street lamps, creating small but regular pools of yellow across the night-darkened landscape of gray buildings and roads. The crack of a gunshot here and there followed by the inevitable screaming. Nothing out of the ordinary, not yet. The streets were still relatively quiet. He knew the emigrants would wait until the eleventh hour. It was in their nature to defy reality by not addressing it.

Scrolling through the pre-sets, he eyed on titled 'The Black Heart'. Curious, he pulled it up and found himself in the middle of a dark, wide open space bracketed by shops on three sides and a sunken amphitheater on the fourth.

It was like being pulled directly back into his childhood. The North Market. A place of filth and vice. A place of disappearance and death.

Mother had only ever uttered one rule, which was to stay away from the market. It was forbidden. But he couldn't help himself. Prohibition without reasoning worked on a child as a magnet would on a metal.

It wasn't like she'd been around to stop him anyway. The gang he'd run around with didn't care that he liked boys and not girls. They just cared that he was tough, that he could hold his own. That he could lift a shiny thing or two and contribute to the group's share, to feed the ones of their number who didn't have homes to go back to at night.

And the opportunities the North Market offered were too rich to ignore, like overripe fruit hanging lazily on the low-hanging limbs, simply begging to be picked by their small, nimble fingers.

Tori reminisced over those days. It wasn't something he found himself doing often; his life in NG couldn't be more different, but he couldn't help but feel some tiny spark of nostalgia. There was something ineffable to it that, despite the riches he could command, he currently lacked.

Freedom.

But that was natural. Everyone, no exception, went through the rite of passage into adulthood that traded freedom, that most cherished and most under-valued of commodities, for responsibility and stability.

His gaze went to the catwalks that soared above the market's main square. Even now children crouched on their black-painted, creaking metal reaches, staring down some thirty feet at the top of the undulating crowd. Spying, just as he had, for clueless tourists wearing their wealth on the outside.

Tori stared at the screens but didn't see them. He was lost in a child-hood he had strived so hard to forget. Swimming in those sometimes exhilarating, sometimes terrifying currents.

Back then they had communicated via hand signal to the others on the ground, directing them on who to hit and when the opportunity to strike was right. He'd been up there the day he quit the gang.

The east side of the market was the low end, where the amphitheater had been dug into the ground. It was little more than a stone pit, but it served its purpose of allowing a crowd reaching into the low thousands to surround the hole and witness and bet on the matches that took place there.

Tori had heard the stories of men, women and modded creatures fighting for money, but he'd never witnessed a battle. That day had been different. The numbers of their street gang had swollen, so that in order to feed every mouth it took twice the time and effort.

He'd been a scout that day. He paced the dim reaches of the networked catwalks above the square. Looking for opportunity. It was getting late and the day-trippers rarely stayed past five. No matter that the market was buried in the vast cavern of the level one and not a single ray of natural light touched it. Five o' clock would come and along with it came the sense of something changing. A dangerous edge crept into the market. The gang could feed off the daytrippers, like parasites glomming onto the bloated riches of fat animals, but the Arc natives were different. Lean, dangerous, sharper. They were more attuned to their surroundings, to incidental contact and the vague brush of light fingers.

Tori didn't like it, it made him nervous, but the boys needed to eat, so they stayed.

As the tourists filtered out, the fighters moved in. Bare-chested men with hands wrapped in the dirty white gauze. Women, some in skin-tight suits, whose musculature rivalled that of some of the men. There seemed to be classes of fighters. Those who fought bare-knuckled always fought one another. Those who modded their biology always fought against other modders or fighters with weapons.

As if following a current, the crowd in the market seemed to shift density and flow to the eastern end where the pit lay. Stepping nimbly above the fray and around the others, mostly lovers, who occupied the catwalk with him, Tori followed.

The fighting began. Tori was fascinated as he watched the men and women beat on one another with flashing fists and bare feet, choking one another and beating heads into the barely padded sand floor of the pit. When the blood flew from the mouth, nose or some other sudden rupture of a contestant the crowd screamed its pleasure.

After a while, Tori shifted his attention from the fighters to the spectators. There was opportunity here. To a person they were almost all certainly of the sharper Arc variety, but for the moment they seemed blinded by their own greed and bloodlust.

Tori made eye contact with one of his confederates and signaled a target close to the upper lip of the pit. The boy—they called him Markle—made a curt nod and slipped in through the crush toward the mark.

When he arrived at the outer perimeter of the amphitheater, he wedged himself in behind the mark, a dark-featured man with a long, black beard and hooded eyes. Tori watched as Markle slipped a hand stealthily into the man's coat pocket.

It was then Tori noticed the man's companion. A woman, standing to his right and back just a little. She did a double-take as tiny Markle began to gently pull his laden hand from the man's pocket.

"Run!" Tori screamed. Markle's head whipped up. They locked eyes and Tori saw an immediate terror. It was too late. The woman grabbed Markle's boney arm just as he turned to run. Pushing the elbow up behind his back, she shoved her sneering face down into his. Markle continued to struggle as she shouted to get the man's attention. As if in slow-motion, the man's face turned so that Tori could see it. A horrifying expression of rage disfigured his features as the realization of the events settled into him. Tori couldn't do anything but watch as he picked Markle up and tossed him, like so much trash, head first into the pit.

He landed head first, arms flailing. From the way his body flopped Tori knew his death was instant. The scuffling in the pit stopped only long enough for two teenage boys to drop down into the pit and haul out Markle's lifeless body. Tori didn't stay long enough to see where they took his friend.

He fled. He hadn't returned to the North Market since, neither in person nor in memory.

Tori closed his eyes a long moment, let the trauma of the memory wash over and through him.

It struck him, then, as he processed the horror of his youth, that fifteen years had gone by and he hadn't changed. He was still that fearful boy living alone in a dangerous world. A heartless place that didn't have eyes for human tragedy.

And he was about to embark on a journey that affected the same treatment on millions of others.

Get it together, Tori. Suffering was temporary. He, of all people, understood this. Just like the day he'd stared his own death in the eyes and then crawled to salvation, those in Legacy would experience the same. NG was just forcing the issue.

Tori flipped the screens off. Tomorrow this place would be crawling with techs and ops people from Novagenica, PerSense and ArcSec, preparing for the exodus. All under one roof, all jostling and jockeying for position. All under his command. Which gave him no pleasure to think about. At least he'd have one day of escape as he gave his last tour before the symptoms started showing and things got *really* busy. Just thinking about it sapped his strength.

He eyed the standing screen in the corner of the room. His sleeping quarters. The foot of a cot peeked out from one side. Tempting. But if he was going to sleep tonight he needed something to shake off the psychic bad taste left by his childhood memories.

He opened a connection to The Bonds of Men.

"Welcome to The Bonds of Men ," the business's system answered. Its foyer, appointed in glossy black and deep red, materialized as an overlay on his vision. The attendant took the form of an asexual teenager, no curves to speak of, short and straight black hair.

"Sato in?" Tori asked.

"Umm, lemme check. Uh, yep. He's free now. Would you like to enter a virtual session with him?"

"No, I want him in person."

The avatar affected a knowing posture, a slight smirk to its lips. "I can't guarantee his availability unless you're here in person. First come, first served."

"Yeah, I know. Just book him now. I'll pay while he waits."

"Absolutely, sir. Done."

Tori disconnected and the overlay dissolved from his vision. He glanced back at the cot. It would have to wait.

MANOLO

MANOLO
AGE: 57
SEGMENT: COMPLEX PROPERTY; NOVAGENICA CPX
POSITION: LABORER, LEVEL 6; LIFELONG
SUBSCRIBER
SOCIAL SCORE: 629

MANOLO WAS AT WAR with himself. One part wanted to sing, to cry out with joy at the prospect of freedom. He'd received the notice that morning, his system pinging with the automated message: *DAY PASS—APPROVED.*

The other part, as he pulled body after lifeless body from their sterile little rooms, bedsheets smeared with bloody-black shit, urine and phlegm, just plain wanted to cry.

He'd been like this all day. Working, trying to keep his down, but having an increasingly difficult time ignoring the fact that the kids seemed to be dying at a more rapid rate.

They'd begun to *expire,* as the Indian doctor had a habit of putting it in her cold, technical voice—like she was talking about cuts of meat that had gone bad—more rapidly. In the last seven days, he'd evacuated twenty rooms.

At this rate, the dorm'll be empty in a week. Maybe that's their plan, he thought. Maybe the actual research was over and this was how they got rid of the evidence. *Evidence,* not people. Not children, full of light and life.

Have the janitor clean it up. Take the bodies to recycling, throw 'em down the drain.

What was once a life, brimming with emotion, thought, potential, was broken down into its constituent parts, used for fertilizer and feedstock for medicine.

And Manolo was a party to it all.

He dragged the body of a massive boy, arms hooked beneath the armpits where stretch marks radiated out like tributaries in the brown flesh, backward out into the corridor. He lay the body on the cool tile. Hands on hips, chest heaving, Manolo looked at the still body, half expecting it to come alive. The boy's afro was the color of milk chocolate and was smashed to one side, from the position he'd ended up in when he'd died. His torso was like a bowling ball that started just below his chin. He was beginning to stiffen.

Manolo hefted the body up once more. "Jesus," he grunted. *Boy has to weight two-fifty.* A hundred pounds heavier than Manolo, maybe more. Manolo scooted backward, foot by foot, toward the recycling closet. The cold light from above shone dully on the boy's graying skin. And when Manolo looked down, the body's puffy, curly hair nuzzled his chin and he was reminded of Sol and Maricel, when they'd come to visit him for dinner, bounding out of the little car and throwing their arms around his middle.

The way they felt. *Warm.* The way they smelled. *Like flowers. Like Esme.*

The memory was jarring, and he had to once more set the boy down. The conflagration within him blazed. How could he be joyful when there was so much pain right smack in front of his nose? *Especially* when he had at least some small measure of power to do something do about it.

That's right. That's exactly right. You can do something about it. Get back to work, that's what. You may be invisible to the AI, but don't push it.

He lifted the body up and shimmied backward down the hall, trying to ignore the thoughts that told him to take action. Once he'd managed to push, pull and prod the boy into the recycling chute, he went back to clean up. *Reset* the room, is what they called it.

They used these clinical terms—*expire, evacuate, recycle, reset* – to mask the truth of what they were doing.

Kidnapping, torture, murder, butchery.

These were words Manolo could understand. He'd never wanted out of something more in his life. And now, finally, he had his chance.

Ignore the noise in your head. Getting out is all that matters, he thought, wiping down the dresser and counters.

He finished cleaning up the mess, bundling the rest of the dirty sheets up in both arms. He exited the room, leaving no visible hint that someone had lived, and died, there. After sending the sheets down the recycling chute, he stood at the door to the tiny closet and its foul-smelling opening.

The hallway was perfectly empty, and perfectly still.

Say goodbye, Manolo.

He checked the time.

4:55. *It's enough.*

He hustled to the supply closet, sneakers squeaking on the floor. He grabbed a couple towels, unsure of what he was doing, but thinking he'd probably need some sort of excuse if he was stopped. *Or seen, you old fool. What are you thinking?*

He didn't know. The nerves hit him, whipped up the acid in his belly to a froth. He tried to move quickly without looking suspicious as he fast-walked down the hall, past the stairwell that led up and out of the sick ward.

Hers was the fifth door down, and he angled toward it. Fingers on the surface of the door, just resting there, he looked over his shoulder one last time to make sure nobody spotted him. The hall was empty. He pushed. The door checked his credentials silently and automatically eased open for him.

The girl sat on the edge of her bed, back to him, staring at the door opposite the one he'd entered. *Probably playing a game* was his first thought, but he didn't know. It was just as likely she could be catatonic or coming down with the illness that seemed to kill them all. Possibly both.

God, don't let me be too late, he prayed. The door clicked shut behind him and her head whipped around, eyes wide and red-rimmed. Her face was puffy and splotched in red and white Rorschach patches from crying.

Thank you, God. At least one prayer had been answered.

She began to stand but he made a settling motion with the fingers supporting the towels, then checked the time again on his system.

4:59. *Shit. No time for pleasantries. Or explanations.*

"What's your name?" Manolo asked.

She looked at him as though she were deaf. When she finally spoke, it was with a wooden numbness. "Kat," she said.

"Your full name," Manolo demanded.

"Kat Merina. Why?"

"I'll be back in a week," Manolo said. "To...bring supplies." It was necessarily vague, and all he could offer at the moment.

"For what?" she asked.

He walked casually to the dresser, set the towels down. "Your friends are dying," he said.

"I know," she responded, sniffling, choking back a fresh round of tears. "We all are."

Manolo ground his teeth together. His jaw ached.

This, he thought. *This is what you're doing, and now you know why.*

He nodded once, almost imperceptibly. She turned her head to the side, confused, waiting.

But there was nothing more he could say. He was out of time. It was the best he could do.

Hopefully it's enough.

He turned and left.

KAT

KAT
AGE: 13
SEGMENT: LEGACY CITIZEN
POSITION: UNDERAGE; GOVERNMENT DOLE
RECIPIENT (REDUCED BENEFIT)
SOCIAL SCORE: 300

"**YOUR FRIENDS ARE DYING?**" Zara repeated. "That's what he said?" She sat in bed, legs crossed beneath her, robed in her hospital gown. The fibers were matted and dirty. A white crust formed at the edges of Zara's eyes, little bits sticking to her black eyelashes, and at the corners of her mouth.

When the man had left, Kat had gone looking for Zara. Immediately bypassed the girl's door, knowing she'd find the social butterfly in the commons or the rec room, huddled around a group of other kids. But she hadn't been there, which had been a warning signal of its own.

Now, looking at her, Kat wanted to deny everything her eyes told her.

She'll be okay. Just...just follow this lead, and maybe we can get out before it all comes crashing down.

"That's it," Kat said. She paced the space between the walls, running a hand over the dresser as she did so. "But..."

"But what?" Zara asked.

"It was like he wanted to tell me more, but couldn't."

Zara rubbed at one of her eyes, the crust flaking, landing on her cheek. It was like she didn't even know it was there.

"You need to tell Antonio. Or Robbie," Zara said. She took a deep, rattling breath, touched her fingers to her downward tilted forehead as though pushing through a monumental task of thought. "Or Robbie," she wheezed, then hacked a great wad of phlegm up into her mouth.

Alarm stunned Kat to silence.

Zara smiled apologetically without showing any teeth. She held a finger up and slid off the bed, went to the bathroom and spit the contents out.

"Sorry," she said, sliding back onto the bed. She pulled the robe tight. "Is it cold in here?"

"Yeah," Kat said, rubbing the goosebumps off her arms. They weren't from the physical cold, but from a sense of premonition. "Hey, you know, you're probably right. I should go tell the boys. Maybe they'll have some ideas."

Head down, her black hair now clumped and shiny from natural grease—that unwashed-for-days look—masking her face. She waved Kat out, fingers fluttering. "Sounds good." She slumped over on her side, not bothering to get beneath the sheets.

Kat pulled open the door that led to the corridor.

Zara's eyes were closed. "Let me know how it goes," she mumbled.

Kat mustered what courage she could find, tried to sound confident, unworried. "Will do, sister," she said, then left.

She went searching for Antonio. Out of her choices, he was the preferable of the two. From their conversations, Antonio had seemed thoughtful while Robbie was perpetually antagonistic. The kind of kid that disagreed for its own sake, that picked an argument not out of spite but because it was his natural rhythm of discourse.

She searched the commons. It was lunchtime now and the most likely place to find the enormous boy at any time throughout the day. Of course, if she found one she'd find the other. The two of them were inseparable. Kind of like her and Zara.

But when her gaze roved the open, monochrome space, she found the tables were mostly unoccupied. There were ten, fifteen people tops, spread around the tables in ones and twos.

She skirted the edge of the large, open room. Headed toward the entrance to the rec room at the far end of the commons.

At a table tucked in the far corner of the room, she spied Robbie, blond head bowed over the table. His stringy hair fell in front of his eyes. A rectangle of food sat on a tray in front of him, untouched. He was all alone. No Antonio.

Kat looked closer, saw his lips moving almost imperceptibly, like he was talking to himself.

Or praying.

His hands were folded in his lap, and his torso swayed in minute motions, back and forth.

That was it. He was praying.

Strange, she thought. *Didn't even know that about him. But how could you? Nobody in this place talks.*

She arrived at the opening to the rec room and was greeted with emptiness. The game tables and couches unoccupied.

No Antonio. No anybody.

She was getting thoroughly creeped out now. She spun, and bumped, hard, into someone.

She made an *oof* noise, stumbled to the side and mumbled an apology. She kept her head down and went to leave, wanting nothing more than to be alone. To process.

Ice cold fingers caught her arm, held her there. She looked up. It was Robbie.

Gone was the customary scowl. Kat looked down at the hand on her arm and he let go hastily.

"Have you seen..." they both asked at the same time.

"Sorry," Robbie said, scratching at his head. "You go first."

An apology, from Robbie? This is getting stranger and stranger.

"Was just going to ask if you've seen Antonio," she asked. "Something...happened. And I wanted his opinion. What was your question?"

Robbie turned his back to her. His shoulders seemed to crumple inward then shake violently.

Jesus. What did I say? She moved around to his front, to find tears dripping from fingers half-covering his face.

Kat gently grabbed his wrists, tenderly pulled them away from his face.

"Hey, it's okay," she said, making *shushing* sounds to calm the boy. With anguish replacing the scowl, he looked about 5 years younger, not like a boy trying hard to be a man but like a hurt, alone child.

"What is it?" Kat asked. She stepped in close, put her arms around him, stroked a hand lightly up and down his back. It was a surreal feeling—she'd always been on the receiving end of this, from her father mostly, but of late the task had fallen to her sister. Kat found it somehow comforting to her as well.

Robbie's sniffle quieted, and Kat felt his body go still.

"All better?"

"Not by a mile," he said, "but a little at least." He stepped backward, holding on to her upper arms with those frigid hands of his. She just now noticed his eyelashes had that white crust to them, like flakes of salt, same as Zara. "I was going to ask if you'd seen Zara."

"I just came from her room. But she looks like shit," she blurted out. "Sorry."

Robbie's eyelids fluttered as he attempted a weak smile. "Yeah, that about sums it up for me too."

"I think..." Robbie started, then got a far-away look in his eye. Kat watched him closely, but it didn't seem like he was intent on finishing the sentence. He slumped into the nearest chair behind him, flopped his arms on the table.

Kat went completely still. *It's like he forgot he was even talking. Like we were talking. Is this what it looks like? When Zara said everyone here dies here? Is this how it starts?*

Regardless, she wasn't going to just give up. She pulled a chair out from beneath the table, planted her butt in it firmly, faced Robbie.

"You think what, Robbie?" she asked.

His chin rested in the crook of his elbow, and only his eyes shifted in Kat's direction. "Huh?"

"You said, 'I think', then stopped."

His narrow shoulders lifted desultorily, then sagged. "I don't know. I'm having a hard time focusing. Can't concentrate on anything for more than a couple seconds." He drew a few ragged breaths, white flakes tumbling from his eyelashes like dry snow. "Antonio's dead, Kat."

She was blindsided, nearly toppled backward in her chair. She flung a hand out, gripped the edge of the table with white knuckles. The shock left her feeling dazed, hollow.

"*What?*"

"Just remembered why I was crying," Robbie said, as though detached from the emotion of it. "Antonio's dead and...I'm probably next." And then, with a suddenness that was jarring, he buried his head in his crossed arms and sobbed. His body shuddered and heaved with the violence of it.

This isn't real, she thought. *This can't be real. I just saw him, eternally round and stuffing his pudgy face. He* can't *be dead.*

Then, in one illuminating flash, she understood the wisdom of Zara's words. Why the kids trapped there in the experimental hospital didn't get close to one another.

She watched Robbie, wracked with emotion. *This is why.*

Kat stood, uncertain of where to go, or what to do. She steadied herself with a hand on the studded plastic back of the chair. Somewhere between dazed denial and complete incomprehension, she left the boy there.

To cry himself out. To fall asleep. And, eventually, to die.

RAY

RAY
AGE: 36
SEGMENT: COMPLEX PROPERTY; PERSENSE CPX
POSITION: REDACTED
SOCIAL SCORE: NA, CPX-SPONSORED

A ROBIN CALLED out into the motionless still of dawn, the lone voice echoing in the frozen morning. Frost limned the surfaces in the low light, turning objects into silvery ghosts of their former selves.

Ray exhaled, captured the video on his system, played it back in magnified slow-mo. Watched ice crystals as they coalesced in the air. He checked the time. 5:54. Getting close. Good thing, too. He was very nearly bored to tears.

He'd been camped out for two hours now, sitting atop the ledge of a ten-story apartment building. Staring down at the proverbial line in the sand. The outer floodgates.

As the first of the snow fell, fat flakes swinging lazily from an iron sky, Ray shifted his gaze to the crest of the hill that rose to the southeast. Through the haze of snow and cloud cover, he could make out the angular shapes of mismatched and stacked freight containers that seemed to radiate out from the center of a circle where an ancient, decrepit radio tower laid claim to the nucleus. He couldn't tell whether it was dark orange from paint or rust.

Fucking homeless, Ray thought. Disgust wasn't a strong enough word for the way he felt about the hobos that laid claim to Signal Hill. Their encampment was a rat's nest, a haphazard warren of addicts that smelled of alcohol, piss, shit and vomit. Ray hadn't actually been inside, only skirted the outermost ring of containers. But he didn't need to see it to believe it. His hatred was enough. The only thing he despised more than the hobos themselves was their ridiculous shantytown, a rusted jumble of freight containers, lean-tos, tarps and scavenged siding. In some places the containers were piled as high as the building he stood on.

Funny, he thought with a wry chuckle. *They still consider themselves homeless, meanwhile I'm staring right at their fucking home, plain as day.* Naturally, it should've disqualified them.

"I'm sorry, sir, but you no longer meet the requirements. The board has decided to revoke your homeless status," he spoke aloud, steam pouring out his mouth into the frigid air. "Please hand in your card."

If it were up to him, he wouldn't be found within a klick of this place. But he had orders. Keep the order. Simple as that.

The whole Legacy population could've emigrated in ones and two, a little bit at a time, at any time since the DS-75 notice had been broadcast. But people were stupid and the vast majority would wait until the very last moment before acting. No planning, just grab and go. Ray knew this for a fact. Thus, today. November 16—DS-75 day. The deadline.

He watched this exact scenario play out before him as people pushed, pulled and kicked their belongings across that line, dispersing toward the Arc, shantytown and the interstitial places in-between.

So far, the exodus had been peaceful, but as the day wore on Ray expected to encounter more than one disturbance. PerSense had access to all of ArcSec's feeds, but witnessing in virtual wasn't the same as being there in the real. Didn't have the same *feel* as meat-space. Ray needed to be here to document, to witness the shuffling herd, to hear the murmur of the mass of refugees and listen in on the details of their individual conversations. To feel the pulse of a revolution before it began.

This last bit was the ostensible purpose of the op, but there was a layered depth to this plan. Something else he wasn't privy to. Before setting up camp atop the apartment building, he'd been ordered to set magnetically-sealing canisters atop the outer floodgate wall. Easy work, considering the wall was still recessed into the ground. But as soon as it began to push upward from the ground, those canisters would start doing their job. Whatever that was. Ray wasn't privy to those details. Yet. He had a feeling, though, that his meeting tomorrow with ArcSec and NG would be enlightening.

Ray stopped his pacing of the rooftop, pulled up the time again. Watched as the second hand rolled over to the next minute. 6:00.

On cue, a loud *clunk* boomed from street-level, followed by an ear-splitting screech, like two battleships scraping hulls. The floodgates—a thick, contiguous wall that encircled all of the Arc and its outlying urban areas—rose out of the ground, inching upward at a glacial pace.

Constructed as the first line of defense against rising tides and catastrophic weather events, Ray knew its real purpose. The same hydrophobic exotics that kept water out were also remarkably good at keeping people in.

As the wall steadily ascended, Ray could begin to make out with the naked eye the wall's vastness. It stretched off into the distance, carving a swath through the concrete landscape, curving gently toward the horizon on either side of him. Past dull gray concrete and ramshackle, faded-brick buildings, a blue-gray line marching east-west.

Ray blinked, rubbed at his eyes. They felt grainy, sore. He was fading when he needed to be alert.

Now is the time to be ready, sir. Alert.

"You always know what I'm thinking?" he asked the AI.

Never. As stated before, we can interpret the signals of a human's bios quite well. For instance, your data suggests you're getting bored, and thus tired. In this state, you could miss a crucial detail. The drug you call Amp can help with this.

"Fuckin' right," he said. "Why're you always fuckin' right?"

Ray accessed his system and bumped some amphetamine. Just five per cent for now. His synthetic spleen could only make so much at a time, so he used it sparingly. That said, he'd toggled his system to full ops readiness the day before, ramping up production of all synthetics he might need, depending on the situation. His reservoirs were at a hundred percent. Slightly more dangerous, running a full inventory of complements. Get punctured or take a concussive hit in the wrong place and a sac could rupture, spilling the entire contents into his bloodstream. If there was a bright side, at least his death would be instant.

The amp hit him, like it always did, with a jolt of alertness that made his extremities tingle. He bounced on his toes, enjoying the feeling of strength in his muscles. Started pacing again, jogging down the length of the rooftop's stone cap. Took a quick gander at the thickening migrant horde below.

The line of shuffling men, women and children extended as far as he could see to the south, a mass of dirty, downturned faces packing the road, making an inching sort of progress. The flow of people narrowed to a point as the floodgates continued their steady march skyward. Prior, the refugees had simply crossed over the top of the wall. Now they were forced to funnel through the lock. Once through the locks, they dispersed in a spray of humanity.

With four locks in total interspersed along the outer wall, Ray knew with certainty that stationed covertly at each of those locks was a PerSense officer, just like himself, with identical orders. *Set up canisters and recording gear, observe the exodus, sort and tag the dissenters using the social database. Most importantly, do not interfere. Not with arguments, scuffles, nothing. Even if it escalates into rioting.*

Ray was pretty sure he knew where the PS top brass were going with this. None of the CPXs could be seen to have a hand in DS-75—The Exodus—despite the fact they were almost without a doubt behind the drafting of the bill. As the outer ring of the Arc progressively went to hell, that chaos collapsing inward to the center of the Arc proper, the Admins of the five districts would feel the fire of the blame.

It'll be fun to see how that plays out, he thought. Not well for the Legacy Admins, that much he knew. And when the fire had died down, PerSense and its partners would swoop in, restore law, sweep the streets clean. Set up a new order. Repeat pattern. In all the Arcs throughout the world. And just like that, Legacy would be history, a poorly remembered thing of the past.

It sounded simple but Ray knew it would be no easy feat. There would be objections, rebellions, death by the thousands, if not tens of thousands. But if there was anything he knew, it was that a good life had to be earned, won through hardship. It was a lesson the vast majority of Legacy needed to learn.

Filled with a manic sort of energy from the Amp, Ray checked the video feed coming from the cams he'd set up. As the people walked through the camera's field of view they were run through the Social DB. Identified, tagged, catalogued. Names flickered to life, locked on the faces. Once a confirmed lock was achieved, the text would turn bold and green, then fade and die. In real time, the process was blindingly quick, too much for the human mind to process, especially with a crowd of this size.

All green, Ray confirmed. *What's next?*

Movement at the corner of his eye caught his attention. It came from the top of the wall. Ray turned to look but saw nothing out of the ordinary. The wall had projected a few feet out of the ground. The masses continued to shuffle through. Nothing special.

Out of curiosity, Ray magnified one of the canisters that was secured to the top of the wall, just a few feet from the lock. It was a squat cylinder, about the diameter of a canned drink. Sunlight shone dully on its blue carbon matte surface. It was featureless for the most part, except for a row of tiny perforations that ran the circumference of the canister near its top. Still nothing out of the ordinary. Ray switched to IR, nothing of note except for the surfaces of buildings already exposed to the sun. And the moving mass of people, of course. Their combined heat was like a sun flare on his vision.

He rarely used the particulate overlay, but it was useful for identifying toxic elements in the air. He switched to it now. The instant he did, he saw a jet of foreign particles emitting from the canister. And not just that one. He pulled his magnification back and saw that each canister was actively streaming its contents, pushing them into the air, creating a cloud of whatever was inside of them. The huddled mass of refugees walked right through the swirling, roiling mist, none the wiser.

For a moment, Ray was lost in the thought of what could possibly be in the containers. Until he realized that whatever it was, it couldn't possibly be good. He tracked individual particles, saw the cloud slowly spreading, drifting along on currents not visible to the naked eye. One such tendril reached its way toward the building he was perched upon. Quickly, Ray whipped his suit's hooded mask down over his head, toggled it to biohazard mode. The mask created a seal with the rest of the suit, filter coming to life with the slightest of vibrations, like a hum in the air.

Bastards. Least they coulda done was told me.

This had to be the work of Novagenica. If it was PerSense, those people down below would be dead. PerSense didn't mess around with gray areas. He looked down at the crowd again. Clueless that their time was coming. Pawns in a game of a grander scale than most could imagine.

Ray watched an element in that game unspool before him. The time was coming soon.

He wasn't happy about being used for another Complex's designs, but then again, PerSense had to have a hand in this as well. One thing was for certain, he wasn't leaving the meeting without getting the whole story.

Then the fun will start.

RIKU "CHIEF" OGUNWE

RIKU "CHIEF" OGUNWE
AGE: 40
SEGMENT: COMPLEX PROPERTY; ARCSEC CPX
POSITION: ARCSEC CHIEF CONTRACTED BY LEGACY
ADMIN: MIDWEST ARCOLITH, LEVELS 1-3
SOCIAL SCORE: NA, CPX-SPONSORED

A RECTANGLE OF SKY, gray and barren, blended into the patchwork tangle of buildings and catwalks above. Normally, the transfer park was a place of vitality. Fountains burbling, children playing, moms sipping steaming drinks and nattering on about things that were, by and large, inconsequential. It had been designed for this sort of thing, for aesthetic pleasure and maximum exposure to the sky above. Not a simple task in the middle of the Arc.

Today it held none of those things.

Just people, queuing up to escape. All these people, heading up the Arc, away from the massive surge of refugees. And me, going down into it. Compliments of Jaim Del Potro.

From where he stood, he could see the Admin's office atop the Rubik's mall overlooking the park. The windows were dark. No sign Mr. Del Potro was there. *Fucking coward created this shitstorm now he wants me to go deal with it.*

He tapped a booted toe, stared at the backs of the heads of the people who stood in front of him. The line snaked its way to the clear wall that stood as a barrier to the elevator shaft. Like him, they waited patiently. But there was a nervous edge to their energy, a restlessness evident in the way they shifted from side to side and surveyed their surroundings regularly.

Ogunwe mirrored their anxiety. A thought, like a hangnail, had nagged at him, growing in urgency over the last several days.

Why won't Olivia respond?

The thought sent a rush of anxiety-inducing chemicals through his system. Scrambled his brains and forced a momentary spike in temperature that caused him to flush and sweat.

It had been seven whole days since he'd called her. No response, no nothing. Absolute silence.

Just the icing on this turd-cake you're being force-fed.

They'd been asked to join forces with Talksmall, to aid in crowd control tactics and offer support wherever they could. Which was ostensibly why Ogunwe and his hand-picked team were heading down the Arc, into the belly of the proverbial beast.

But in actuality Ogunwe had other priorities. Plans he couldn't reveal to his team. Namely, to find Olivia and Stef, then drag them back to 3, through the Golden Road if need be.

Morales and Sweet stood to his left, inseparable, Carpenter to his right. A meeting like this, it would've been useful to have a frame like Reichmann's along, but this was a mission of delicacy. It required a more subtle intellect, which Reichmann most definitely did not possess. A little extra muscle never hurt, but where they were going it wouldn't matter much. They'd be so outnumbered that one extra man, known for being a hothead, would be more a liability that an asset.

Add that to the fact that the new rep for Talksmall, Jayne Moon, had a growing reputation for cunning, an ability to outmaneuver his opponent, and this meeting was a chess match where every word was a move in itself. Which made Carpenter and Morales the people for the job. Sweet was along for the ride because, well, he wasn't sure. It just seemed she wouldn't let Morales out of her sight and managed to bully her way into any mission that involved her long-time partner and wife.

"Lemme just say, Chief, thank you, a million times over, for not bringing Reichmann along," Sweet said. They shuffled forward a step as the line moved.

Before Ogunwe could respond, a voice chimed in on the shared feed they were using for the op.

"Tut, tut, detective. Not exactly living up to your namesake, are you now?" Reichmann chided.

"Can the moron please stay off the line if it's not operationally critical?" Sweet hissed under her breath.

"Quit, you two," Ogunwe said, low and stern. "Or I'll mute you both."

"Ten-four, Chief," Reichmann said brightly. "I've got eyes on Riverside Park. It's like a hobo orgy-slash-carnival down there. Want me to patch you guys in?"

"Not yet. Just stay quiet for now."

"Over and out."

The queue moved steadily toward a clear box the size of a small closet, seated within the wall. These were the measures Ogunwe had helped put in place, what the PASS directive had mandated. People were calling it a lightbox. It was the first time he'd experienced them firsthand.

The line they stood in was for the way down; it moved at a regular clip. The downward direction was default mode and always left in the open position. Heading up was another story. Ogunwe watched the line to their left for the elevator that climbed up the Arc. They'd only been there five minutes and already he'd seen three people refused passage, the entry door to the clear box opening back up instead of the exit door on the opposite side. The system had weighed the person, analyzing their Social Score, and determined it didn't pass muster. The clear box lit up with a red glow.

Not high enough, he thought. It was like a walk of shame; the rejects hung their heads, slunk back out not making eye contact with the hundreds of other faces there to witness their perfidy.

After a few minutes, Ogunwe came to the checkpoint. It was comprised of clear composite, like the barriers to the Free Drug Zones down on 2. But instead of open passageways manned by live guards, the transfer stations were now fully barricaded and automated. Ogunwe could see through to the other side, where travelers entered and exited the giant elevators that would transport them to the other levels.

Ogunwe looked over his shoulder briefly before entering the box. He could see the stairs, a hundred yards behind him. The barrier extended to the stairs, which were packed solid with people, moving just a few feet every minute. The free route was always the slowest route. And anyone on 3 wasn't heading down, only up. Already there was a growing fear that the lightboxes would reject someone erroneously.

He went before the others, stepped inside the box. The door sealed shut behind him with a soft click. He could feel the eyes of everyone in line fixed on him. It was unnerving, like being on stage.

The sensor array began its work. From all around the box a rapid-fire clicking sounded. The system started with his ID first, pulling up his Social Score, compiling it with his meta-score—pending warrants, debts and legal decisions that hadn't yet been handed down. All of the stuff that hadn't yet been dumped into the Social Scoring system. The algorithm ArcSec had devised then compiled a weighted transport score and compared it against averages of the inhabitants at each level. Either way, up or down, if the individual's score was anomalous up against the destination score, that ID was tagged for further observation.

Simultaneously, the sensor array was busy scanning his body for weapons. Both external and embedded.

Ogunwe came equipped with the standard host of self-defense gear. Projectile repulsion, wound-stasis and quick-healing complements. The array would see these but would also know that he belonged to ArcSec and was licensed to carry a lot more than what he currently had on him—his built-in self-defense suite, a mini pharmaglove that was stashed in the pocket of his black jacket and the hilt of a protracting shortsword secured by print to his belt for easy access.

Five seconds had gone by when the clicking stopped. The lightbox glowed green and the exit door, opposite the entry, swung open. Ogunwe passed through, turned to wait for his cohorts.

Morales went first, then Sweet, then Carpenter, all without issue.

"That was a breeze," Morales said as Carpenter joined them.

"Going down's the easy part," Carpenter said. Sweet grunted her agreement as they all fell in step, heading toward the elevator doors.

The elevator was a holding bay, twenty to a side. Ogunwe knew if they packed in as tightly as possible, the hold could carry almost two hundred passengers.

More like a ferry than an elevator, Ogunwe thought. It brought him back to his days outside Cleveland, at ArcSec Great Lakes. Taking a boat over to Canada, one of the few remaining unsplintered nations. *Give it time,* he remembered thinking at the time, as though all Legacy governments were the enemy. The Canadian government maintained tight control over its social policies but that couldn't last forever, not up against the mounting pressures and burdens of the rapidly changing world.

The four of them filtered to the back. The doors slid shut on 3, a lone shaft of sunlight beaming in from between a break in the buildings cut off as the elevator hummed to life. Ogunwe closed his eyes on the sterile interior of the elevator, held the image of the transfer station on 3 in his mind. People going about their business patiently, peacefully. All was as it should have been. It gave Ogunwe pause. Maybe PASS *was* the solution, the right path to take. After all, the idea that scoring someone based on their social choices and behaviors was incentive for them to make better choices, was it not? Maybe behavior modification wasn't the evil he perceived it to be, if this was the result. His military mind couldn't help but appreciate an orderly society, people behaving respectfully, following the rules. Maybe both physical and social mobility was the right motivator for Legacy civilians to better themselves. Make better decisions that, when piled up, person after person, became a force that could transform society.

Ogunwe felt the elevator descending, opened his eyes. His gaze roamed the faces of his fellow passengers. Just people, same as him. Distinguishable by their many different features, colors and shapes, but otherwise unremarkable. Strip them all down to nothing and you couldn't distinguish a bottom-feeder from a Level 7 Plutocrat.

Except that isn't quite true, is it?

Not when wealth could buy better health, looks and, ultimately, a longer life-expectancy. It was immensely unfair, Ogunwe had to acknowledge, and PASS was realistically the final nail in the coffin of humanity's aspirations to equality. Especially when the levers for access were manned by those in power.

The elevator lurched softly, coming to a halt. The doors separated, opened onto a different world. Passengers filed out, Ogunwe waiting his turn. As he passed through the doors, the first thing he felt more than noticed was the lack of natural light. Combined with the lack of current made the air feel stagnant and subterranean. They exited the elevator and into the holding area, built identically to the one on 3, and it felt like walking into a cavern. The next thing that hit him was the stench. It smelled like a portable toilet, the overwhelming, heavy scent of waste combined with a slight chemical overlay that was supposed to repress the smell but simply sat beside it.

Through the clear walls, Ogunwe eyed the line to the elevator heading up from Level 1. Watched as one man, in a filthy, ragged orange shirt, was hauled out of the box and thrown to the ground by two other men when he refused to budge after the lightbox flashed red. In the span of minutes, it happened again and again. Person after person was rejected for upward passage. They jostled and shouted at one another in line. In the short span, not one had been admitted.

Ogunwe observed dispassionately, not allowing any sign of emotion to touch his face. Right now, he knew that the transport algorithm was supposed to work without judgment as to what level a person currently resided on. But he also knew that it was just one iteration away from rejecting all passengers that called 1 their home.

Hell, maybe that update's already been pushed. By the look of things, it likely had.

They exited the waiting area, onto the streets of 1. It was far more crowded than he'd ever seen, their group forced to scrunch together with Ogunwe, the tallest of them, as their nucleus.

Sweet jostled at his left elbow. "You know where you're going?" she asked.

"North. Straight on from here to the Riverside wall."

"Jesus," Carpenter whispered, appearing on his right. "This place is a mess."

Ogunwe didn't respond. Somewhere in this "mess", Olivia and Stef wandered, shuffling along with the horde. Fingers of guilt touched his soul, and he asked himself the same question that dogged him when confronted with failure.

What more could I have done?

Control was the issue. And for a man who was used to being in control, to be able to affect the outcome within his sphere of influence, the entire scenario seemed to be slipping from his grasp. Devolving into chaos and lawlessness.

No better example of that than down here on 1.

The law down here was governed and meted out by the parties, with very little oversight from Legacy's Admins. Subbed out to thugs and it showed. Just like this little mission was subbed out to Ogunwe and his team. At the request of one of those Admins—Jaim Del Potro. Doing his dirty work for him because he knew that if he set foot below 3 he'd likely be taken hostage and strung up for his part in passing DS-75, the forced exodus law.

Flanking either side of the street, Talksmall Enforcers wore white helmets and navy jackets with the bolded *TS* on the front and back. Each guard held him or herself erect and alert, shoulders back, eyes vigilant. It was a posture that clearly stated *Do not fuck with me, or anyone else for that matter, while I'm on watch.*

The group fell into single file, walked in silence. More of the same. Packed streets, hungry mouths, vacant eyes. It filled him with disgust. Deep pockets beneath hollow eyes made even darker by the lifeless, artificial light that brayed from storefronts. He passed an alley, peered into its depths to see a refugee down on her knees, performing fellatio on some scumbag with his back pressed up against the building wall.

Surrounded by other refugees who pretended not to notice. Ogunwe choked back the anger that surged within him.

That isn't the mission, it's a symptom. He had to set his aim higher, needed to stay focused.

He stopped anyway, looked closer. Morales bumped into him from behind.

"*Oof*...Sorry Chief. What's up?" she asked. He said nothing, as anger boiled within him. It was just a girl. Barely a teen.

No older than Stef.

Ogunwe turned and charged down the alley, a bull seeing red. He bulldozed through the people on a straight path, ignoring the cries of protest. In a moment he was on the asshole. Ogunwe didn't bother addressing him, and he didn't stop. With a fluidity that spoke of practice, he used his momentum to throw a right cross, snapping all his weight and force through the perv's head in one violent motion.

The man's head snapped to the right and his body tumbled backward against the wall, then flopped to the greasy, littered pavement below.

The girl performing the act fell backward on her butt, scrambled away in an awkward crabwalk. She didn't scream, just looked up at Ogunwe with wide eyes. Beneath the layers of grime she couldn't have been more than thirteen.

Ogunwe sunk to his knees, held out a hand to the girl. *Just a kid,* he thought. *Is this what childhood has become?*

The girl took his hand, then scooped up a red pack and fled toward the mouth of the alley.

A little bit of the anxiety that weighed so heavily on him slid off. He wasn't sure he'd made any appreciable difference, yet he felt lighter, better.

Violence doesn't solve problems, but sometimes it sure does help.

Ogunwe left the alley, trailed by turning heads and awe-struck stares.

"Holy shit, Chief! You knocked him out *cold*!" Carpenter said. Ogunwe rounded the alley, the other following, and inserted himself back in the flow of people moving down the street.

"Fucker got what he deserved," Sweet said.

"What?" Reichmann chimed in over the shared line. "Who got what?"

"You missed the fun," Morales virtually sang. "Chief just laid some pervert out. Almost took his head off."

"Aw, dammit Chief! You know the rules, no fun without me!" Reichmann shouted with boyish exuberance, the tiny speaker in Ogunwe's ear crackling with distortion.

"Alright alright, calm down," Ogunwe said. "We're just about there."

Before them, a narrowing of the street brought the traffic almost to a halt. Beyond the bottleneck, Ogunwe could see another transfer gate—this one just a staircase—off to the left, and a wide plaza, like an open-air market, directly ahead.

It was known as the Black Hole, a market where pretty much anything illicit could be had. But the Black Hole wasn't their destination. Today, their target was the only thing farther north on 1 than the Black Hole—Riverside Park.

Ogunwe had been to the park once before. If memory served, it was less a park and more a broad field with more homeless tents than trees. The park butted up against the curve of the Riverside wall, the northernmost barrier to the Arc.

They made their way to a narrow street that skirted the market's western edge. It was bordered by a row of 2-story buildings on one side and a hulking parking structure on the other. The parking garage had been retrofitted to serve as a hostel, and from its different levels catwalks jutted out into open air, vaulted above the street to connect in a network with the buildings surrounding it. Some projected into the blackness back toward the market.

Carpenter craned his head back. "You can tell the hobos've been here."

"Looks like their work," Morales chimed in, voice lilting with a touch of amusement.

Past the parking garage, through a makeshift arch formed by the catwalks strewn above the street, the road terminated into a wide, open space. A soft rise choked with people prevented them from seeing much beyond it. Above the rise, the gray vastness of the Riverside wall, framed by the garage on one side and a squat stucco building on the other.

"Have to at least admire their ingenuity," Sweet said. "Don't approve of their way of life, but you gotta have some respect for people that just figure..." Sweet choked on the last word.

A low hum grew to a dull roar as they topped the rise. They looked down into a kind of shallow bowl of land that was wider than it was deep. At the far end, up against the massive wall, a cleared space where black buildings roughly the size of freight containers were being constructed.

Ogunwe had never seen so many people. Packed shoulder to shoulder, a sea of humanity spread before him. Men, women, children, infants. He surveyed those closest to him—they looked every bit the refugee. Dirt-lined faces, graven in concern. Screaming of children, squalling of babies pierced the din.

More worrisome were the astonishing number who lay on the ground, inert. As much as one in ten, it looked like.

"Chief." He felt a tug at his right elbow. Ogunwe turned back to see Carpenter looking worriedly at a child, a girl, maybe ten, balled up on the ground, shivering as though the ground were vibrating.

"There something we can do about this?" Carpenter asked plaintively, as though he already knew the answer.

Ogunwe bent at the knees, forearms on his haunches, to look at the girl more closely. Despite her shivering, tiny droplets of sweat beaded her forehead. A yellow-green crust of mucus lined her nostrils, which nobody had bothered to clean. Her short black hair was caked with mud and through her brown complexion he could see the signs of poor health, of illness taking root.

He stood up abruptly. "Flag the location, point her out to Talksmall. Do it now, and make sure you get a promise of immediate aid."

"On it," Morales said.

"Sweet." Ogunwe addressed the graying, stern-faced woman. "Stay with her til they come."

Sweet's attention flickered briefly to Morales. He could see her biting back the protest that wanted to claw its way out of her throat. The last thing she wanted was to leave her partner out of sight.

Her back stiffened, but all she said was a perfunctory, "Sir."

"Reichmann?" Ogunwe called out.

"Here, Chief." He sounded subdued, not his usual ebullient self.

"Where we headed?" Ogunwe asked.

"Ah, OK, eyes down the hill. Center of the park, up against the wall."

Ogunwe did as told, looking back toward the black, blocky structure he'd noted earlier as they'd topped the rise.

"See that thing they're building down there? That's your final destination," Reichmann said.

"Ten four." Ogunwe started down the hill, others in tow.

"Uh, Chief?" Jonas again.

"What is it?" Ogunwe asked as he stomped his way down the hill, mud squelching beneath his boots.

"Well, I tried to warn ya, but whatever's going on here, it ain't good." Reichmann said.

"What do you mean?"

"That little girl on the ground isn't alone. I've been checking feeds for the other encampments inside the gates and turns out we've got ourselves a bona fide pattern here."

Ogunwe almost allowed his shoulders to sag. *Another goddamn problem.* He couldn't carry the weight of the world on his shoulders, but increasingly it felt like it was his responsibility to at least try.

"Pattern for what?" Ogunwe asked. He kept his voice even.

"Dunno, some kind of illness. Looks like a current affliction rate of about two percent. Well, that's just counting the people that've been knocked down so far. Who knows how many have it but haven't started showing symptoms yet."

"Put it on your watchlist," Ogunwe said. He would deal with it later. He needed to stay focused. "We've got a job to do."

They kept moving forward, weaving between tents and people, down the gentle slope, when Reichmann spoke up again.

"Uh, Chief?"

"What *is* it?" Ogunwe asked, unable to keep the growing irritation out of his voice.

"Yeah, well, I know I'm an annoying SOB right about now, but since I'm just sittin' here with my thumb up my ass I decided to run some epidemic scenarios through ArcSec's riot-prediction model."

"And?" Again the sinking feeling in his gut, the tension ratcheting the muscles in his back.

"Current affliction rate is two out of a hundred. That number climbs to fifteen and you've got an old-fashioned stampede on your hands."

"Fine, keep tabs on it. Update me when the percentage changes. Whole integers please."

"That's not all, O,"

"What?" Ogunwe growled.

"Hey, don't kill the messenger," Reichmann whined. "I looked at the history of the feed, computed the rate of affliction since yesterday."

"Why yesterday?" Ogunwe asked. He continued the trek downward. It was slow going as he had to sidle between and step over people. The low-light grass below their feet had been stamped into mud, become indistinguishable from sewage. The stench was terrific.

"DS-75 day," Morales said over their feed.

"You got it sweetheart," Reichmann said. "Over the first 24 hours this thing has been knocking people out at a pretty low rate. But over the last few hours, since this morning really, the rate's accelerated. Started at under one percent at six AM and now we're at two. Chart shows we'll hit that fifteen percent mark somewhere between twenty-four to thirty-six hours. But...I really don't know for sure."

"What are you saying, Jonas?" Ogunwe asked. His irritation had given way to serious concern. And not just for the lives of his team. For all of Legacy. If this, whatever it was, moved as quickly as Reichmann suspected, it'd spread like wildfire through the Arc. And when it did, people would panic.

And when people panicked, people died.

"Say what you need to say to Talksmall and get your asses out of there. I don't need a front-row seat to *The Burning of the Arc*, starring Chief and his Golden Road Minions."

Ogunwe stopped, felt Morales bump into him again. The moment threatened to overtake him. There was that issue of control again. The threads seemed to be pulling away, slipping from his fingers. There were too many unknowns and too many decisions that could lead them down the wrong path, to disaster. PASS, the exodus, Olivia gone and disappeared. And now this...this disease. It stank of a plot, but of what he wasn't sure.

Genocide? What's the point of that? It wasn't like most people could point to one single ethnicity and say this was what defined them. Everyone was a mix, a hybrid. Maybe a war against the ninety-nine. But even then, it didn't seem like mass-murder served a purpose.

So what, then?

Whatever it was, the realization that there seemed to be a greater plot at work only served to fortify his resolve. Ogunwe steeled himself for what he knew he had to do.

They were close to the structure that stood up against the wall. Around it, Talksmall engineers worked to add on to the containers that had already been constructed, using a system of pullies and ladders to hoist individual wall pieces up on top of the existing domiciles. They then fastened each wall, working quickly, putting up a new domicile every ten minutes or so. Already the structure was three stories tall in some places and spanned the several-hundred-yard width of the park.

No less than a hundred Talksmall Security, different than the Enforcers, in white helmets and navy jackets, dotted the circumference of the area. It was clear from the density of people inside the circle compared to those on the outside that their mission was to keep people out, allow the engineers to do their job.

Stepping cautiously through the muck, Ogunwe stepped up to the closest guard, a man of average height with square shoulders and a short, red beard. The man's expression didn't change but Ogunwe was trained to look at the details, noticed the slightest stiffening in his posture, the skin over the knuckles that gripped the standard issue fiber bat tightening.

"Chief Ogunwe here to see Mr. Moon," Ogunwe said calmly.

Much more quickly than expected, the guard pivoted, his body swinging backward on the one foot like a door opening on its hinges.

"Thanks," Ogunwe said to the man as he passed through. The rest of his contingent followed in single file.

Here they'd put down gravel to combat the mud, and Ogunwe moved cautiously across it, scanning the faces of the engineers and other workers for Moon's. He'd never met the man before, and this meeting had been engineered by the Admin on 3. But, like anyone else, his information was widely available.

Sharp looks, sharper intellect. Ivy-league trained but instead of heading into the world of Corporations and Complexes he's here, in your Arc, heading up a pseudo-political revolutionary party. Just in time for the whole place to go tits up.

Ogunwe was trained to sniff out plots, and this certainly smelled of one. He just wasn't sure what part Talksmall played in it.

Ogunwe's black boots collected a chalky white layer as he gravitated toward the center of the clearing. He was supposed to be looking for Moon, but his attention was drawn to the structure they were building.

Homes, he thought. Anchored to the river wall, it would be the first node in a massive network of small, rectangular rooms that would eventually cover the entire volume of the park, filling the space from the ground up to the belly of level 2. Ogunwe could almost envision it unfold in his mind, another shantytown, albeit more organized, with planned sectors and walkways dividing them up.

Lined up in front of the boxes, a small army of men and women toiled alongside a host of printers. Ogunwe could feel the heat coming off the machines, accompanied by a sweet, burning smell. One in the group, a man with his back to Ogunwe, looked over his shoulder. Did a double-take before stopping what he was doing and turning around.

Ogunwe knew what Moon looked like from his ArcSec file. The man beamed beneath casually swept back black hair as he recognized Ogunwe.

"Chief Ogunwe! I see my friend Mr. Del Potro followed through on my request to send you as his emissary," Moon approached, stretched his hand out.

Interesting. Moon asked for the meeting? Not the Admin?

"Mr. Moon," Ogunwe acknowledged, taking Moon's hand. "This is my team," he said, motioning as his people spread out to the left and right.

Sweet jogged up and joined them, puffing. Ogunwe looked to her and she simply nodded.

"I'm afraid you've got me outnumbered," Moon joked as a young woman and a man detached themselves from a contingent of engineers and joined him. The woman was short, cute, with tight, curly hair that spread like a lion's man out from her head. "This is Estelle." She nodded and smiled. No teeth, a hardness, maybe worry, in those eyes.

"And this is Ian. He runs our network." The man, taller than even Ogunwe, was pale and gangly. Shaggy brown hair framed a face that betrayed no emotion, showed no expression.

"A pleasure," Ogunwe said to the both of them, then, to Moon, he uttered a soft, "Are we private?"

"We are," Moon replied. "But follow me, if you would." He turned and led the way to the interior of the printed domicile that stood in the center of the structure. Workers were busy assembling more rooms above and on the sides, but when they entered the eight by twelve rectangle all sounds became hushed, as though the outside had been stuffed with cotton.

"I know it doesn't seem like much," Moon gestured to the container, "but once it's complete it'll seem like heaven to those people out there."

"It's a little tight," Sweet said, bringing up the rear, "but it'll do."

"Well, for our purposes what matters most is that it's secure," Moon said. He knocked on the wall and it made a hollow, muted ringing. "Not even ArcSec can see what's happening in here."

The double-threat wasn't lost on Ogunwe.

We can't see in.

"You don't have to worry about any dirty business from us. We're aware that any harm done to you here would result in us not making it beyond your enforcers," Ogunwe said matter-of-factly. "What you see is what you get."

Moon studied Ogunwe's face, evidently looking for something there. Ogunwe wasn't sure what, but he knew what the man would find. Nothing. Ogunwe was a master at this game.

Despite Moon's assurances of security, Ogunwe needed his own. He wrote a quick message on the shared feed.

[OGUNWE]—
REICHMANN can you hear or see us? fact check

"Rest easy, friends," Moon said, an easy smile playing on his lips. Ogunwe felt something pull at his emotions, making him want to relax and enjoy this man's company. An alarm bell rang inside of him. There was something unnatural about the feeling, like his emotions were being pulled at with strings outside his body. It put Ogunwe on full alert. He would have to watch Moon's every little action. "We know what the stakes are here. Same for us as they are for you," Moon continued.

[REICHMANN]—

> CO I'm blind and deaf from the outside, 'cept your optical feed of
> course. lemme try and read the vibrations off that tin can you're
> sitting in. keep talking but be still

"Very well," Ogunwe said, as though making his mind up just then. In fact, this was mostly planned out in advance, but Moon couldn't be allowed to know that. "I suppose ultimately trust is a leap of faith," he paused for a moment, looked around at those who'd accompanied him to the meeting. They returned his solemn gaze and, on cue, nodded their assent.

Ogunwe turned back to Moon. "I was asked by the Admin to come and help you coordinate your response to the refugees, which I most certainly will do. But there's another reason I'm here. To make a trade."

Moon raised his eyebrows as if it say *Go on.*

"I can appreciate what you're doing here, helping all these people. PASS is anathema to what we believe in. We want to help."

"Mm-hmm, let me stop you right there, Mr. Ogunwe. PASS is but one brick in the giant wall of resource warfare. What we're doing here isn't just because of PASS, or DS-75, or any one particular policy instituted by Legacy. And it isn't anything different from what we do day in and day out."

[REICHMANN]—

> CO picking up a few vibrations from the walls but nothing
> translatable into words. you're safe

"You see those fabricators out there?" Moon asked, pointing toward the container's entrance. "That's our whole fleet of black boxes. Normally we put them to work every day, all across Legacy. But this isn't something you'd see, or know. We're silently doing the work to keep all these people alive and well. The ignored ones. Every day we turn trash, battered shoes, threadbare sweaters into food, medicine, clothes, shelter. That's the only payment Talksmall requires—bring us a bit of worthless trash to use as stock for the printer and we'll give you something you need."

"And it makes me wonder, are we just keeping Legacy in business? Are we just delaying the inevitable? If we were to stop all our aid activities, would whole-scale revolution ensue?" Moon asked with a far-off look.

"I don't know, but I do know that I took an oath when I joined Talks-mall. The welfare of people means more to me than hollow words of change." He pointed to the letters on his jacket. "You see that? It means something. Talk small. Let your actions be louder than your words. I believe in that. People see these letters and they know, whether they agree or not, to respect the purpose of the person wearing it."

"Today was made for our purpose, Chief. This is why we have all eighty-two engineers and fifteen fabricators in one place today, grinding away at building a small city. Because we believe in a different way of life and our actions, every day, speak louder than anything we could say. So when you say you don't agree with PASS, I can admire that, but I have to tell you—and it's important to me that we're very clear on this—your words don't mean shit to me. Until you prove otherwise with your actions, you still work for ArcSec. Complex, Legacy...they're all the same in our eyes.

[OGUNWE]—
SWEET he talks a good game but does he mean it?

"How is a Complex like ArcSec comparable to Legacy?" Ogunwe asked, careful to sound earnest, not mocking.

Moon's eyes narrowed a touch. "You could look it up in the handbook, but I can also quote it for you. 'Gears of an oppressive machine whose sole purpose is to lock away, from the people, access to resources.'"

[SWEET]—
CO no tics or tells thus far, conviction seems to be there

Ogunwe almost reacted to the internal dialogue he had running with the team, but caught himself before he nodded to Sweet.

[OGUNWE]—
ALL agree so far, odds of betrayal low. MORALES where to from here

Ogunwe scrambled for something to say, anything to stall. "Just playing devil's advocate here," he started, putting his hands up palm out, "but some, most probably, in the CPX world would take offense at your comparison to Legacy. Legacy is haphazard, feckless, lawless. CPX life is none of those things."

Moon scoffed. "You're missing the point. A Complex does the same thing but in a different way. They say 'Join us! The greater the commitment the greater the access to resources!' But if you decide not to, you're locked out. Whereas Legacy's social policies lock the people into a loop of apathy, a CPX's policy maintains rigid control over permissions. Indifference on one hand, avarice on the other. They're just opposites on the same spectrum."

[MORALES]—
    CO empathize, compare his oath to yours. ArcSec is making you
    break that oath with PASS, then specify the trade

"Point taken," Ogunwe said. "Listen, I didn't come here to argue with you. I *do* admire what you're doing." Ogunwe paused, resisted the urge to scratch at his head. This was too important to seem false, or worse, nervous. He clasped his hands in front of him. "Fourteen years ago I joined ArcSec. Oath was the same as it is today. Simple, three words. *Viribus Ad Populam.*"

[CARPENTER]—
    CO got a different take here. attempts to emotionally maneuver will
    be painfully transparent to this guy, just shoot it straight

"Strength for the people," Moon said.

"Yeah," Ogunwe agreed. "Seems to take on a whole new meaning now."

Moon chuckled. "Couldn't agree more. Which people, right? PASS is choosing for you."

A calm sort of silence ensued. Ogunwe's gut was telling him Carpenter was right, that this guy was too sharp a knife to toy around with. He leaned one palm flat against the side of the container. The surface was cool, smooth. Outside, he could hear the murmur of the crowd interspersed with banging as engineers slotted and hammered components into place.

"Hey, I have a novel idea," Ogunwe offered.

"Let's hear it," Moon said.

"How does this sound...I tell you what I'm looking to give, what I want in return, and then we talk. You okay with that?"

"Sounds like the start of an honest relationship," Moon said, his face unreadable.

Ogunwe's thoughts drifted, as objects pulled in toward a huge mass, back to Olivia. She was the planet inside his soul, the thing around which all thoughts gravitated.

If there's anyone who can help find her, he's standing right before me.

"We've created pathways to circumvent the PASS checkpoints. We're calling it the Golden..."

The girl, Estelle, interrupted, her face lit up with revelation. "The Golden Road," she said. "That was you? The tip came from you guys?"

Ogunwe put his hands up as if to say *guilty as charged*. "That was. Just a few of us. Disenchanted Complexers. We put it out there because we believe your organization can find ways to exploit these pathways to subvert the PASS checkpoints and, ultimately..."

"Subvert the system responsible for putting PASS in place," Moon said, finishing his thought.

"Yes, that's right," Ogunwe said, a little deflated. Moon had finished the thought for him as though he'd known what Ogunwe would say.

"Which would be, what? PerSense? Your parent Complex?" Moon asked.

Ogunwe's thoughts seemed scrambled. It was such a simple question, but it cut to the heart of the matter in such a clear way that it seemed to shine a light on all their little cabal's suppositions, exposed all the faults in its foundations. He fumbled for words. "The...directive...came from my boss, Briggs, but...to answer your question...I'm not sure. It seems too coincidental to have been implemented just before the pass of DS-75."

"Hmm," Moon ruminated, hands steepled in front of his face, "I agree with you there. It *isn't* a coincidence. But let's get back to *you*. It seems to me you're operating on assumptions, without a clear idea of who your actual opponent, or oppressor, is."

A lump formed in Ogunwe's throat, prevented him from speaking. He knew that not having a clear answer to the origin of PASS was a black mark against his credibility. More than that, it felt like the question had been designed in advance as a trap, that Moon had bided his time to spring the question on him. He felt like a child caught in a lie, in way over his head.

[OGUNWE]—
ALL some help here? re-direct?

"Before we get too far into this, it's important you see how little time you've actually spent thinking about this. A group like Talksmall, these are the very types of questions that form our ideology. The questions that birthed a movement. You've been laboring under a separate set of tenets for...fourteen years, did you say?"

Ogunwe stared. He was stuck in the trap and had nowhere to go. Could do little else but wait and see where Moon was taking him. The silence on the feed was like the calm before a storm. He could almost feel the anxiety emanating, like steam off his team.

"That creates a completely different set of expectations. And now, after fourteen years of being in enforcement, you want to get in the revolution game?"

Moon paused. Ogunwe knew Moon wasn't expecting an answer. This was for the drama, for emphasis. A feeling of horror welled up inside him, like he was on a plane careening out of control, spiraling downward through the atmosphere. He could no more pilot a plane than he could wrest control of this conversation away from Moon's iron grip.

"That's a leap you're woefully unprepared to make," Moon said.

[MORALES]—
CO double down, admit your faults, go with it

[REICHMANN]—
CO disagree, I say get your butts out now, I don't like the way this is going

Ogunwe considered both options briefly, but deep down he knew the answer before his mind had time to process it. It was too late to abandon what they'd started. They were committed now, completely at Moon's mercy. Besides, he wasn't leaving before extracting some sort of aid in the hunt for Olivia and Stef.

"You're right," he said gravely. "Which is why we need Talksmall. My team can get you access to run ops beyond 2. Get intel, spread your message. In return, we need a safe harbor to operate. I have five people that are going to need your protection."

Moon seemed thoughtful, then looked at the woman. There was a whole conversation that occurred there, not just in the look, but via their systems as well. Ogunwe could tell they were discussing something heatedly. Almost a minute passed before Moon looked back up, addressed Ogunwe.

"I can get you protection, but I don't need access." The words hung in the air, a caveat that rang out and set Ogunwe's nerves on edge. He could feel goosepimples rise up on his flesh.

"What...*do* you need?" Ogunwe asked, afraid of the answer.

"Well, I should clarify. I don't need access to 3 and up. That's why I have Ian here," Moon patted the tall man on the shoulder. Ian didn't react, continued to observe impassively. "Running a network is akin to running security, so he helps with both. What I need more than anything are your resources combined with your access. In short, what I need is *you*."

"Me?" A strange mixture of both trepidation and relief flooded Ogunwe. Relief that there could be a deal and this wasn't just a trap; fear of what Moon would say next.

"I mean I need your eyes, your ears, every little bit of information you can provide. And I need your commitment that this isn't a flash in the pan, that you're committed to doing what's right for the people not just for today and tomorrow, but for life," Moon said, the implication clear as day. There was no mincing of words in his message, and Moon's steely gaze held no humor. "I need you to take the Talksmall oath."

The floor dropped out from underneath Ogunwe. It was like each step of the conversation pushed him farther down the rabbit hole. All their preparations had been pointless.

[REICHMANN]—
     ALL GET OUT! YOU TAKE THAT OATH YOU'RE AS
     GOOD AS DEAD

[SWEET]—
     REICHMANN then what? Go back to your cushy job as a transfer
     guard? Babysit the addicts at the FDZ?

[REICHMANN]—
  SWEET better than your life not being your own. you want to hook
  up with these distrocialists be my guest, just leave me out of it

[SWEET]—
  REICHMANN you committed yourself the same as the rest of us,
  no backing out now

Ogunwe blocked out the furious back and forth of his team. It was all
happening too quickly. He needed time to clear his mind and think.

"Okay," he blurted out, "let's say I take your oath. What then? Am I
just your puppet, til death do us part?"

Moon chuckled, a gesture that, strangely, reassured Ogunwe. "You
don't know much about Talksmall do you? That's not how this works, not
how *we* work. TS is about distribution of power through open access to
resources. Hierarchy isn't our thing. I'm not technically anybody's boss.
Communist and socialist regimes of the past have proven that concept is
untenable. What TS believes is something closer to social anarchy."

"You're a free man and will continue to be a free man after the oath.
The oath simply requires your commitment to share for the good of all.
If you have something useful, share it. Naturally, this means different
things for different people. For example Ian here," Moon angled to his
left, "shares his technological and computing expertise. Just today he got
wind of a little birdie selling root access to Novagenica and its forward
base here in the Arc. Now, the chances that birdy came across that information honestly are zero, so we didn't hesitate to accept when Ian said he
could copy the access without the seller's knowledge."

"In the case of someone in your position, the most valuable thing you
possess is your knowledge, your information, your connections with those
in powerful positions."

"If I commit, what about my team? Can they decide for themselves?"
Ogunwe asked, but he was just following the forms, doing it for the sake
of appearances. He knew the answer.

Moon's face fell. "Sorry, this is a package deal."

"If we decline?"

"Well, I can't say I wouldn't be disappointed, but this is a commitment
each man and woman must make for him and herself." Moon looked at
Sweet, Morales, Carpenter as though trying to gauge their interest. "But

if you're asking if you can decline and still be allowed to leave, then the answer is yes, of course. We're not monsters. We don't kill people in cold blood."

Ogunwe nearly scoffed. But he had to admit, Moon was a damn good liar.

"Give us a moment," Ogunwe said to Moon, who nodded in response.

Ogunwe turned around, ushered the three of them to the lip of the container. Outside, he could see the continued bustle of activity. The team of engineers had set up floodlights as the scaffolding of the project had now reached several stories into the air, clinging to the wall like the cliff-homes of the Pueblo. The excited animation and purpose of the group seemed to contrast with the heavy weight of decision that lay on Ogunwe's mind.

"Tell me your thoughts, quick," he said.

"I'm in," Sweet said, searching Morales' face with questing eyes.

"*We're* in," Morales corrected her.

"Don't see any other way, Chief," Carpenter said softly, looking down at the floor then back up. "But what about Sophin?"

"Leave him out of it for now," Ogunwe said in a low, gruff tone. The silence from Reichmann over the feed was as clear a rebuke as he'd ever made. Ogunwe knew where he stood, just hoped it wouldn't be a problem.

So it came as a complete shock when Reichmann chimed in over the feed.

[REICHMANN]—
    ALL I'm in.....fuckers

Carpenter grinned. "He does have a way with words."

"It's done then," Ogunwe said, giving each of them one final, sobering look, peering down over the rims of his glasses into their eyes. The resolve, and faith, gave him heart.

Hope I can live up to it.

"Alright," Ogunwe said, turning back to Moon and his team. "We're in. What now?"

An excitement lit up Moon's features. He looked at the woman, Estelle, who seemed to mirror his excitement. She was near to bouncing on her toes.

"Now," Moon said, "you take the oath. After that's done, we'll talk. Simple as that. You share what you know and we'll do the same. Beyond that, there's just one thing I need you to do for me, and it has to be done today."

"Oh?" Ogunwe said, unable to contain his surprise.

"That forward base we were talking about earlier? We've already taken a peek at the building. Turns out you have a meeting today, with some people tied to powerful Complexes."

It took a few seconds for Ogunwe to recover. He hadn't realized earlier that when Moon had mentioned the *Novagenica* forward base that it was the location of his meeting later that day. "Okay," he managed. "So you want to see the meeting? Or a report after?"

"Oh, not that. I'll be looking in, but you won't be able to see me. What we need is something more subtle."

"What then?"

Moon turned to the woman, Estelle. "You're up," he said, taking a step back and put an arm out against the enclosure, body tilted against it.

"We need you to tell a little white lie," she said with an evil little grin that sent a tingle of premonition jolting through Ogunwe. The words were innocent enough, but the way in which she said them made the goosepimples stand up on his arms. A seed of fear took root in his chest, spread a growing sense of dread throughout his body.

He'd started the day assuming the meeting would be a blip, then he could wander off and go look for Olivia. Now, he'd found himself bound to strangers, thugs, his will in this whole thing spinning wildly out of control.

"Fine," he said. "But I need something too."

Moon's face was calculating, his lips pursed in a thin line. "Tit for tat. What is it?"

"I need you to find someone."

VAL

VAL
AGE: 19
SEGMENT: LEGACY CITIZEN
POSITION: UNEMPLOYED; GOVERNMENT DOLE
RECIPIENT (STANDARD UNIVERSAL CREDIT)
SOCIAL SCORE: 478

VAL GAZED out the window. A view onto a cold, gray morning. A wet mixture of snow and drizzle fell, the flakes heavy enough to descend straight down, as though in a vacuum.

Or maybe it's the lack of wind, Val ruminated absently. It was hard to tell from inside the warmth of her apartment. Secluded behind her window, Val felt the stillness of it all. A feeling of safety, of being wrapped up, deaf to the outside world.

She sat on the edge of Kat's bed, one leg tucked underneath her, eyes glued to the somber sky outside. The bedsheets beneath her were still pulled back, disheveled, the result of that night that seemed years distant.

She'd been there for some time, uncertain of how long, almost numb to the passage of time. She felt a dull rumbling in her belly, ignored it. She hadn't eaten much in the past few days, not since the...revelation.

Dad had been right.

Never join a Complex, he'd said. And in that brain of his, back when he'd still been alive, were secrets and information he'd possessed that might've been able to help Val. To help her keep Kat safe. But all he'd said was not to join them. Had he somehow foreseen what would happen to his children? How they'd get wrapped up in the designs of the world's most powerful Complex? One kidnapped, the other forced to debase herself for information?

The questions circled like vultures, occasionally swooping in and taking vicious stabs at the carrion. At first, Val diligently waved them away, trying to wrangle her thoughts in an optimistic direction. But even that became tiresome, and eventually Val's mental state devolved into something akin to shock.

She and Trevor had parted ways after Ian's place on 3, the last communication between them a solemn promise of aid. He'd said he was going to see what favors he could call in. And they should both start thinking about a plan.

But his silence was all the evidence she needed. Three days and counting now. The mountain was too great, the task too insurmountable to their feeble skills. He'd abandoned her. Now all that was left was to simply let the vultures in and allow them to feed.

The solitude had not served Val well. She was beginning to have thoughts, the kind that, left unchecked, drove a person to self-destruction in one form or another.

The snow was mesmerizing, held Val in a trance. But something kept nagging at her, begging for her attention.

A thought? Something she was supposed to remember? It rang like a chime in her head, nagging to the point of irritation.

Her system. It was her system, beeping at her. Telling her she had a message. Sighing as though it took a monumental effort of will, Val checked the message.

Trevor. Her pulse quickened. She opened the message with a sort of voraciousness, born of a starving need to communicate.

It was a video. He sat on a narrow cot in a dim box of a room. He spread his arms wide, smiling even wider. "Hey, you're late. Figured you'd call yesterday to say you were on your way. Anyway, welcome to my kingdom. It ain't much, but it's home. This whole exodus thing is the shits," he scratched at a thigh, "but we'll make the best of it, put our heads together. Sending a pin now so you know how to get here. Oh," he leaned in close to the camera, "don't bring much. As you can see, my quarters are kinda...basic. I don't have much, but consider it yours. See ya soon."

The video went dark.

Exodus?

Something was attempting to pry through the fog in her brain.

Exodus...exodus. *Why does that sound so familiar?*

And then it hit her. She rushed to the window, put a hand up against a cold pane, peered down over the edge.

The floodgates, a massively thick wall that curved its way around the city, thrust up through the snow. Three stories high, the top was blanketed in white.

The street below was packed, people shuffling through a cut in the wall, in one direction.

Toward the Arc.

Oh shit.

It was like she'd been living in a cave. She pushed open the entrance door to her building.

Into a warzone.

In the span of three days the streets had gone from merely crowded, with their own natural ebb and flow of people and a variety of transit vehicles, all moving in a mostly orderly fashion, to a scene on the verge of raging bedlam. The Arc was like an overcrowded slave ship; humans packed in the smallest of crevices, just waiting for that one spark to organize the bloody slaughter of the slavers. The sidewalks and streets alike were choked with refugees who milled about, no place to go and nothing productive to do.

She'd been so wrapped up in her own cocoon of worry she'd forgotten about the emergency order. DS-75. The forced exodus.

The snow had stopped falling; without it the day was simply cold, depressing. Val struggled through the flood of dirty faces. Inched her way toward the Arc, where she could reach the nearest transfer gate and take the stairs to 2, where Trevor lived. She was shocked at how many people, children mostly, she witnessed lying on the ground, in the wet slush. Some covered in blankets, family members huddled around them in a protective circle.

Then again, none of it was shocking. This was exactly the sort of thing she'd come to expect. She was just grateful she had somewhere to go.

I don't have much, but consider it yours.

At this point in time, she didn't have much of a choice. She could hunker down in her apartment and lose all but a measly portion of her Legacy dole—not enough to pay the rent, much less feed herself—or she could join the hundreds of thousands on the Arcward migration. And live to fight another day.

Fight. The thought weighed on her like a lead suit. Three whole days had passed and they still didn't have an in to Novagenica, a way to free Kat.

She trudged on, laden with her pack full of meal bars, part of the herd. Looking back at her apartment building, the place she'd lived her entire life, it seemed a cruel joke—Grand Place overlooked the floodgates from the southern edge.

Literally the first building outside the floodgates. Just deposit this building on the other side of that wall and problem solved.

Sadly, it wasn't as easy as all that. The disenfranchised didn't get to make the rules.

Val inserted herself into the flow of people, the *Misery Shuffle,* as it was already being called on the feeds.

Like it's a dance.

She laughed, briefly, a miserable sort of mewling that forced the people around her to inch away.

The optimism at having an actual destination was false and faded quickly, violently. The walk inward, toward the Arc, was a death crawl. She shuffled in the shadows of tall buildings, the Arc looming ever closer, through the slush, feet wet and cold, fingers pink and raw. The snow continued to fall, and steam rose from the masses of refugees that congregated together in a pack. Inward. Nowhere. Just one out of the many thousands, but feeling the most alone she'd felt since the night Kat had been taken.

She blew hot air into her hands, but the feeling of warmth was fleeting. The tips had tingled for the last half hour and were beginning to go numb. She pinched them to try to get some feeling, then shoved them in the pockets of her pants. Her teeth chattered and she hunched forward, folding inward to try to preserve warmth.

In the span of an hour, she'd gone from worrying about how to jailbreak her kid sister from an all-powerful Complex, to worrying about where she'd sleep that night. And the next.

God this blows, and in so many fucking ways. Why me? She looked around at her fellow walkers. *Why us? Why now? What the fuck is the purpose of this?*

The thoughts aroused that familiar feeling, of injustice, of the *wrongness* of it all, of being swept up in vast waves, currents driven by powerful machines that would just as soon dash her body against the rocks than relent or alter their course.

She was just meat, eventual feedstock to be sent to recycling.

A unit of work. And what happens when you work against *their plans?*

The instant her face and name were known, to Legacy or the Complexes, or whoever was behind this, she was as good as dead. But none of that really mattered anymore. The only thing that mattered was her little sister.

So whatever you're feeling, stow it. Think of Kat. Imagine what she's *enduring right now. I guarantee it's worse than a little cold.*

The thought was enough to break down the reality of her situation to barest essentials. If it came to it, she'd trade her life for Kat's. She knew that.

She ground her teeth together, forced them to stop vibrating in her skull. *You were already disposable, but you can be more than that. You can be a threat. They just don't know it yet.*

Up ahead, a half mile through the gray gloom, she spied the horizontal beam that marked the beginning of the Arc. Despite her most recent memories of Level 1, of standing on a street corner selling herself to pervs and pederasts, she was actually looking forward to passing below that beam. Her body needed its shelter.

She pulled her hands from her pockets, alternated between clenching and splaying her fingers to get the blood moving, rubbed them together rapidly. The friction generated some warmth. She touched them to the sides of her neck, shocked at how cold they still felt.

Keep your eyes on the prize. Almost there.

She repeated it like a mantra. The Arc grew in her vision until it consumed everything in her northern field of view.

The walk took another ten minutes, but finally she passed below the structure's massive outer beam and into the shadow of Level 1. Like passing through an invisible curtain, the air temperature shot up thirty degrees, while the lack of natural light forced her to strain to see up ahead. She sighed at the warmth, and the walkers around her sighed collectively, in response.

Eyes adjusting to the dark, she called up Trevor's location on the map.

Ok. From here, she thought, *I just go to the transfer gate, take the stairs to two, and...*

Her system pinged with a message.

That's probably him, she thought.

But when she checked the sender, it wasn't from Trevor. From a stranger. Not marked as spam, but not unheard of.

Eh. Whatever you're sellin, I'm not buyin mister.

"Can't afford to," she mumbled, checking the ID card of the sender. *Marco. Mister Marco. Sorry dude, no SUC for you.*

Val went to discard the file, but as she perused the contact card, something seemed off. Random. The sender was middle-aged, a resident of 3. Well-to-do. His ID wasn't a sales profile, not tied to a business or Complex of any sort. And Val's personal node wasn't listed, so she didn't get all the *sugadaddy* and *dom-seeking* offers other girls her age and looks normally received.

What the hell, she thought, *not like your schedule's all booked up.* She opened the message.

It was a simple voice message, no video. The man's voice was deep, friendly, warm. A halting tinge of uncertainty.

As she listened to the contents, she stopped walking. The glow of neon bathed her pale skin. Her hands began to shake, lower lip trembling. The crowd flowed around her, a stream around a boulder.

Her addled mind could only generate a single thought: *Is this real?*

She listened to the message again.

"Hello, this message is for Val Merina. I realize this may not be a good time for you, with the exodus and all, but I have information regarding Kat Merina. Please contact me today. And, please, it has to be *today.*"

RIKU "CHIEF" OGUNWE

RIKU "CHIEF" OGUNWE
AGE: 40
SEGMENT: COMPLEX PROPERTY; ARCSEC CPX
POSITION: ARCSEC CHIEF CONTRACTED BY LEGACY
ADMIN: MIDWEST ARCOLITH, LEVELS 1-3
SOCIAL SCORE: NA, CPX-SPONSORED

A RIBBON OF LIGHT delineated the edge of the Arc, where the streets led out from underneath the megastructure and into the gridded city beyond. He'd traveled this way countless times before. From 3 down to 1. Up and back down with Olivia and Stef. Pre and post-treatment. Then back up to ArcSec's offices on 3 where he lived. Happier circumstances than this, all of them.

He exited the Arc walking a straight line down the middle of street, a steady influx of refugees *swish-swishing* around him in their plastic coats, heading the opposite direction.

The sky above was a sheet of dull metal. Not bright, but still he squinted against the light. His eyes felt like they'd been taken out and roughly sanded down, then shoved back in his skull. He rubbed them with dry knuckles, trying to massage some moisture back into them. The headache he was just going to have to ignore. It had come halfway through his meeting with Talksmall—Moon and his goons—and hadn't relented since, throbbing dully. He could feel his every pulse through the nerves in his temples. Not even Mamajean's tincture had helped.

He sniffed the air. *Storm's coming.* He thrust a hand out, palm up, collecting flakes before they melted into tiny puddles. The snow was getting thicker, picking up. *Might already be here.*

Before him, a sea of humanity, heads bent against the snow. He barreled downhill, shimmying left and right, inserting his body here and there against the waves of people. An unease, like a hint of impending sickness, soured his stomach. There was too much on his mind to think clearly and the feedback his body was giving him just added to the noise.

You need to sort it out. Write it down then pick it apart, one problem at a time.

The problem was, all his problems seemed to intertwine. He couldn't address one without fiddling with the state of the other.

And, he thought, *they're all competing for priority in your head. That's the real issue here.*

He had an influx of potentially hundreds of thousands of refugees to worry about—their safety and, ultimately, their food and shelter as that was directly tied to how tenable this situation would end up being. On top of that was the nagging worry of his now permanent allies in this whole mess: Talksmall. He wouldn't be able to do a thing without their assistance, their *partnership.* For all of Moon's high-minded talk of ideals and duty to help the people, Ogunwe got the distinct feeling there was a subtext he was missing. Moon seemed to be the type that only looked out for number one.

And then there's the last thread in this grand tapestry of fuck-ups. The most worrisome of all, if Ogunwe was being honest with himself.

It had been a week since he'd first attempted to contact Olivia. Seven whole days he'd been trying to get in touch with her, and in return he'd gotten nothing. He'd had all his arguments prepared. It would be the last-ditch attempt to relocate her and Stef to 3, before DS-75—the Exodus—went into effect and they'd be left with the decision to be either homeless or fundless.

It isn't like her. She's smarter than this, and she'd at least call or drop a message saying she was okay, but busy. The silence doesn't add up.

And it was eating away at his sanity, in tiny little morsels, day by day. He knew hers was only one life in the face of thousands. *One of these thousands,* he thought, looking down the hill at the multitudes thronging up the street, toward the Arc.

But to Ogunwe hers mattered so much more.

Which was why he was here, fighting against the flow of the exodus, to search her apartment.

Down at the bottom of the hill, through a scrim of falling snow, the floodgates jutted up out of the ground, bluish-gray and surprisingly tall.

He hadn't realized how tall they were, had never before seen them raised up.

Behind the floodgates arose Grand Place Apartments, looking forlorn and forgotten through the flurry of snow. He was close now, and the anticipation juiced him. He raised his voice and shouted, "*Comin through!*" Squared his shoulders and plowed through the crowd, putting his hands together in a V, pushing people aside like water before the prow of a ship.

Ignoring the angry cries, he fought his way to the building's stoop. Punched in the code, flung the door open and took the elevator to the eleventh floor.

The elevator *dinged* softly and the doors, clunky and battered, slid apart erratically. Ogunwe couldn't wait. He slammed the doors open, ran into the hall.

1106. Third door on the right.

The door was closed, as it should've been. He rattled the knob. Locked.

Again, as it should be. But if she'd left under force then would she, or whoever, stop to lock up?

Ogunwe kicked the door open. It was ridiculously easy to do. The frame around the knob splintered and the door banged backward against the wall.

The apartment was one bedroom, with a couch that pulled out for Olivia. She gave Stef the bedroom, God knew why.

Because, he thought as he entered the space, *that's who she is. Kind, almost to a fault.*

"Olivia?" he called out. No response, and nothing out of the ordinary. Lights off, couch up against the far wall, beneath the window. He poked his head into the bedroom. Nothing. Stef had even made his bed.

Back into the living room, Ogunwe rounded the corner to the kitchen. Clean and tidy. He pulled open the cabinet drawers, where he knew she kept her food.

Empty cabinets stared back at him. And there it was, all the evidence he needed. She'd left, taken what food she had, and joined the exodus. He'd never see her again.

Anger, white and hot, blinded him. He slammed the cabinet door shut, a hinge coming loose and falling, clinking to the cheap countertop.

The door hung on by its bottom hinge, crooked and dangling like a loose tooth ready to be pulled.

Of all the selfish motherfucking things you could've done, Olivia, this is the worst.

He exited the apartment fuming, but still thought to close the door. At least as best he could. Out of the corner of his vision, a rectangle of natural light. He glanced over, saw that the door to an apartment across the hall had also been busted down. Something tickled at the back of his brain.

A door busted down, why does that remind me of something?

But the thought was blown away, like dandelion puff on a wind, when out of the doorway he heard a murmuring of voices. Ogunwe crept closer, loosened the heavy baton from its sleeve on his hip.

At the lip of the frame, back against the wall, he eavesdropped.

"Cozy'n here, ain't it?" said one voice, high and pinched.

Another person grunted in response.

"Findin' those codes was a blessin'. Nice to be lookin' at a storm for once. Instead a bein' out in it."

"Storm dudn't bother me. It's the people," the other one spoke like he had rocks in his throat. "Ain't one a them knows how to live like us. They all soft, and when they start dyin' they gonna take everyone else with em."

Ogunwe had heard enough. He rounded the corner, flicked his wrist. The baton telescoped with a satisfying *snick*.

The two of them, both men, were swaddled in gray rags and had sunken as far as they could into the carpeted floor. The instant they saw him they pushed back, scrambling away until their backs were up against a couch.

"How'd you get in here?" Ogunwe demanded.

"Uh," the one with the pinched voice looked to the other. "We sorta found a code. To the building."

"These are people's homes. Someone lives here," Ogunwe said, pointing the baton at them.

They both looked at him like he was crazy.

"Look, man" the one with the gruff voice said, pulling himself up a little, "this whole building'll be empty in a couple days anyway. You kickin' us out only makes room for someone else to set up shop."

They were right.

Everything's changed, Riku, he could hear Olivia say to him. Her voice lilting like a song. *And that's alright. You'll learn to live with it.*

No.

He wouldn't just *live with it.* He wouldn't accept that she was gone, just like he couldn't accept that PASS was the right thing for everyone.

He left the room, and the building, not looking back. Into the snow, the cold, the crush of refugees, not feeling it, just focusing on what had to be done. He had business to tend to first, but one way or another he was going to find Olivia and Stef and bring them through this. Kicking and screaming if need be.

The building was on a lonely stretch of road between the Arc's major thoroughfares. Lacking the garish lighting typical of storefronts, the area looked to be abandoned. Ogunwe approached the building, what Moon had styled Novagenica's forward base. Devoid of windows, it was a faceless giant that seemed to deter the eye from looking at it directly. Which wasn't too hard considering this stretch of sidewalk was more like a desert wasteland compared to the packed confines of the rest of the Level 1.

Walking right into the lion's den, aren't ya, Riku? He could feel his back tensing up, the muscles tightening in a gradual path up to his neck. He tried to breathe through it, but the noxious fumes typical of Level 1 weren't helping. He envisioned a green gas slipping into his lungs, infusing his blood with toxic elements.

Not helping, he thought. *Well, if it's a trap then at least I can go down fighting.* He stuffed his hands in his pockets, fingered the weapons residing there. Strangely, it gave him some comfort knowing they were there. Less like a caged animal.

Per his instructions, Ogunwe entered through a foul-smelling alley on the building's right. A heavy, plain gray door clicked open onto a scene in contrast with the abandoned exterior. Ogunwe stepped inside into an open area that hummed with activity. The door clicked shut behind him and a momentary panic flitted through his nerves, a flaring image of his imprisonment in his mind, then fled again as he took in the details of the activity around him.

Stacked pallets were organized in a grid, filling up most of the floor. A central shaft, probably an elevator, took up the center of the building. Around it, several printers stood arrayed, diodes blinking. None appeared to be operating at the moment.

What wasn't taken up by pallets or printers was occupied by people, maybe forty in total. They bustled about the first floor, some climbing on scaffolding to reach the tops of the pallets, zipping utility knives down shrink-wrapped exteriors, handing boxes down to their floor-bound co-workers, who then removed the contents of the boxes, what looked to be one-foot white cubes of some compressed powder. They inserted the cubes into waist-high rolling machines that had NARK printed below a red cross emblazoned on all four sides.

"Mr. Ogunwe!"

Ogunwe snapped his head up, eyes going to a balcony that ringed the central area, some forty feet up. A man stood at the railing, waving to him.

"Stairs in the corner," he pointed to the corner opposite Ogunwe. "Come on up."

Ogunwe weaved his way through pallets, scaffolds and workers, who reminded him of ants at work, going about their business, completely oblivious to his presence. Dust in the air tickled his nose.

He looked back on the scene for a minute before heading up the stairs. *What is going on here?*

Another clue, but he couldn't seem to fit the pieces together. He had a feeling he wouldn't have to wait much longer until it all started to come together. This meeting was one giant mystery and clue rolled into one.

He jogged his way up the stairs. Emerged onto a landing that was all open space surrounding the central shaft, a balcony allowing for a line of sight onto the main floor. On what would've been walls, screens displayed scenes from all around the Arc. Ogunwe's stomach cinched up as he noted a view looking onto Riverside Park.

Were they watching me? Do they know about my meeting with Moon?

The man who had called out to Ogunwe walked over briskly, smiling broadly. He had an attractive, sculpted face that matched the rest of his features. Like a throwback to a surfer from a previous generation. A sun-tan that looked genuine, matching slightly sun-bleached hair.

"Chief Ogunwe, so glad to meet you," the man said, touching a hand lightly to the back of Ogunwe's arm. A gesture that clearly meant *Come along.*

"Tori Fanning, NG," he said, ushering Ogunwe over to a table planted in a dark corner of the floor. Another man sat there, partially masked in shadow. One side of his face was lit up with the blue glow that emanated from the wall-screen nearby, the other cast in shadows that morphed his features as the lighting on the screen changed. Mother nature had gifted him with a bland, fleshy face but hard eyes that stared at Ogunwe with a cold, unwavering glare.

"This is, uh," Tori started, then faltered.

"Call me Ray," the man said flatly.

"Ray," Tori finished.

"Call me Chief," Ogunwe said, matching the man's tone. There was something in his posture that was immediately, viscerally adversarial. Ogunwe pulled a chair out from underneath the table, took a seat, stared at Ray. Two could play at this game.

"Great," Tori said uncomfortably, trying to gloss over the tension in the room. "Now that introductions are done, let's dig into the details. I'll be heading up this op, so it's..."

Ogunwe whipped his head to Tori. "Hold up," Ogunwe interrupted. "What op?" he asked, forehead furrowed.

Tori looked to Ray, then back at Ogunwe.

"I'm sorry, I don't understand." Tori finally managed to get out through his confusion.

"I mean" Ogunwe repeated slowly, voice laced with irritation, "what op? I was told my teams would be needed for support. Nobody said anything about an action."

"Nobody told you *anything*?" Tori asked, dumbfounded.

Ogunwe shook his head slowly from side to side. "Just here for support."

Tori huffed in frustration. "Well, as you're already aware, DS-75 has pushed a lot of people into a limited amount of space. When overcrowding happens, we know with one hundred percent certainty that disease follows. Infection rates skyrocket. You follow?"

Ogunwe nodded. He followed, but it smelled fishy.

Like a setup.

"That's about it, really," Tori said. "The details of the op, from our perspective, are to wait for the inevitable then extend a helping hand when needed. It's a win-win. NG has the opportunity to extend humanitarian aid *and* gain market-share.

"So where do my people come in?" Ogunwe asked, trying to keep the disgust out of his voice. The pieces of the puzzle were falling into place, rotating of their own accord. He could start to see it now, the image those pieces made when combined, and he didn't like the look of it at all.

"While we do have security teams, something of this magnitude requires strength in numbers," Tori said. "Our security isn't designed for an Arc-wide event, but your teams are. Thanks to PerSense," he gestured to Ray, "Your teams at ArcSec are on loan."

Ah, Ogunwe thought. *So big man over there works for PerSense. Military man. Probably thinks he's my boss.*

"So, what, I just wait around for something to happen? That the plan?" Ogunwe asked.

"Uh, no. Not exactly," Tori stammered. "We, uh, we have a map of dispersal points. Each one is close to a warehouse where we can store product stock."

"Medicine, you mean," Ogunwe said.

"Yes," Tori said. "When we've identified a pathogen, we can then identify the correct product to manufacture..."

"Medicine," Ogunwe interrupted again. Tori just stared at him as though he were stupid, continued. "Then we'll send our NARKs out to the dispersal points to distribute the...medicine."

Ogunwe leaned back, laced his hands behind his head. "So you'll need my teams to escort these NARK units to the dispersal point, keep them safe. Babysit the machines until they toddle on home to their warehouses."

"Correct," Tori said.

"What's the time commitment for each dispersal?" Ogunwe asked.

"Fifteen, twenty minutes each way to the dispersal point. Another hour to distribute the payload. So maybe a couple hours total."

"And how many times a day?"

"There's no way to estimate that. Depends entirely on the severity of the outbreak."

"So you just want my people to be on call, twenty-four seven," Ogunwe said flatly.

"No, no," Tori waved his hands, "it wouldn't work like that. We would operate between the hours of eight AM to eight PM."

"How many dispersal points?" Ogunwe asked.

"I'll do you one better," Tori said.

Just then, Ogunwe's system pinged. Tori had sent him an image file;

he opened to find an overview map of level 1. Interspersed in a grid were red dots. A key in the lower right corner defined the red dots. *Recruitment Site.*

He counted up the dots, twenty-five total. Which meant that he'd have to put together twenty-five teams, each with its own command officer, corps of engineers and infantry unit. A hundred men and women per red dot, twenty-five hundred total. He didn't have that kind of personnel. But he knew who did.

Ogunwe nodded in confirmation to Tori, to at least give the appearance of thinking it over.

"I'm not going to tell you I can't do it, but it's gonna be a stretch. And I'm gonna need help."

"What kind of help?" Ray asked from the corner. His words came out slow and quiet, menacing.

Ogunwe locked eyes with the man. "Like the manpower kind." He looked back to Tori. "You're talking over two thousand people. That's more than I have. I can do it, but I'm going to need more people to protect those machines of yours."

"Let me guess," Ray drawled, voice dripping with cynicism, "you want to use Talksmall Enforcers."

A spike of ice drove through Ogunwe's mind. *He knows. About Moon, about the meeting. Shit, he could know about everything.*

"The thought crossed my mind," Ogunwe said.

Tori looked hesitant, unsure. "I don't know."

Ogunwe collected himself. He could've gotten the soldiers from other ArcSec offices, but Talksmall had asked for this favor and he'd granted it, in exchange for a favor of his own. Whether or not this Ray character knew all the details didn't change the fact that Ogunwe was committed. It wasn't just opposing the injustice of PASS, or creating the Golden Road safe routes to move freely up and down the Arc, or even the broad, magnanimous and often unwieldy goal of making the world a better place. It was about this. Right here and now. It was about combatting the evil before him, veiled in the schemes of Complexes and the people that did their bidding.

Whatever these two are planning, it can't be good. And it's your job to sell this, make them think you're on their side, so when the time comes you can fight it. From the inside.

Ogunwe elevated his chin, pointed it at Tori. "Let me ask you a question. How did you get here?"

"The rail to the station on 3, a few days back," Tori said as if this were the most obvious thing.

"And then?" Ogunwe asked.

"Um, down the elevator to 1. Main routes from there."

"And along those main routes, did you notice anything unusual, other than the amount of crowding?"

"Well, I was going to say the crowding was heavier, but, no, other than that, nothing I guess."

"You didn't notice that the routes you walked were relatively peaceful, devoid of violent outbursts?" Ogunwe pressed.

"Well," Tori started, "I guess belonging to a Complex has inured me to the insults of Legacy. Civility is just a way of life, something to be expected."

"For you and me, safe in our guarded towers, yes. But not these people. Down here, the parties are the law, and the main party is Talksmall. The reason you *weren't* accosted, beaten to death and stripped of all your possessions is only because of the presence of Talksmall Enforcers on the streets."

Tori scoffed, clearly unconvinced.

"Don't believe me, take a stroll down an unprotected route. See how far you get."

Ogunwe leaned back again, looked expectantly from Tori to Ray then back again. The ball was in their court now.

Tori looked to Ray, clearly hoping for help. All he got was pursed lips and the slightest, almost imperceptible shake of the head.

"You don't have *anybody* to spare?" Tori pleaded to Ray.

"Not my department. That's why we have genius here," he cocked his head toward Ogunwe, "to do the mindless stuff for us."

Anger spiked in Ogunwe. He was a headache away from slamming his fist into the pudgy bastard's mouth.

"Alright, fine," Tori said, scowling, flopping his body down in a chair. "Do what you have to."

Ogunwe leaned back. "When do you need us to be ready?"

"Two days? Maybe tomorrow? I really don't know. I think we have to plan for worst-case," Tori said.

"Which would be today."

"Well. I guess." Tori put his hands in his back pockets, elbows jutting out. "Yes."

"That's a pretty damn tall order," Ogunwe said.

"But," Tori protested, "I was told you could make it happen. Is that not right?"

I'll make it happen, Ogunwe thought. *And it'll be the end of you.* He looked at Ray. *The both of you.*

Outwardly, Ogunwe grunted, said, "I'll make it happen." He pushed back from the table, stood.

"Where are *you* going?" Ray asked. There was something in his voice that irritated Ogunwe. Something like implied ownership, as if he had any say in what Ogunwe did.

"Did you just sleep through that whole exchange?" Ogunwe asked, irritation bubbling up to the surface. "I've got work to do." He pushed away from the table, headed to the stairs in the corner.

Ray's tone was mocking. "Alright then. Just don't get too buddy-buddy with your guy at Talksmall."

Ogunwe had his hand on the rail, foot on the top step. He froze.

"Moon, right? Jayne Moon?" Ray asked.

Something stuck in Ogunwe's throat, made it hard to swallow. He turned back, eyes narrowed.

"What about him?" Ogunwe said.

"Hey, no need to get defensive," Ray said, maintaining a light tone, like a sugar-coated dagger. Threat of force below a sweet exterior. "I just notice what I notice. Thought the timing was awfully convenient."

Ogunwe had to think quickly. It was unlikely Ray was privy to the contents of his and Moon's conversation. They'd tested Moon's assertions of privacy. Which meant Ray only knew that they'd met and was fishing for details.

The best lie contains a kernel of truth.

"You think I have a choice?" Ogunwe asked. "Talksmall's our only hope of maintaining control down here. I don't have the men, and any idiot can see Level 1 is a heartbeat away from burning. Moon and I have a mutual interest in not allowing that to happen. While you at PerSense may have the luxury of, god knows what, probably only thinking about how to end people, my job is to protect them."

Chew on that, dick, Ogunwe thought.

Ray narrowed his eyes but said nothing. Tori stood in-between them, head swiveling from Ogunwe to Ray and back again.

"We...done then?" Tori asked.

"I'm good. What about you, pitbull?" Ogunwe asked Ray.

Ray looked to Tori, nodded.

Ogunwe turned and walked to the stairs. As he went to descend, he was surprised to find Tori by his side.

"I'll walk you out," he said, following him down the stairs.

"Appreciate it," Ogunwe said. They reached the ground floor, the NG workers going about their business without sparing a glance at the two of them as they crossed the floor. Dodging the pallets, Ogunwe ran his hand over one of the black machines as he walked past. Stenciled in the side in white block letters, was the word NARK.

"Out of curiosity, what does NARK stand for?" Ogunwe asked Tori, who walked at his side. They reached the door that led to the alley.

"Novagenica Acute Recruitment Kiosk."

A wry, knowing smile formed on Ogunwe's dark lips. "So *there's* the market-share piece. It's an army of recruitment machines."

Ogunwe looked at the spare, angular machines. *Fuckers are using the opportunity to recruit more people to their Complex. Don't want to join? You can go to hell, no medicine for you. Shit, they probably* created *this fiasco in the first place.*

"Naturally," Tori said, zero hint of shame evident in his expression. "The machine samples the DNA as one part of the contract then disperses medicine upon completion of the agreement. You don't approve?"

"Oh, it's not that," Ogunwe said, putting his hands up. "I'm just not usually privy to this level of scheming, er, planning. My obligations to ArcSec require me to manage people. You're on a whole other level."

I manage 'em, you manipulate 'em, Ogunwe thought.

Tori stared at Ogunwe as though trying to read his thoughts. Ray hadn't followed them down the stairs, but he'd be damned if he couldn't feel Ray's eyes on him.

Tori sighed, deflated. "Just let me know when your preparations are done," Tori said. It was obvious he wasn't used to this type of conflict. It seemed to wear on him. He had greeted Ogunwe with so much energy and now here he was, shoulders slumped, scowling and dejected. Some people just couldn't handle the pressure. Not everybody had it, that was something Ogunwe knew intimately. This sort of thing was baked into Ogunwe's fiber—it was part and parcel of what he did every day.

"You'll have your men," Ogunwe said and turned to leave. It looked like he'd be talking to Moon sooner than expected.

KAT

KAT
AGE: 13
SEGMENT: LEGACY CITIZEN
POSITION: UNDERAGE; GOVERNMENT DOLE
RECIPIENT (REDUCED BENEFIT)
SOCIAL SCORE: 300

K AT TOSSED in her bed. The dawn lights hadn't yet come up, probably wouldn't for at least another hour, but she couldn't continue to
lie there, utterly dominated by the thoughts in her head. Thoughts that
were gradually leading her, step by awful step, to a state of total despair.

Doesn't help that you feel like a sloppy turd, she thought. Her body felt
weak, loosely assembled, and her breathing was becoming more labored.

*But hey, at least you can still piece two thoughts together. Not drooling
out the sides of your mouth. Not yet.* Unlike Robbie. The day before he'd
barely been able to remember what he'd been thinking. Was that the next
step for her?

*Must be a symptom of whatever they've given us. Right along with this
sore throat, body ache and chills. Your average cocktail of death. Merry
Christmas, kids!*

She found herself thinking of where Val was. What she was doing.
How she was doing. And then something she hadn't felt in a long while
rushed her, tackled her to the ground.

I miss her. God, I miss her.

Then came the shame. At the way she'd behaved, the horrible things
she'd said during their last conversation. She'd accused of her being
incompetent when Kat knew Val was just trying to do her best. That she
hadn't chosen to be a parent at the age of 18, but that was the hand she'd
been dealt.

So she dealt, while you whined and cried like a little baby.

Kat rolled out of bed, feet hitting the cold floor. She pulled her robe on. She'd go talk to Zara. She was like a surrogate big sister in this place. She'd know just the right thing to say to make Kat feel better.

Maybe better wasn't the right word. *Wiser.* She seemed to have a knack to make Kat reconsider her assumptions.

Kat entered the bathroom that joined her room to Zara's, knocked on the door.

"Come in." Zara's voice was barely audible through the door. Kat swung the door open and was hit by the stench. It smelled like bad breath amplified by ten.

"Hey chica," she said, trying her best not to betray her revulsion. She smoothed the covers, sat on the foot of the bed.

"Hey," Zara replied weakly. She was laying on her side beneath the covers, facing away from the door. Her raven hair spilled in a shower over the pillow, framing her face in a halo. Her eyes were red and gummy with pus and crusted tears.

Kat didn't bother asking her how she felt.

"Sorry, not really feeling up to talking," Zara said, almost as if she'd read Kat's thoughts, then coughed up a wry laugh that seemed to wrack her for minutes on end. When she finally managed to stop, her breathing was labored and uneven. Her eyes remained closed.

Kat started to push up from the bed when she felt a hand on her wrist.

"Don't go," Zara muttered. "Just sit with me. If that's okay."

Tears began to stream without warning down Kat's cheeks. She managed to utter a weak *Okay* without sobbing.

Kat sat there for what must have been an hour, gently stroking Zara's hair as though the girl was her daughter and not two years her elder. She watched the blankets rise and fall, an abrupt panic erupting every so often when Zara's huddled form seemed to remain at rest.

When finally Zara's breathing evened out and her eyes started to shift back and forth beneath their lids, Kat gently got up and retreated to her room. Closed the door behind her with a soft click.

She lay down on her bed, stared at the blank canvas of the ceiling. Something inside threatened her as she lay there. She wasn't sure what it was, but suspected it was hopelessness, or something like it. It lurked there at the edges of her mind, waiting for the right time to invade. To take over and tell her to just give up.

But fuck that. She wasn't the sort to give up and she wasn't about to start now.

It happened quickly. One day there. The next, gone.

When Kat didn't hear the customary sound of morning bathroom usage—toilet flushing and faint static of running water—she threw back the covers, ignoring the ache that had spread up her torso to the base of her neck, tip-toed into the bathroom and up to the door to Zara's room.

She scratched a fingernail at the door.

"Zara?"

She waited a minute. No response.

Quietly, she eased the door open, acid and anxiety flooding her gut. She closed her eyes and uttered a silent prayer.

Please let her be okay.

Kat poked her head around the corner and, once again, the first thing she noticed was the smell. A bright, clean citrus scent greeted her. The bed was neat, freshly made. Everything was different, off. Zara had rarely made her bed, and never so crisply as this, the cover pulled tight so that not one wrinkle showed in the white fabric. The white floor tiles gleamed.

Gone.

No girl, no robes strewn about the chair, none of the sweet-smelling flowers she liked to steal from the commons area.

Zara was gone, disappeared, and this room made to look like she'd never existed in the first place. But Kat knew the truth, even if none of the patients ever spoke of it. Zara was dead. She'd died in the middle of the night and then been cleaned up and thrown out like so much trash.

Kat felt her throat tightening up, thunder pounding in her ears. She banged through the door that led to the dormitory hallway, scrambled down its empty stretches, right hand dragging along the walls. All she could hear was the rushing sound, like a waterfall, in her ears. The wide double doors at the end of the hall were open and she rushed to them like she was in a race to the finish. A race her life depended upon.

She came upon the doors quickly, halted at the threshold as her mind tried to process what her eyes took in.

Emptiness.

The same as Zara's room.

Kat turned around. A sick of sense of dread welled up within her, attached to a horrible hypothesis that she knew needed to be tested.

One by one, she twisted the knobs to the dormitory rooms. One by one, the doors swung open easily, revealing rooms identical to Zara's.

Pristine and empty. Sanitized, wiped clean of evidence.

Not one child was left. Except for Kat, of course. She was the only one.

That night she awoke in her bed, body on fire, sweating profusely. The ache now had its grip on her entire body, crushing her into a state of total weakness. It felt like she'd been dropped from fifty feet up, every part of her body making contact with unyielding ground all at once. Her head throbbed with an ebb and flow that worsened with every miniscule move.

Her mouth was dry and pasty, tongue bloated and sore. She winced as she ran her tongue over her lips; the flesh was dry and cracked, shot through with fissures. She groaned. It was about all she felt she could do.

Shit. So this was how Zara had felt as Kat sat there and stupidly petted her like she was some cat, thinking she was offering solace and comfort. Nothing could offer comfort for what she was feeling now, except for maybe a blow to the head to knock her out.

Kat had a pretty good idea of how much time she had left. Another day, two at the outside, and she'd be in the same place as Zara. Fertilizer for the farm.

Same as Antonio. And Robbie. And every single other soul that had been trapped in there with her. Through the haze in her mind she realized dimly, again, that she was the only one left.

It should've been terrifying, but she was too exhausted to project the timeline into her own future. Too weak to think straight.

The world went dark.

She dreamed of the church, Deus Ex. Of walking its endless halls, which were the same but not the same. Darker, wetter. More ominous. They held a strange, pulsating life all their own. The lights grew dim and, in that way you simply knew something to be true in a dream, she knew she was being watched. She was prey in a maze, taking turns at random, corridors replicated at every junction, a seamless sameness to them. The tapestries on the walls depicted historical scenes of prophets—Jesus, Siddhartha, Vishnu, Mohammad. And now the prophets were administering relief to a giant shark that lay beached on a sandbar, laying eggs, moaning as each egg plopped out onto the hot sand. The buddha blessed each egg with an outstretched palm. Without transition, the cavernous atmosphere was replaced with blinding light, which faded into a single long corridor. White walls, white floors, white light overhead allowing for no shadows. A voice on an intercom, deep, foreboding, muttering something incomprehensible. Yet she knew with certainty the low, guttural, indecipherable babble contained an imminent threat, chilling her insides and terrifying her to the core.

And then she was running down the hospital corridor, doors blurring by as she picked up speed. A dark presence chased her, pushed her down the hallway, passing door after door. All the same. She told her legs to pump faster but they wouldn't respond. Frustration built in her chest as she tried harder and harder to escape the amorphous, hungry thing behind her. To Kat's growing horror her body slowed. She willed herself to move faster, but to no effect.

She tripped over her own feet, her body spilling to the white-tiled floor. The darkness, a smothering fog of black, consumed her vision.

Kat let out a silent scream and it swallowed her up. She felt as though her lungs would implode, but nothing came out. The fog slipped down her throat and blocked her airway.

Her chest convulsed once. Twice.

On the third convulsion Kat awoke, eyes snapping open. She gasped, drinking sweet air, and clutched at her blankets with stiff, clenched fingers.

Her hair was matted, body drenched in sweat.

The foul aftertaste of the nightmare lingered like a poison in her psyche. She took a few long, deep gulps of air, intentionally pushing away the remnants of terror that clung to her mind. She battled there silently in her sweat-soaked sheets as the images tried to force their way in.

No, she thought forcefully, defiantly. *This is MY body.*

With one final effort, she swatted the nightmare away and the despair that seemed to be tangled up in it. She focused on her breathing. In. Out.

After a few minutes of doing nothing more than counting her breaths, her mind stilled.

She was met with the sound of a throat clearing. Tinny. From the intercom.

Kat opened her eyes to find the lights up on a dimmed setting. Orange sunset glow. Comforting.

Through the screen on the door, the Indian doctor locked eyes with Kat, a knowing smile hanging on one corner of her mouth.

"Congratulations. You made it through the long night." The doctor looked at her as though studying a bug. "You're going to save a lot of lives," she concluded.

The absurdity of it all struck Kat then, reminded her of that time she'd seen a young mother encouraging her toddler to take a dump in the middle of a packed roadway, just inside the shade of the Arc. Legacy could be crazy like that, sometimes a random jumble of shit that didn't make sense. This felt the same.

"Yay me," she tried to shout, but it came out as a hoarse whisper. She barely had the energy but still managed to let out a long, hacking cackle that sounded evil, destitute and manic all at once.

She closed her eyes and fell into a deep, dreamless sleep.

VAL

VAL
AGE: 19
SEGMENT: LEGACY CITIZEN
POSITION: UNEMPLOYED; GOVERNMENT DOLE
RECIPIENT (STANDARD UNIVERSAL CREDIT)
SOCIAL SCORE: 478

Hello?

Hi, uh, Marco? Mister Dimayuga?

Got it right on the first try, I'm impressed. You must be Val.

Yeah, that's me. You left me a message. About Kat? My sister?

My father belongs to Novagenica. He works in Research. That's where, uh, she's staying.

Held. You mean held.

Yes, I suppose you're right. My father's seen her.

Wait, your father is one of the sickos that kidnapped my sister?

No no no. He's a low-level subscriber. Mops the floors, cleans the messes, that kind of thing.

Oh, right. Sorry. I don't mean to come off like a total bitch, it's just...

Please, don't apologize. Why I called is he left me this insane message. Said he wanted my help, to smuggle out one of the patients.

My sister.

Your sister, yes. I...I told him no. I'm sorry, but I have children of my own.

Can you call him back? Ask him to wait? Can you at least do that?

I can.

SECONDS SLIPPED through Val's fingers. She felt like each one that somehow eluded her grasp was directly tied to her odds of rescuing Kat.

She'd made it to Trevor's earlier in the day, riding high after the call with Marco. They'd set a plan in motion, and now here they were, road-blocked.

Roadblocked, bottlenecked, generally fucked, she thought as they moved one agonizing inch at a time. *Fucking misery shuffle is right.*

She'd been the one to suggest the church, *Deus Ex Machina*, as the rendezvous point. Now she was regretting it, hardcore. She understood it would take some time, coming from Trevor's coffin on 2, but it seemed logical. After all, she and Trevor were coming from 2 and Trevor's friends from Talksmall were coming from Riverside. *Deus Ex* was about the same distance from both. It made sense at the time.

But now here they were, descending one miserable step each minute, just trying to get from 2 to 1.

Hey, at least you can see Level 1 now.

And she could. They'd just rounded the landing that switchbacked to the final stretch down to 1. Steam rose in lazy tendrils from the shoulders and heads of the congregation below. The people were jammed between buildings, in the streets, in dark doorways. Everywhere, below, people and more people. Humanity compressed.

She tried to stave off the self-defeating thoughts that wanted to intrude, but it was no use. *How the hell are we gonna get through all these people, much less get out to Novagenica and spring Kat?*

It was a good question, one that didn't have an answer. Not yet. Hence, the meeting at *Deus Ex*.

Trevor stood to the side, holding her hand. He gave it a squeeze.

"How you holding up?" he asked.

She briefly considered a snappy, snarky response but wasn't feeling it.

"Worried." She dipped her head in the direction of the crush at the bottom of the steps. "How do we get through that?"

He chewed at a lip. "Inch at a time, I guess."

As they descended the last step, the question answered itself. In front of them, a young man with a shaved head and a face the color of a beet snarled and shoved the woman next to him. The tightness of the crowd, like particles in a solid mass unto itself, held the woman locked in place. Someone else threw a punch at beet-face, kicking off a chain-reaction.

They found themselves in the eye of a twisting hurricane of bodies. The knot of people writhed, shifted back and forth, knocking Val forward and back. She spun in a rotation of human flesh as arms, feet, hands pushed, pulled, kicked. Something clipped the back of her head and her vision flashed as her head was knocked forward.

She threw her hands out, braced herself against the bodies around her. *You fall, you die.* The words a whisper of instinct.

Like a black hole, the scrum sucked more people into its mass as it swirled and rotated ferociously. People fell, got trampled on. A hand grabbed her roughly, a painful grip on her upper arm, and another on her ribs.

She looked up and saw Trevor's face set in a grim mask. "Come *on*," he snarled. He pulled, hard, yanked her away from the swirling knot and to the side, shouldering his way out of the middle and onto a sidewalk, up against a building, not looking back.

Val couldn't help herself. She ogled like some rubbernecking bystander as Talksmall Enforcers charged into the eye of the storm, hauling people backward by their hair, heads whipping back, necks exposed to the under-belly of Level 2, like supplicants to a dark god. The Enforcers wielded their bats indiscriminately. The dull thud of dense material to flesh was inaudible save the sudden cries of pain and shock.

Val watched with horror. All she could think of were all the people, mostly children, still lying on the ground, being stomped, kicked, crushed. She willed herself to look away, focused her eyes forward, quick-ened her pace.

What should've taken fifteen minutes stretched into an hour-long trial of forced passage, squeezing with narrowed shoulders through spaces occupied by a sea of people robed in multi-colored plastic ponchos, where elbows consistently jabbed into her ribs and back.

When they finally reached the black and gray, water-stained steps of *Deus Ex Machina*, a line of Enforcers stood waiting.

The streets were just as crowded here, but an aura of safety surrounded the church's doors. She and Trevor pushed through a wave of people that surged up against the steps, testing and breaking against the line of guards.

Trevor approached one of the Enforcers, whispered something, and slipped through the TS line unaccosted, tugging Val along. He pulled open a brass door, allowed Val to enter before him. He stepped inside behind her, the heavy door easing shut in its frame, sealing out the chaos of the outside world.

Inside, it was calm, quiet. A robed Brother Anthony waited for them.

His robe looks different. Simpler. Not the usual cloth, it was made of what looked like burlap, a coarse brown fabric caked with mud along the sleeves and bottom hem.

"I apologize for my appearance," Brother Anthony said. "I've been...traveling."

"Traveling? In this?" Val said.

"Yes, well," he looked to the side, wouldn't make eye contact, "the universe needs its warriors." Something struck Val as off. It was uncharacteristic of the monk, who'd always been straightforward in his dealings. It was like he was avoiding something, sidestepping some issue she couldn't see.

Maybe he's just tired. Certainly looks it.

His complexion had yellowed, soured, and his face was shiny with grease. Lines radiated in starbursts from the corners of his eyes. Fissures plowed through the tissue on his forehead, deep and broken. The beatific air had been deleted from existence. He was a frail, disembodied head poking out beneath an ugly brown robe.

He looks old. Weathered.

Nonetheless, he clasped Val's hands in both of his, which were surprisingly warm and soft. As he slipped his hands back beneath his robes, she noted the dirt lodged beneath his fingernails.

"Val." His eyes were stressed, but kindly. Somehow seeing him like this erased her dislike for the man. Made all her previous interactions with him seem...

Childish. That's exactly what they were, cuz that's what you were acting like. A little baby. But he's still here, still caring for people. Helping people. Helping you.

"Thanks for making room," Val said.

"I was surprised, happily so, by your call," the monk said. "I see a lot of your father in you. He wasn't one to ask for help either."

She waved it off. "Special circumstances and all that. Besides, it's for Kat," Val said, looking down at her feet.

"This is hard on everybody, but it's what we do. Our purpose now is clearer than ever," Brother Anthony said. "The first move has been made, and now the hands of God must unite. To fight the oppressor."

"Uh-huh." *Still batshit crazy I see.*

He led them through the hallways to a staircase barely wide enough for one person. Past the first landing, the walls transitioned from concrete to stone, painted white. Something was bugging her about what Brother Anthony had said. Something she'd been thinking about a lot lately, given her situation.

"The oppressor," Val said. "You mean the Complexes? Or Legacy? Or whoever started this mess?" Her voice rang in the narrow space.

"Both, somewhat. Guilty by association I suppose. Both are two sides of the same coin. They both serve to advance the goals of the oppressor," the monk said.

"So, by oppressor, you mean, what, the devil?" Val asked.

From ahead, she could hear him laugh softly at the suggestion.

"No, nothing so ephemeral," he said, and left it at that.

I will never understand you people, Val thought.

They continued down. The air smelled different, heavy, and she could feel it on her skin.

"We, uh, going to the seventh level of hell or what?" Trevor whispered behind her.

From the sound of it, *Deus Ex* didn't believe in hell, at least not in the traditional sense. Val just shrugged. She'd never been down here before.

Step by step, a murmur grew until finally they came out onto a laminated hallway cacophonous with noise. They walked on, passing doors on the left and right to rooms jammed full of refugees. Val peered in as they passed. People of all types and ages were crammed in the rooms, talking, sipping at steaming cups of liquid. Children pried through spaces between legs, chasing each other, laughing, screaming.

The laughter was bright and beautiful, a reminder that life persisted, and should be cherished.

"You're housing refugees?" Val asked.

Brother Anthony ignored the question, stopped at a doorway near the end of the hallway and gestured to a room that was only half full.

"This will be your space. When your friends come I'll send them down to join you. And remember..."

"No use of internal systems," Val intoned. "We got it." She turned to Trevor, patted him twice against the chest, her fingers *tap-tapping* against the brown vinyl. "Turn it off," she told him.

He screwed his face up in bewilderment, but rolled his eyes and said, "Sure."

With that the monk left, robes swishing, and Val pulled Trevor into the room. The space was a square, bathed in that familiar sickly green light. Stone walls, like the corridor, were painted white. People were clumped in threes and fours, talking quietly, the space notably devoid of children. Val went to the far corner, a place visible from the doorway, and took up residence there.

"What's the deal with not using systems in here?" Trevor asked. "I feel naked without it."

Val leaned against the wall, a sharp angle of stone pressing into her arm. "I don't get it either. But it's religion. I'm not sure anyone gets it except the people making the rules."

"How can we make a plan if we can't pull anything up?" Trevor asked.

"I have an idea about that." Val said. "Maybe Ian can help."

Estelle and Ian came into the room, an oddly matched couple. Her with her lion's mane of hair and he, a foot taller, aloof and impassive.

Estelle crossed to them quickly, wrapping Trevor up in a tight hug, said something in his ear Val couldn't hear.

Trevor turned, with an arm behind Estelle's back, and introduced Val.

"Hi," Val said.

Estelle stepped forward and pulled Val into a hug. Val's arms hung limply at her side as the woman squeezed.

We've got a hugger, Val thought. The woman was small, but strong.

"Thank you," Estelle said, barely above a whisper directly into Val's ear.

"Uh, you're welcome?" Val said.

Estelle released her, offering no further explanation. She looked back to Trevor, then to Ian who had finally managed to saunter up behind her.

"You already know Ian," Estelle said.

Val offered a perfunctory royalty-wave at the man. He looked exactly the same as the last time she'd seen him, down to the green button-down and khaki pants. He blinked, which seemed to take an eternity.

"Hello," he said in his curt manner.

"Shall we?" Estelle asked.

"Let's," Val said. "We don't have much time."

They huddled up in the corner, the four of them, as best they could.

"What do you know?" Estelle asked. "Other than what Trevor already told me."

"She's being held in their Research area," Val said. "The guy who sent me the message, Marco, his dad lives there. Belongs to ."

"Sounds like he grew a conscience," Estelle said.

"From the sound of it, he's not really in a position to make those decisions. Low-level subscriber was how his son put it," Trevor said.

"Okay, so how do you plan on making use of him?" Estelle asked.

Val cringed. "Uh, I'm not exactly comfortable with the idea of using people. That's kinda why we're here in the first place."

Estelle stared blankly, then reached out and stroked Val's arm once, softly. Val tried hard not to recoil. She was still getting used to this whole touching thing.

"I understand your hesitation," Estelle said. "But he's offered himself up. Call it what you want, however you want, but you'll need to use him, one way or another, if you want your sister back."

Estelle was right, of course. The man, Marco's father—*Christ*, she didn't even know his name—had reached out first, had contacted them. He was willing to risk his life to rescue Kat, which meant they would use him. However they needed.

"Well, we kinda have a plan," Val started, "at least a way in, thanks to our buddy Ian here. What we need help with is the getting out part."

Estelle raised an eyebrow.

"This is where the big man comes in," Trevor said, slapping Ian on the shoulder. Ian looked at his shoulder, then at Trevor, expression unreadable.

"Nothing? Anything? Okay then," Trevor said, "here's the deal. Novagenica holds daily tours for people that might want to join up."

"And that's how you get in," Estelle said, nodding. "That's good. Smart.

"And you, my friend," Trevor said to Ian, "are going to help us get out."

"How?" Ian asked. Coming from anyone else it would've sounded like a challenge, but that was just how Ian communicated. In the shortest form possible. He and Trevor began to discuss the logistics and technical details of how they'd get Kat out without being noticed. At the risk of missing some minute detail, Val tuned them out.

Pick your poison—miss out on the tech-speak or be bored to sleep by it. She directed her gaze to the floor and discreetly pressed her fingers together, prayed silently.

> *God please let this not be a mistake. God please let this not be a mistake.*

It felt absurd, but she continued to repeat the prayer. It was about all she had, and Kat's life hung in the balance.

She felt a hand on her arm, warmth through the material of her thin jacket.

She looked up to see Estelle, a sort of angelic glow mixed with a determined set of the jaw, confidence in the eyes. That halo of hair framing the brown skin of her face.

"It'll be okay, Val. We'll help get your sister back."

"I hope you're right," she said.

MANOLO

MANOLO
AGE: 57
SEGMENT: COMPLEX PROPERTY; NOVAGENICA CPX
POSITION: LABORER, LEVEL 6; LIFELONG
SUBSCRIBER
SOCIAL SCORE: 629

MARCO:
Hey Dad.

MANOLO:
Son. It's good to hear your voice.

MARCO:
Yours too. You sound...better.

MANOLO:
I *feel* better.

MARCO:
I'm supposed to pass a message along, but this whole thing...I don't
know. And with the exodus...it's just *nuts*.

MANOLO:
It's okay to be worried, son, just save it 'til you see me next. Then we
can talk about it face to face.

MARCO:
So you got a day pass then?

MANOLO:
Proud owner.

MARCO:
How'd you manage that?

MANOLO:
Just your average rule-following, Complex-loving subscriber.

MARCO:
Now I know you've gone nuts.

MANOLO:
Never saner, son. I've lived. I want to continue living, but now I need to serve something that means more. More than just scraping along, day by day, shoveling shit left one day then right the next.

MARCO:
You're sure then? I can't convince you otherwise?

MANOLO:
Couldn't *live* with myself otherwise.

MARCO:
I'll cut to it then. They want you to bring her to the tour. Tomorrow, as the group leaves.

MANOLO:
That works. Tell them I'll try my damnedest.

MARCO:
Okay dad. I love you, and be careful.

MANOLO:
Love you too, son.

JAIM

| JAIM
| AGE: 35
| SEGMENT: LEGACY CITIZEN
| POSITION: LEGACY ADMINISTRATOR: MIDWEST
| ARCOLITH; DISTRICTS 35-134
| SOCIAL SCORE: 880

JAIM JAMMED HIS HAND into the flimsy foil of the bag. It made a crackling sound as he rooted around for the choice pieces, pulled out a fistful of chips. Crammed them into his mouth and chewed noisily. Watched, on his system, aerial footage of refugees cramming the roadways, cut with protests at the transfer gates on 1 and 2.

Shit, he thought, *I could probably see it with my own two eyes if I just hopped on down to the park, took a peek down the steps.*

The unrest was on every feed, local *and* global. It was a shitstorm in the making. The bugger of it was he couldn't make it go away just by ignoring it. Wouldn't work, he'd already tried.

He wasn't the first person to ever wish he could turn back time, but it didn't stop him from thinking it. Given the opportunity of a do-over, he'd take that first suggestion from Moon—to pass the DS-75 budget bill— and shove it.

He chomped loudly, little bits escaping his mouth as he flipped from one news feed to the next. A tightness gripped his chest, was making it difficult to breathe.

The lights in his office were all down. Kara had gone home for the night. Jaim sat there in the dark, alone behind his desk with his feeds and his bag of *GROSERO'S*™ brand potato chips. Hidden from the world. He wanted it that way. If he could stay hidden he would, just shut the office down until the bad weather stopped raining down and blew on through.

Deep down he knew that wouldn't do. The forms had to be observed.

And the pact he'd made with the other Admins would fall to pieces if one of them buckled. And that wouldn't be Jaim. He'd do what was needed. Put on a brave face, project an outward façade of power, control.

But the bad news wasn't helping. Level 1 was like a pig-farm without its human handlers. Soon it'd be overrun. Talksmall was doing what they could in building new shelters, but they couldn't put them up quickly enough.

And you have to admit, a part of you wants them to fail.

All the feeds he'd been watching showed drone footage of hordes of people packed shoulder to shoulder in the streets as if it was some kind of festival of suffering. Most of the footage showed Talksmall Enforcers in their black helmets lining the thoroughfares. They'd managed to maintain an uneasy peace thus far but Jaim knew it wouldn't last. It couldn't. It was a powder keg that need just the tiniest of sparks. Sooner or later, that spark would be provided.

And then, the shitstorm.

He'd played out the scenarios in his mind so often that they'd begun to invade his dreams. Panic-laced sequences of mobs rioting, rushing the transfer stations, armed with kitchen knives, gardening trowels, whatever could be printed. He could hear the stomp of their boots on the catwalk, feel the rumble of it through the flimsy framing of the Rubik's mall before their incoherent shouts resonated through the air and filtered through the office front. His body would only ever move in slow-motion as the mob bottle-necked in his office doorway, then burst through like a jet of water through a levee. Fingers would rip at his face, tear at his clothes, pulling, pushing, pummeling, lifting and carrying him outside the office to the catwalk. Flinging his seemingly weightless body over the black railing. Falling, flailing, screaming, he'd awaken with a gasp.

Jaim shut his HUD down. This wasn't helping.

He stood, rubbed his aching knees, walked slowly outside. The air was dark and cool, smelled like vanilla. Below, sweeping machines sprayed and cleaned the streets. The lights at the transfer park glowed yellow, illuminated the odd couple strolling through the artificial trees, near the playground, which stood still, silent.

He made his way down the metal stairs, out into the roadway. It was split up with wide lanes for pedestrians and wheeled vehicles. Another narrow line accommodated the maglev stripe.

Jaim meandered north along the pedestrian strip, the occasional scooter buzzing by. Not many people out this late, which was perfect. It allowed him the space, peace to think. Everything was closed. Bars and cafes, restaurants, clothing and tech boutiques, small offices—all were shuttered for the evening, giving the streets a lonely, deserted feel.

Almost haunting, he thought. *If only 1 was more like 3.*

Had he been on 1, he'd most likely have been dead within an hour.

Yeah, but if 1 was more like 3 you wouldn't have these problems to worry about. The only thing stopping them from remaking 1 in the image of 3 was...the people.

Jaim snorted aloud, looked around to see if anybody heard him. The streets were still empty. *Yeah, if you could only get rid of all the people it'd be a great place to live. Frickin' paradise.*

He walked on in silence, the thought echoing in his mind. Dim light from high hanging globes mimicked white moonlight, glistened on the pavement.

If you could only get rid of the people...

Seemed like the situation was primed for that sort of thing.

The thing that made 1 "1" was its culture. It had a reputation for theft and violence because the people that lived there perpetrated it.

Hollow out the den of thieves and murderers, but leave the infrastructure.

His system chimed, startling him.

Moon. Jaim opened the connection.

"Y'know, at one in the morning, most people have the decency to just send a delayed message," Jaim said.

"Normally I would but we've had some developments down here. It's kind of urgent. Thought you should know."

Jaim sighed. "Yeah, it's okay. Wasn't sleeping anyway. Got a lot going on."

"That might be the understatement of the decade," Moon said darkly. "It's been tough, but we think we're getting a grasp on it."

"Good, good, that's good," Jaim said without feeling. "So what's the urgent news?"

"You see the Riverside shelters?" Moon asked.

"Mm-hmm. Pretty amazing what you did there. You parties have the luxury of doing what we can't. You see a need, you address it. Doesn't have to go through committees, polls and all that other bullshit."

"Ten tiers of stacked housing, good for about fifty thousand. She's my baby, I'm proud of it," Moon said.

"Quite the accomplishment," Jaim agreed. "What else?"

A slight hesitation.

"You been watching the feeds? I mean, lately?" Moon asked.

"Sadly. What I've seen today isn't exactly encouraging. It looks like level one's about to blow."

"Not gonna happen," Moon said.

"Jayne, you know I respect you, and I do get the feeling that you can move mountains when you want to, but that statement seems just a touch ignorant," Jaim said. "What if you're wrong? What's your plan then?"

Jaim could hear Moon taking a long, deep breath. When he spoke, his voice was surprisingly calm. "We have people in place, as I'm sure you've seen on the feeds. We're on all the major routes, where most of the refugees are congregated. Well, that and the parks. Plus we've got more shelters going up. But to your point, we've instituted rules. A sort of Martial law if you will. No soapboxing and no fighting. All conflagrations are dealt with swiftly.

"You mean...violently?" Jaim said.

"I mean we do what's necessary. But...there will be no riots," Moon reiterated. "What I was calling about will help prevent that from happening. We need to disperse the crowds, give these people a place to go, a home. But it requires your help."

Over my dead body, Jaim thought, but said, "What do you need?" He tried to keep the wariness out of his voice. If he'd learned anything from his earlier dealings with Moon, it was that every conversation was a chess match, a strategic battle for position. *Whether* to trust was not so much the question, but *what* to trust.

"We need you to do a press conference. It doesn't have to be anything fancy, in fact it can just be a statement released by your office."

Alarm bells rang in Jaim's head that seemed to scream *PROCEED WITH CAUTION.*

"What kind of statement?" he asked.

"Two things. First, an informational piece on where free housing can be found," Moon said.

"Pointing them back to Talksmall and your Riverside project," Jaim concluded. "Okay, that makes sense. And two?"

Moon hesitated again.

Here it comes, the big ask.

"You said you've been watching the feeds?" Moon said.

"Not in the last couple hours, but other than that pretty much all day," Jaim replied.

"Have you seen anything about a virus?"

Jaim's heart went still. "What virus?" he asked quietly.

"We don't know. Not yet. But people are dropping. Just a handful of fatalities so far, but..."

"*What*?!" Jaim's panicky shout echoed through the quiet, empty street. A rising tide of panic threatened to overwhelm him. He stopped where he stood, feet on wet cobblestones.

"Only a few," Moon continued. "But we expect it to get worse, and soon."

Jaim could feel his chest being squeezed, a band that ratcheted down, tighter, with every second. Every successive breath became more shallow, quick. He looked up, almost as a supplicant to God. *Knock this shit* off, he told himself. *Look for solutions, look for solutions. There's gotta be one somewhere in this dungheap of a situation.*

"Jaim? You there?" Moon asked.

Jaim ignored him. A part of him wished he'd never even met Moon, recognized the fact that this whole scenario had played out mostly against his initial instinct, which had been to stay the course. Keep Legacy running, alive at least. Nothing could be done about the Complexes; they'd take over the world in time, but not immediately. But Moon had sold him a story, had sung it with lyrical beauty and cunning efficiency, like a siren luring him to within striking distance. And Jaim had fallen for it, walked of his own volition right into the swirling vortex—a conflict that could very well burn Legacy, and Jaim along with it, to cinders.

But you're here, safe on 3, while Moon is down there in the muck. That means something. And you can still exert some control over the situation. Over Moon. Grant the favor, but play it his way. With half-truths.

"Alright," Jaim replied after a minute. "What do you want me to say?"

"It's really pretty simple," Moon explained. "Just tell people to keep calm and promise them that aid is on the way."

"How can I do that?" Jaim asked, flustered. "How can I promise something I don't have?"

"Lie," Moon said, his voice containing a threat and a command all at once.

A revelation hit Jaim then. He was dumbstruck for several moments, unable to speak.

"Don't get me wrong," Moon said. "You don't have to follow through on acquiring medicine. That's not *really* the point. We've got engineers working on a genetic map and a solution but this one is proving trickier than normal. It'll take time. I need *you* to buy me that time."

Jaim wasn't hearing Moon. All he could think was...*It's the perfect scenario.*

Moon had played into his hands. Jaim could wipe himself clean of Moon *and* his own involvement in the debacle on 1. Start over with a clean slate once everything died down.

This was the fire he had daydreamed about, that he needed, to save his own ass and to recreate Levels 1 and 2 in the image of 3 and up. It came with the added bonus that Moon would be caught up in the middle of it. Burning, along with the rest of them, because he was too dangerous to let live.

"Okay," Jaim said grudgingly. "I'll do it."

VAL

VAL
AGE: 19
SEGMENT: LEGACY CITIZEN
POSITION: UNEMPLOYED; GOVERNMENT DOLE
RECIPIENT (STANDARD UNIVERSAL CREDIT)
SOCIAL SCORE: 822

"CHANGING YOUR ID ISN'T NECESSARY."
That's what Ian had said the night before, face like a stone in a cramped, dim corner of Deus Ex. He'd said he could adjust her and Trevor's social scores so they could board the elevator up to 3, where the rail station stood. Trevor had agreed. Said he knew firsthand it could be done, but that he trusted Ian to do it better than he could. Val knew she had no choice, but still...it was hard to trust what she didn't understand.

She and Trevor stood, waiting in the long line that fed into the lightbox. Beyond that clear wall, the elevator had surprisingly few souls waiting for it. A few minutes of observation told her why. The lightbox continued to admit people to its tiny, clear closet and, within matters of seconds, it would flash red. Only once out of the thirty or forty instances had it flashed green and permitted its occupant to pass beyond the barrier.

She looked with trepidation at the crush of people that milled about at the transfer gate. They were a mass of downtrodden, awaiting their turn to see whether or not their life decisions had warranted a high enough PASS score to allow them to travel upward, away from the chaos of Level 1.

Trevor stood to face her, slipped his fingers in-between hers. His hand was dry and warm. Her thumb pressed up against his wrist. Through the skin, a faint pulse. A vestige of the warmth from the previous night hummed through his skin, transmitted to hers. A muted simpatico. She felt a tiny echo of what they'd experienced the night before.

By the time they had returned from the meeting at *Deus Ex* to his coffin on 2, fighting the thickening crowds, it was late and she was exhausted. Not that you could tell what time it was either inside the windowless coffin or out on the streets of Level 2. It all looked the same, like living underground.

The space was cramped, but he'd tidied it up at least.

"So...wanna share the bed with me? It's more comfortable than you sleeping on the floor," he said, grinning. He'd lobbed the suggestion lamely, as a softball meant to be easily deflected, swatted away.

"Okay, sure," she'd said, taking a little pleasure in the surprised look on his face. She sat on the cot with square shoulders turned inward, narrowed. She was acutely aware of the vulnerability within her and, for once, she felt safe enough to show it.

Never had sex, but it'd be okay if that changed tonight, she thought, silently admiring Trevor.

It felt like the time she'd see this sort of scenario play out in movies and interactive dramas. *The night before the big heist-rescue-escape-whatever, the heroes bunk up for one last hoorah.* But maybe the trope existed because it was true. Art imitating life. They were one good night's sleep away from complete success or total failure. There was no in-between.

Trevor's lop-sided grin remained frozen. "Really?" he said, unblinking.

"Really," she said quietly.

"Well, I mean, *I* can sleep on the floor. It's no big deal," he said, the words coming in a rush.

"Yes, it is. I don't want you on the floor. I want you right here beside me." She scooted back on the bed, stretched out, patted the mattress next to her. "Come on. I won't bite, promise. I just want you to lay beside me."

He obliged, tossing his jacket on the floor and lying down beside her, hands tucked into his sides.

"Jesus, Trevor, you're stiff as a board. Here, turn over," she pushed and prodded until he lay on his side, facing away from her. She started to work on his shoulders first, where the muscles met the neck, her hands kneading the tightly bound cords. He emitted non-descript sounds of pleasure as she worked her way down his back. Suddenly, he jumped up.

"What?" she asked.

"Just had a thought," he said as he went to rummage through one of three drawers in the tiny kitchenette. "Ah, gotcha," he said, slamming the drawer shut then turning around. In his outstretched hand, two tiny yellow cubes with red roses printed on the sides.

"What is it?" Val asked, wary.

"Ever Tangoed before?"

"Uh, no," she responded flatly. She figured her tone would be sufficient in relaying how she felt about drugs, but Trevor remained oblivious. His face was alight with excitement.

"Ok, well, first *you* take this," he pinched a cube and held it out to her. Reluctantly, she accepted. "And *I* take this." He popped the cube in his mouth, started to suck on it. "Go ahead," he said, urging as he sat back down on the edge of the narrow bed. She eyed the cube up close. Then looked up to Trevor.

The real question is do you trust him.

"It's safe?" she asked.

"Oh yeah, the safest," he said.

"That's can't be accurate, but what the hell," she said, then pinched the cube with thumb and forefinger and popped it in her mouth. It tasted faintly of fruit, a little sweet with the vaguest floral hint of the roses printed on the sides.

"Don't chew it. That'll overload you and desensitize you to the experience."

"Uh, okay."

"Now, I need you to open your system. Total access to me. I'll do the same for you. We're gonna link up nervous systems so you feel what I feel and vice versa."

She did as he requested. He eased himself onto the bed beside her, dimmed the lights. Now it was her turn to tense up. She didn't feel any different yet but was afraid of what the drug would do. She didn't want to lose control.

He put a hand on her arm, brushed the skin lightly with the back of his hand. "I'm gonna keep doing this until I get the settings right. You'll know it when the tango hits."

The touch shot tingles up her arm, sending a thrill of goosebumps all over her body. She felt a heightened awareness of her own body as the touch aroused something else within her. Then something happened, instantly, that made her feel as though the world had just flipped. The vertigo was so sudden she had to shut her eyes to prevent from getting sick.

"Oh yeah," Trevor muttered. "Shoulda told you to shut your eyes. Makes the transition easier."

The brushing on her arm continued, but she could also feel the hand that was brushing it, as though it were hers. One and the same. But her arm still felt as though *someone else* was touching it.

"Whoa," she breathed.

"Try this," Trevor said. Then she felt the lightest of fingers caressing her cheek, down the jaw line, tiny drops cascading over her neck then encircling her ear. She gasped, then reached out for Trevor in return.

"Start slow," he advised. "It's like learning to walk with four legs at first, but you'll get it."

He was right. She reached out, tentatively. At first, the motion was mechanical as she had to think about what limb was what. Was that his hand or hers? Her inputs had just instantly doubled, so parsing which was which took a few seconds of deliberate thought.

Time passed, hours maybe, as they delved one another's depths with something as simple as touch. They were completely and wholly intertwined, physically, spiritually, without the actual physical act of making love. Neither of them had wanted it. Not yet. As they tangoed, they discovered something that felt more intimate than sex.

Maybe not more intimate, Val thought. After all, she didn't really know as she'd never physically had it. Only in virtual. *Maybe a precursor. A requisite to a deeper connection.*

That sounded right, felt right.

As time passed, the drug wore off as exhaustion took root. They fell asleep on the tiny cot, her body tucked inside his arms, head pressed to his chest.

Val smiled at the memory, looked down at Trevor's hand, the flashbacks from the night before lingering lightly as they took another step forward in line at the transfer gate.

"It'll be fine," he whispered, his warm breath tickling her ear. She used her other hand to rub it away. Up ahead, the line for the elevator moved at a crawl. They had budgeted for this, but still...if they missed the train everything snowballed. They'd miss the tour and their only opportunity to get Kat out, away, alive. The thought made her tense up. They *could not* miss the train. They would not.

"It'll be fine," Trevor reiterated, squeezing her hand. "Just...relax."

She almost glared at him but thought better of it. No reason to take out her nerves on him, he was just trying to help. Regardless, she did not *want* to relax. She wanted to move, to *do*. To kick ass. To just get on with it. Not stand and wait in this infuriatingly slow line. She groaned loudly, earning a backward glance from the woman in front of her. Val glared at her instead.

The line continued to crawl, seeming to take an hour to move fifty feet. Finally, it was her turn. She went before Trevor, stepping around green glass shards and dirty, trampled plastic bags that had somehow found their way inside the booth. The door eased shut behind her, muting the noise of the crowd that either waited in line for the elevator or shuffled slowly up the stairs. The box proceeded to click and whir.

She couldn't help but feel anxiety, the kind derived from helplessness, a total lack of control. *Please work please work please work,* she thought. *What's taking so long? Just decide, dammit, is my score enough for me to pass or what?*

And then it was over. A green light washed over her and the exit door opened.

Oh thank God. THANK GOD, she thought, relief spilling into her bloodstream, loosening the cords around her middle. *And thank you Ian. Even if you are a bit weird, I could still shower you with kisses right now.*

Changing your ID isn't necessary, he'd said, and he'd been right. She'd been foolish to doubt him.

She stepped through into the near-empty boarding area, turned to wait for Trevor. He stepped into the box, gave her a careless sort of wink. The door eased shut behind him and the seconds clicked by. She felt a relaxed sort of confidence, a world away from how she'd felt just moments ago.

If it worked for me, it's gotta work for him. He's the one that knows about this kinda stuff.

She counted in her head. When thirty came and went she felt an instant anxiety turn on, as an electrical current flowed, as if by switch. She looked to Trevor's face, the worry in his face mirroring hers.

What if he doesn't pass through?

She dismissed the thought. That wouldn't happen. It couldn't.

The lightbox flashed red.

Shock and numbness like paralysis hit her. A dismissal of reality. It wasn't right. Furthermore, it wasn't *possible*. She needed him. They were in this together. He couldn't just leave her to do it all alone.

Behind him, the people in line shouted, pointing at Trevor and advancing on him. He locked eyes with Val, a certain sadness in them.

It's okay, he mouthed, before grasping hands hauled him backward violently, out of the box and flung him into the crowd.

And like that, he was gone.

"*Trevor!*" She rushed to the barrier, slamming up against its unyielding surface, splaying her hands there, searching for his face through the multitudes of onlookers. Every face was a stranger. Set, rigid, hostile.

An alarm wailed behind her, a soul-grating sound that spiked her adrenaline, made her jump. She looked back to lights flashing above the wide frame of the elevator doors. They were closing.

No no no no. She scanned the crowd for Trevor's face, his hair, anything that looked like him, sensing the elevator doors creeping shut behind her, inch by inch, as an opportunity lost.

What do I do?

But she knew. She couldn't miss that train.

With one last glance beyond the barrier, she turned and fled, into the elevator.

MANOLO

MANOLO
AGE: 57
SEGMENT: COMPLEX PROPERTY; NOVAGENICA CPX
POSITION: LABORER, LEVEL 6; LIFELONG
SUBSCRIBER
SOCIAL SCORE: 629

MANOLO GAZED OUT the front windows of his home. Beyond the frozen, snow-dusted field where dull green cornstalks shivered in the wind.

The white menace. The wall. It had become a symbol, something he could pour his loathing into.

His hatred was visceral, started in his gut, fueled the rest of his body. If he could, he'd take a sledgehammer to that wall. It wasn't about keeping the bad elements out, or even keeping acceptable ones in. It was about maintaining control.

A controlled environment, like a culture in a petri dish, couldn't be counted on to behave properly, as predicted, unless it was perfectly sealed.

That's what Novagenica was. A hermetically sealed, controlled experiment where the subjects were forced to live a proscribed existence. Manolo felt his hands trembling, the pressure in his chest increasing. A dull throb in his ears.

This won't do, especially right now. He didn't think the machine would revoke his day pass, but he also wasn't sure if an individual, an actual person, could peer inside Novagenica's artificial mind and parse the logic of its discontent algorithm. If it could decide Manolo was finally worthy of a pass, he was sure it could also decide to rescind the privilege.

And who knows how long Maria's amnesty will last, he wondered. It seemed like he had more time before the AI was fully aware of him again, but he didn't want to get lulled into a false sense of security.

Manolo practiced his breathing exercises, mindfully slowing down his respiration. Forced his mind to go blank.

He stood with his legs pressed up against a console that was situated just below the front window. A bamboo-framed display rested atop the table. It only ever held one photo. Manolo turned it upright, revealing a small rectangle of dust-free surface on the table. Over the past year, the picture had spent most of its time looking at the tabletop, turned over on its face. The image only served to exacerbate the bitterness, the caustic anger, the *blame* he directed at Novagenica.

Manolo studied the image. He and Esme, in front of the tiny hovel of tin and stucco they used to occupy on 2. The light was gray and dead but Esme was shining, her brown cheeks suffused with warmth. She always did like her bright colors. She wore a vibrant yellow blouse and an even brighter smile. Joy radiated from eyes the color and shape of almonds. Their arms were wrapped tightly around one another.

Now the photo brought a sense of peace. Calm. He shook his head. So much time wasted when all he needed was a purpose. He was going to right the scales. Bring balance, justice.

Hands on the frame, his gaze went to his system. Checked the time. 9:20.

He gingerly picked the frame up. "Now or never, dear," he said, touching his lips gently to his wife's smiling face.

Manolo put the screen down for the last time. This time, he stood the frame upright, letting Esme's joy shine outward and illuminate the space.

VAL

VAL
AGE: 19
SEGMENT: LEGACY CITIZEN
POSITION: UNEMPLOYED; GOVERNMENT DOLE
RECIPIENT (STANDARD UNIVERSAL CREDIT)
SOCIAL SCORE: 822

THE RIDE TO 3 was uneventful, which was completely at odds with the way she felt. The elevator was crammed with people like Val—a step above destitute. Twice during the debarking process her cap had been knocked off by the passengers jostling, pushing their way out of the wide elevator doors. They all seemed to be fleeing something.

Unlike yours truly, the hero riding her white horse into the belly of the beast.

She exited the waiting area, another clear-walled corral, into a different world. One full of light, life and, most notably, space. The streets on 3 were busy, but a good sort of busy. Organized, clean. Everything seemed to move in straight lines.

In other words, the complete opposite of 1, which is ramshackle as fuck.

She swiveled to take in the multi-tiered, multi-colored array of shops, offices and housing. It was all constructed in a way that seemed to make sense. The aesthetic value, while understated, was enough to invoke some dull appreciation within her—it was less something she thought about and more something she simply enjoyed. An energy, a feeling to the place.

She stepped into the street, the pedestrian lane, and headed west toward the train station. Her system chimed. It was Trevor.

"You there?" his voice came through directly in her ear. She couldn't remember hearing a sweeter sound. Relief suffused her.

"Thank god," she said. "You okay? What happened?"

"I'm good," he said, "but Ian goofed. He missed a hold on my credit. Not prosecuted, but pending, so..."

"Guilty until proven innocent," Val said, finishing the thought.

"Pretty much. I'm sorry Val. Shit I'm *so* sorry. I should've looked it over before we went. I fucked up. Bad," he said.

"It's okay," she said, her voice calming, soothing. "I'll be okay."

Silence.

"Trevor?"

"Uh," Trevor said. "Yeah. I mean, okay."

"What does that mean?" she asked.

"It means yes, no, I mean goddamn I hope so, Val, because I don't want anything to happen to you and now I can't be there to make sure."

"This isn't about me. It's about my sister. Don't forget that."

"Yeah? Well if you get caught she sure as hell ain't gettin' out of there, so it's about both of you now. Permit me to worry, if only just a little, about your safety. After last night, I'd like to think we can be honest with one another. So please don't make this out to be a business transaction. I care about you, and that's okay."

Trevor's words clarified the muddled emotions within her. *He cares about me...and that's okay.* It made sense then and, despite her circumstances, she felt comforted, embraced, *loved.*

"Okay," she agreed.

"Good. We need to talk about the plan," he said.

"What about it?" she asked.

As far as plans went it was fairly simple. At least Val's part in it was. They were relying on Ian for all the hard bits. All Val had to do was walk in, pretend to be interested in joining the stupid Complex, and walk out, sister in tow. Give the girl a cap, maybe change her hair a bit. It was up to Ian to get Val access to the tour, then add Kat to the tour's registry *after* the tour had started. Which meant they had to have access to Novagenica's intranet. Which, apparently, they'd managed to acquire on the black market.

From Saul, no less. That weasel had managed to profit *twice* off Val and her sister.

He's not your concern anymore, she thought. *The only thing that matters is getting Kat out.*

"I don't know. I'm worried," Trevor said. "At least with me there you

wouldn't be all alone. Everything has to go just right at the right time."
She could hear the anxiety building in his voice. "Too many things can
go wrong. What happens if Ian can't get you in? I mean, just getting to 3
was supposed to be a breeze, but that didn't exactly work out, did it? He's
smart but obviously not super-fucking-man." His breathing had become
ragged over the connection. "I don't know. I'm kinda freakin' out here,"
he said.

"Yeah, you are dude," Val said. She could almost see his green eyes,
creased with worry. She needed to dispel it before it infected her. "Hey,"
she tried again, this time more gently. "It'll work out. I promise."

"Yeah, you say that, but what if it doesn't? What if," his voice turned
thick, "what if something happens to *you*?"

It happened, every once in a while, that somebody said something in
the course of conversation that Val didn't just disagree with but, like a
severe allergic reaction, her body revolted violently at the words.

"I give you permission to care about me, but *not* to whimper and cry
like a little baby," she spat.

"Sorry?" Trevor sputtered.

"You heard me. It's bullshit. I don't accept your worry over me. You
don't *get* to worry about me. I'm here to get Kat, that's all. You can either
get on board with that, and only that, or step aside. That's all I have room
for right now. If you want to help, start by throwing the rest of that shit
you're thinking away. I don't want to hear any more of it."

Trevor grunted as though punched in the gut. "Harsh," he said.

"That mean you're in?" she asked.

"I'm in," he said. She heard the return of his typical cool confidence.
"And thanks. I was spiraling hard there for a minute."

"No shit," she said, and hung up.

The station was a large platform that fronted a rail 100 yards wide by
50 deep. It stood near the edge of the Arc, cutting in and out of the northwest corner of Level 3. There wasn't much real estate on the far side of the
rail. Beyond, a wide field stood in the shadow of the buildings that demarcated the edge of the Arc. It was the first time she'd ever seen such a vast
expanse of unused, untouched land. A place allowed to grow wild. More
than anything she'd seen, it seemed to be the biggest luxury of all.

The platform was covered by a giant canopy that billowed upward like
a sail full of wind. A train sat dormant on the rail—bright white, sleek,

full of potential energy. It reminded Val of a bullet loaded in the chamber of a gun.

She stepped up onto the platform, entered the train, found her seat and waited. The seats were a fake leather, firm yet worn. Reminded her of Trevor's jacket. She closed her eyes and waited. A few minutes passed and her ears picked up a quiet susurration.

That would be the doors. Without warning, the train shot forward, pressing Val into her seat. Through the windows, a blurry canvas of gray and black that eventually melded into a dirty brown.

A message flashed on her system:

YOUR FRIEND TREVOR HAS A SPECIAL HOLD ON HIS CREDIT. UNDER INVESTIGATION BUT NOT YET PROSECUTED. I MISSED IT.

THE TRAIN IS ON TIME. THE SCHEDULE HOLDS.

8:30—TRAIN DEPARTS

8:45—TRAIN ARRIVES NOVAGENICA

9:00—TOUR BEGINS

9:40—TOUR ENDS

10:00—TRAIN DEPARTS NOVAGENICA

She nodded to herself internally, eyes paying special attention to the space between 9:40 and 10:00. That was when Ian would insert Kat into the tour register and presumably some unknown Samaritan would miraculously appear with Kat in tow. Marco's father. Mr. Dimayuga the senior. It was all pretty thin. But if she didn't have Kat, she didn't have anything. She'd just have to go with what she had, thin or no.

Another message blinked:

GOOD LUCK

The message made her smile. She knew it should've been creepy that Ian could patch directly into her system, but for some reason she found it soothing. Comforting, she guessed, to know someone was looking out for her.

The train shot out from the side of the Arc, speeding hundreds of feet above the city's expanse, like some foreign ejecta from the belly of an artificial beast. The day beyond the Arc looked cold, gloomy. Featureless weather. Val watched blankly as the landscape gradually altered. Like the ends of a bell curve, the buildings and landscape flattened out as they sped away from the Arc and its outlying territories.

Without meaning to, Val found herself praying.

God, whatever you are, please...please help me. Help us. Help Kat. Help me get my sister out safe.

And so the prayer went. In variations of roughly the same text, over and over. She ran the tip of her thumb over the side of her index finger, worrying at a rough patch of skin.

Another message:

TREVOR TOLD ME TO SAY "YOU GOT THIS", "YOU'RE A COLD-BLOODED BITCH" AND THAT HE'S HEADING TO 3 USING THE GOLDEN ROAD. HE'LL MEET YOU AT THE TRAIN STATION WHEN YOU GET BACK.

She laughed out loud in her seat, earning her curious glances from her fellow passengers.

The ride was short. The suburbs, which were mostly overgrown and abandoned save for a few enclaves that operated like mini-Complexes, gave way to open, fallow fields interspersed with sparse, sickly forest growing like malnourished weeds out of poisoned ground. By the time the train glided effortlessly to a halt, Val found herself eager, almost bouncing in her seat.

She exited the train onto a platform that stood, isolated, in the countryside. Surrounded on three sides by fields of what looked to be native grass. The dry, golden stalks swayed with the wind.

The fourth side faced the colossal white walls of Novagenica, CPX.

They were close enough that it was hard to tell where the wall's corners were, where it turned back to encompass the rest of the Complex. Beyond and above the wall, the top of an oblong sphere was hazed by distance. To the left, the tops of a few buildings were visible. To the far right, a forested hillside hid the occasional glimmer of a building or home.

Behind her, the doors of the train whooshed closed. Without fanfare, the train raised up on its track and shot out of the station. Val looked around at the people on the platform. Without the train there, they looked aimless and lonely, lost. Roughly twenty of them had exited along with her. All shapes and sizes but united by one thing—poverty. You could read it in the lines in their faces, the frayed bits of clothing, the filth beneath their fingernails.

In ones and twos they stepped down from the platform, followed the only path, which led straight to the wall, adjacent to a wide, unblemished road.

"Alright," she whispered to herself, "let's go."

She followed the others to the base of the wall, which loomed, intimidating, when you stood beneath it. Next to where the path terminated, a beveled square was framed, at about head height, set in the wall. Within the frame, a message blinked.

> *Tour Participants Wait on Path.*
>
> *Tour Begins at 9:00 AM.*
>
> *Be Prepared for Movement.*

The words were accompanied by an animation of a person on a walkway that was thrown to the ground when the walkway started up. Below the message, the current time blinked. 8:57.

As the clock ticked down to 8:59, Val's system dinged. She pulled up the message.

> *ONCE YOU PASS THE WALLS I WILL REMAIN SILENT. DO NOT FEAR. ALL NORMAL. I WILL COME THROUGH.*

The time clicked over to 9:00. Where the path terminated in the wall, an arched entrance pulled back of its own accord, the material melting back into the wall, revealing a long, unlighted tunnel. A circle of light, like the promise of salvation, beckoned at the far end. Small, distant, yet not without hope.

The path jerked to life. Despite the earlier warning, Val was still caught off-guard. She flung her hands out, touching the sides of the wall. They were cool and seemed to hum with energy, vibrating in rhythm with the walkway as it steadily advanced the group toward, and through, the black hole.

OGUNWE

OGUNWE
AGE: 40
SEGMENT: COMPLEX PROPERTY; ARCSEC CPX
POSITION: ARCSEC CHIEF CONTRACTED BY LEGACY
ADMIN: MIDWEST ARCOLITH, LEVELS 1-3
SOCIAL SCORE: NA, CPX-SPONSORED

T HE WHITE MASK contrasted with Ogunwe's deep brown skin. He wore it against his face, the elastic digging into his flesh. The area beneath it was moist, warm. The only exposed portion of his face was a narrow swath that included his eyes, which were dry from the cold. He resisted the urge to scratch at them. He wore the mask knowing full well it wouldn't do much to stop the virus and wondered how long it would take before he and his men started showing symptoms. But it was an abstract thought, one he didn't bother giving much weight. They were, after all, distributing medicine. Surely he and his troops would be the first to receive it.

Ogunwe lived in the concrete, in the here and now. Still, every time he was confronted with the overwhelming masses of refugees, his heart ached as he wondered where Olivia and Stef were. And whether they were safe.

But right now there was a job to do, so he wore the mask, and watched.

Ogunwe's men formed a cordon around Novagenica's secret warehouse on 1, with himself at the center. His back up against the cold, smooth material of the building, he watched the thickening stream of refugees pack into the street in the dim light, press up against the edge of the cordon. The mass of humanity had pushed outward from the main streets, squeezed out into the byways and alleys, like water flowing to its lowest point. Here, on a street that had on any given day been lucky if it saw more than a small handful of travelers, the people now had filled in and begun to carve out little bits of territory, setting up sheets of plastic connected to poles and the sides of buildings. Yellow, white, blue, black—wrinkled, torn, rigid. Whatever scraps of poly they could lay their hands on they erected.

Their own little slice of heaven.

The sardonic thought took on a darker note as his eyes roved, picked out the bodies amid the mass of people—some lying inert, skin blue with the cold, others racked with convulsive, coughing fits. The telltale rings of dried mucus collecting around their nostrils, dripping down like stalactites on their upper lips. Some of them would meet their maker all too soon.

Did Olivia have it? What about Stef? Were they, like so many others, mere hours away from falling face down in a gutter?

Ogunwe felt his chest tighten. Just inside the building behind him was medicine that could save their lives. He itched to do something.

You are doing something, dimwit. You're organizing the medicine convoy. It'll save countless lives. He snorted derisively at his own thoughts. *You can lie to all these people, but you can't lie to yourself. You know the price.*

NG had held a gun to the whole of the Arc's head and effectively said, "*Your life or your freedom.*"

He dismissed the thoughts. They weren't getting him anywhere.

"Vander!" Ogunwe shouted over the din. His next in command stood at a corner of the cordon where he was speaking with one of his men. He turned, two black eyes peering out from above a mask, which hid his narrow face and jutting chin. Ogunwe waved him over.

Vander strode over. "What's up, Boss?" he asked, voice muffled through the mask. "We ready to get the show on the road?"

A curt shake of the head. "Not yet."

"Hmm," Vander mused. "A shame. Troops are gettin' a touch antsy."

"Yeah, me too. Soon as I get the word, you'll be the first to know."

Vander nodded. "Appreciate it boss." Wisps of steam curled from around the side of his mask, proof of their ineffectiveness. *How long do we have?* he wondered again, then dispelled the thought. They had work to do.

"For now, I want you to open up a corridor down the middle of the street, heading first to Vine, then 12TH, then Baltimore where you'll set up a ten-by in the intersection. Use the AWOLS, riot protocol."

Vander's eyes wrinkled up in a smile above the mask. "You got it Chief. Be relieved just to be doin' somethin'."

Ogunwe nodded, watched as Vander got to work. He was good with the men, efficient and well-liked without being their buddy. Respected. The men bustled as Vander issued orders Ogunwe couldn't hear over the crowd noise.

Each autowall—or AWOL as the men called them—started out as a tiny square of black, rigid material that was light to carry. Once expanded to its full size of twelve feet high by sixteen wide, an AWOL could bond to any material, including another AWOL. They were lightweight and used mostly for crowd control.

About a quarter of the hundred men he'd brought belonged to the engineering corps, and each engineer wore a pack on his back that could hold as many as fifty of the AWOLs. They began to unsling their packs, pulling stacks of the little black squares out. The rest of the men created an outline of the wall using their bodies. Once in formation, they set their chemguns to *Wide Euphoric* – known simply as *yoop* – and started spraying outward, away from the line they'd created. The gas was clear and odorless but diffused quickly. Standard riot protocol. Ogunwe and his men had been modified to be immune to the chemical's effect.

Wouldn't mind a little sniff though, he thought, then imagined the outcome and barked out a laugh. The *yoop* took away all concern for anything, to the point of recklessness. If he took a sniff now the medicine would never make it outside of the building. They'd all die face down in the streets. But hey, at least they'd do it with a smile on their faces.

At least we'd have a smile on our faces, he thought, letting out a grim chuckle.

But that wasn't going to happen. Ogunwe watched his engineers lay the squares down in the open walkway between the soldiers, unfolding them so that each one was about the width of a man's shoulders. They continued to work, following the line of men as it went out into the middle of the street then curved eastward out to Vine. The process was remarkably smooth. Watching the engineers, it was like they were laying train tracks, following the line of soldiers that allowed them enough space to put down the AWOLs in parallel lines six feet apart. No scuffles either. The *yoop* did its work quickly; the crowd softened, looking somewhere between a mellow high and a happy drunk. His men herded the squatters with open arms, gently pushing them out of the way. The refugees regarded the soldiers with complacent, bovine eyes, the slightest of smirks.

Once the line reached Vine, they stopped, held position. Vander sent a team of ten scouts up the center of the street, chemical emissaries to soften up the crowd. The lead engineer, a tall and lithe, dark-skinned woman by the name of Green, consulted with Vander in the distance. Vander issued a command over their local net, a countdown to build time. Thirty seconds.

The time counted down backward on Ogunwe's system. When there was just two seconds left, all of the soldiers that formed the line down the center of the street took one giant step over the AWOLs so that they were on the inside.

The two seconds disappeared and, like a translucent flower blossoming, the AWOLs unfolded in unison, erupting upward to create a clear wall that stretched twelve feet tall. Ogunwe had been told by engineering that the individual wall segments chemically bonded with not only one another, but also with whatever substrate they were laid upon. This was what gave them their strength. They weren't just sitting atop the ground, they became part of it, an extension.

They'd have a distribution node set up in the intersection of 12TH and Baltimore in another ten, fifteen minutes at the outside.

Then the Emergency Broadcast cut in, hijacked his system.

The image overlaid his view, a man behind a lectern against the backdrop of what looked to be a Transfer Park. It had to be an Admin, but Ogunwe had a hard time reconciling what he knew of Admins to the image of the man standing there. He looked slovenly, like he'd slept in his clothes for a week. An overgrown beard, like a tangle of brambles, covered his face and crept down his neck. His hair was probably the most put together part of him, but that wasn't saying much. It looked like it'd been finger combed. His eyes were red-rimmed and glassy. A lone ray of early morning light streamed through the giant buildings behind him, touched his hair, lit it the color of a dull fire.

The man cleared his throat. "Ahem," he started uncertainly, casting his gaze about like a lost child. "Hello. My name is Jaim Del Potro, Admin of Districts thirty-five through one thirty-four. I come to you via emergency broadcast today to put to rest the rumors of a public health crisis. Over the last three days my office has been fielding reports of a spreading illness. I'm here today to confirm the existence of this illness."

"Oh, *shit*," Ogunwe said involuntarily. The floor seemed to drop out from beneath his legs. He put a hand up against the wall of the building to stabilize himself. He sucked wind through the facemask, the resistance seeming all of a sudden greater.

Vander's voice, laced with urgency, came over their local connection. "Boss?" he drew the word out, voice rising in pitch. "You seeing this?"

"I wished to God I wasn't," Ogunwe said lowly. "Keep your men in position but double-time it. Send another squad of scouts for wider penetration of the yoop. I want the route to the intersection sealed in 3 minutes, not one second more. You get me, Green?" Ogunwe said, directing the question at the lead engineer.

"Clear as a whistle, loud as a bell. Thirty-second countdown in two minutes, thirty seconds, ready or not," came the woman's voice, which was more raspy than usual. It sounded like she'd been screaming at a concert the night before.

Or like she was getting sick.

"Get moving," Ogunwe reiterated.

"On it, boss," Vander said firmly.

The Admin, Jaim Del Potro, was still talking over the Emergency band.

"While we're not sure of this disease's root cause," he said, posture growing stiffer, voice bolder, more confident, "we *do* know that as of nine AM it reached epidemic proportions at its epicenter, which is Arcside, level one.

"In order to stall the spread of the disease, this office is declaring a local emergency, applicable to all those on the interior of the floodgates. All locomotion between levels is currently terminated until further notice, beginning at level 3. All transit in and out of the Arc will be confined solely to levels 3 through 7, where no sign of the infection has yet been detected. This is subject to change."

The Admin scratched at his reddish-brown beard, creases of concern shadowing his face.

"We realize that this is a difficult time. Rest assured that we have discovery teams working on the issue. For residents of levels one and two who are," he paused, "*uniquely* impacted by this situation, please know that the situation is in the hands of a most trusted proxy of the Arc government. The Talksmall Party is working diligently on distribution channels for medicine, and crews will be ready to distribute that medicine as soon as a viable solution is ready. For those on levels one and two, please direct *all* inquiries regarding location and timing of that distribution to Jayne Moon of TS.

And then he was gone, the window in his view dissolving. Macabre fascination transitioned to outright horror as the Admin's words sunk in. "Holy God," Ogunwe hissed through clenched teeth. "These people are gonna go *berserk*."

He barely had time to complete the thought before his system buzzed with an incoming call.

Moon. Ogunwe opened the connection.

"I just s –" Ogunwe started.

"Secure the building now," Moon commanded.

"What?" Ogunwe said, dumbfounded.

"We had an agreement," Moon reminded him. "To help one another. Secure the building, and all of its contents, right now."

"Fuck our agreement," Ogunwe bristled. "What about the tens of thousands that'll die if this city blows up?"

"I'm well aware, Chief. Believe it or not, we're on the same side. Listen," Moon said, his voice softening. "This isn't about you and me, governing philosophy, none of that. It's about an emergent need. Like you said, it's about a hundred thousand people or more dying in the streets. If we don't mobilize that medicine this very instant, we won't have to worry about a virus decimating the population. Riots and fires will do the job much more effectively."

Ogunwe listened, the logic of Moon's argument sinking in. *So this is it, time to commit. No more working in the shadows.* He sighed. "How do I get in?"

"We've unlocked all the warehouses *and* we've unlocked the machines to distribute without ID sampling, meaning..."

"No contract," Ogunwe said. "Holy balls, that's a coup if I ever saw one."

"That's right. Once you control the NARKs you'll be able to distribute medicine without requiring the user to subscribe."

"How the hell you manage that?" Ogunwe asked.

"It's not important," Moon came back quickly, voice flat. A little too quickly. There was something there he didn't want Ogunwe to know. Somehow they'd acquired high level access to Novagenica. *Or stole it,* Ogunwe thought with distaste. *Oh get over yourself. You've got a dying city to deal with. Methods don't matter at this point in time.*

Moon continued. "What *is* important is utilizing those machines while we can. The only way we can diffuse a riot is to start distributing medicine within the next ten, fifteen minutes tops. You'll need to get in, locate the lead engineer and enlist their help."

"How do I do that?"

"Just get me their name, and an image. I'll use my access to spoof authorization from their boss."

"What happens if they have troops in there?" Ogunwe knew the answer, but he wanted to hear Moon say it. If only to share the responsibility for what he'd be forced to do.

"Immobilize them." Moon's voice was devoid of emotion, of humanity, the verbal equivalent of a stone. "Kill them if you have to. It's the devil's bargain. If you have to kill a few to save thousands, it'll be worth it."

He didn't like it, but he understood it.

"Okay, say I get the machines. What then? After I secure the NARKs it won't be long before ArcSec or PerSense send their troops to take it all back. Or better yet, my orders'll be countermanded by Briggs and I'll find myself at the end of my officer's barrel."

The connection was silent for a moment. "Let's break this down then. One, secure the medicine. Two, order your men to begin distribution immediately at all the nodes. Get as much out there as quick as you can. Three, come to me at the new Riverside tenements."

Ogunwe mulled it over. Getting the medicine out there as soon as possible made sense, but what would happen with Ogunwe gone? Someone else would take his place, stop the distribution.

"I appreciate your concern for my safety, but what purpose does my leaving serve?" Ogunwe asked. He knew he wouldn't have much time once word got out that the medicine was being distributed without contract signatures. Ray, or some other CPX assassin, would be on him in no time.

"I can protect you."

Fat chance, he thought. It was a simple enough equation—the longer he stayed in command, the more lives he'd save. The only variable was now a matter of *how many* lives he could save.

"Appreciate the offer, but I'm staying."

"They'll kill you," Moon stated.

"They can try. It's a free world after all," Ogunwe responded, "but I like my odds."

Besides, I've got something to live for.

"Hey Moon," Ogunwe said, "what's the status on that favor?"

He wanted to allow himself to hope, for even the tiniest glimmer of divine light in this hellscape. Just one positive thing. But he quashed the urge. Hope wasn't good for a damn thing. It just got in the way of people when they needed to make difficult decisions.

"No," Moon said. "I'm sorry, but we haven't picked up any facial scans of your friend or her son. Are you sure you won't change your mind?"

Ogunwe steeled himself. All he could do was control what was right before him. "No," he said.

"Have it your way," Moon said. "You know where to go if you change your mind."

"Ten-four."

"Goodbye, Chief. And thank you."

"Mm-hmm," Ogunwe said and the connection went dead.

He looked back at Novagenica's building, craning his neck back to see where its upper stories disappeared into the darkly opaque underbelly of Level 2.

"Alright you son-of-a-bitch," he said to the building. "Time to give up your secrets."

VAL

VAL
AGE: 19
SEGMENT: LEGACY CITIZEN
POSITION: UNEMPLOYED; GOVERNMENT DOLE
RECIPIENT (STANDARD UNIVERSAL CREDIT)
SOCIAL SCORE: 822

V AL HAD BEEN SURPRISED to find that once she passed through
the giant white walls of the Complex, her fear ebbed away, replaced
by wonder.

Despite her familiarity with the monks' garden at Deus Ex, she had
never seen so much green. The moving walkway took them through a
copse of pines, the ground littered with a bed of browning needles. She
looked to either side. Through the multitude of tree trunks, Val spied
buildings dotting the campus, each one surrounded by neatly manicured
fields and hedges.

Up ahead, an enormous crystalline structure monopolized the landscape. It resembled an egg standing on its fat end. Despite the dull, overcast light, it seemed to shimmer through the trees.

The copse of trees gave way, widening her view of the Complex
grounds. The walkway carried them onward, snaking through a precisely
cut lawn that held only a trace of the snow that had fallen the previous
day. They moved past a circle drive that fronted the giant egg and were fed
into the building itself.

The walk spilled the tour participants into the foyer, some of them
bumping into one another as they forgot to step clear of the end of the
moving walkway. Val stepped off to the side, avoiding the cluster of bodies. She looked up and felt suddenly very tiny. It wasn't unlike standing
at the edge of the Arc. She almost got a sense of vertigo, the scale was so
huge. The foyer was at the building's center, and the open space extended

up to very top of the building. The cavernous space gleamed, seemed to emanate light of its own accord.

"Hello, everyone!" a chipper voice shouted from the center of the floor. Its owner was a casually dressed man with sunkissed skin and a perfect, brilliant smile. "Welcome to Novagenica, CPX!" He clapped his hands together, looking around the tour group. Val could tell he wasn't looking at them. *Just another bunch of filthy subs from the slums, eh buddy?* she thought. It was in the eyes. His smile seemed genuine, but the eyes spoke of detachment.

"My name is Tori Fanning and it's my honor," he touched the fingertips of his left hand to his sternum, "to walk you all through the ins and outs of the life you could have here at NG. As we say, only forty minutes and one signature away! Now," he turned briskly and began to walk, "follow me and try to keep up. We'll see a lot in the next forty minutes."

They took the visitors through a door, one by one, never to be seen again. Probably. Or not. Val didn't really know, she just imagined it was the sort of thing a CPX would do. After all, she'd already seen sufficient evidence they were capable of it.

The room was a waiting room of sorts, closed off, distinctly unlike the rest of the egg-shaped building, which was defined by its very openness. It was positioned up against the curved exterior wall, facing west. Val suspected it was a not-so-subtle reminder of the beauty, the perfection attained by those who decided to join up.

Their tour guide was inviting people through a far door, one by one, to some uncertain place. *Probably where they sample and you sign your life away.*

This was the part that Val dreaded. Though she and Kat weren't identical, separated by years and slightly different features—Kat was softer, still a little awkward in her pre-teen body whereas Val was athletic, strong-featured—a genetic sampling would be a dead giveaway to her real identity, as opposed to the ones Ian had planted in their systems. She had no intention of allowing herself to be sampled, but what if she had no choice?

People kept disappearing through the doorway until it was just Val, alone in the room with her anxiety. Looking out the window, she couldn't deny the beauty of the Complex. In the foreground, a large field lay beneath the faint shadow of the egg. A few low buildings dotted the landscape, which had paths wending around and between them. Far off to the northwest there was what looked like a small town with a central street that was lined with buildings. Probably shops, entertainment. All the distractions people needed to stave off the questions, the dangers of introspection.

From where she was sitting, it looked like the ideal place to live.

Never give up your independence. Never join a Complex, she could hear Dad's words ringing in her ears. His low, gruff voice reprimanding her as though she'd already made the wrong decision. But that was just Dad. She suspected she understood him better now. Being a father of two girls in the Arc was enough to turn the softest of people into the hardest of iron. If it didn't kill you first.

You won't know the price of your freedom until you've given it up. His voice had always been clear, resounding, but the cancer had made it watery, weak, filled with phlegm. *By then it'll be too late.*

Dad's admonishment rang truer than ever before. She looked over the Complex grounds, thinking it was remarkably like the story of Hansel and Gretel. Just supplant the house of sweets for the Complex and the evil witch for the machinations of a human-AI hybrid system that ground up and ate souls.

Val chuckled lightly. It sounded like something Trevor would come up with.

The door opened and Mr. Tori Fanning's perfectly tanned face smiled at her. "Last but not least," he said jovially. "Come on in."

Val gave him a polite smile, steeling herself inside. She stood, pulled the brim of her ballcap down low on her forehead. *Here we go.*

She walked through the doorway into a dimly lighted cell of a room. No windows, a small rectangular table, two chairs. A long, shiny cylinder affixed to the right side of the table. The door clicked behind her, which gave her a start.

"No need to be frightened, Miss." The guide walked up to the table and pulled out the chair closest to Val. "Please, make yourself comfortable," he said as he went around the table and took the other chair.

Val sat, but did not touch the arm rests, much less scoot in beneath the table. She needed to be able to get up quickly. Besides, she didn't want her DNA to be harvested. The futuristic looking cylinder on the table seemed the most likely piece of sampling hardware, but realistically any surface in the room could serve as a collection point. Conscious of that fact, she tucked her elbows in, folded her hands in her lap.

The guide rested his elbows on the table, clasped both hands together. "So!" he said in a bright voice, "if you're ready, go ahead and insert your arm into the tube. That'll get us started."

Val simply looked at the tube and, after a short silence, uttered a terse, "No thanks."

"I see," Tori ruminated, rubbing at his cheek. "Tell me, what did you think of the tour?"

"It was nice," Val said obtusely.

What's the quickest way out of this hole? she thought. *Play along, make him think you'll go home and consider it.*

"Oh? Just nice?" Through the polite veneer, he seemed almost offended. *Jeez, you'd think he took the rejection personally.*

"Well, yeah. I mean, don't get me wrong, this place is beautiful. I'm just not sure yet. I need some time, to think it over." Val began to stand but Tori motioned for her to sit.

"Hold on, hold on. Humor me for a minute. Tell me, what *exactly* is there to think about?" She was right. The politeness was mostly gone, replaced with a hardness. She sighed internally. *Typical dude,* she thought. *Can't handle rejection.*

"I'm not sure I'm ready to give up everything I know to come live here," she said sweetly, playing the part. She looked Tori in the eyes and she saw an instant shift to genuine caring, concern. *This guy is good.* It made her all the more wary. He could be playing a game way more dangerous than she expected. *I just need to get my ass out of here.*

"I completely understand," he said slowly, "that can be a hard choice to make. The good news is, doesn't force you to make that choice. Just because you've joined up doesn't mean you have to give up everything—everyone—you know outside the walls."

"It doesn't?"

The guide threw his head back, let loose a full-throated laugh. "Heavens no! If that were the case nobody would join! No, we're very sensitive to the needs of our members. Everyone is allowed passage outside the walls."

Val didn't know where to take the conversation from there. This was taking too long. Fed by this thought, a trickle of anxiety began to claw its way up her gut.

"Y'know, you look familiar," Tori said, and with the remark the trickle suddenly turned to a flood. Heat flashed through her body and she could feel sweat begin to bead on her forehead.

"Oh...yeah?" she uttered stupidly. *Jesus, get a hold of yourself!*

"Have you been here before, on a tour maybe?" he pressed, perfect face scrunched up in consideration.

"Nope," Val tried to sound nonchalant, but it came out stiff, wooden.

Tori stared at her another moment, eyes searching her face.

"I'm sorry," Val said, words spilling out of her, "I have to go."

Instead of accommodating her, he leaned back in his chair and studied Val, eyes narrowing.

"Do you?" he asked.

Oh shit.

MANOLO

MANOLO
AGE: 57
SEGMENT: COMPLEX PROPERTY; NOVAGENICA CPX
POSITION: LABORER, LEVEL 6; LIFELONG
SUBSCRIBER
SOCIAL SCORE: 629

MANOLO SHIVERED. He wore two sets of clothing, normal work scrubs over a smaller, tighter set of civilian clothing. Despite that, the cold cut through as he waited.

He checked the time. 9:34

The blank face of the building loomed before him. Its gray material blended into the gloom of the day.

He took stock of his state. He wasn't anxious. He had channeled his anger into intent, a belief of righteous action. His breathing was steady, heart rate felt normal. All of the things he knew Novagenica's autopanopticon was watching he felt he had under control.

Control wasn't the right word. It wasn't like he was wrestling with his emotions, fighting an inner turmoil. It was more like Judo. He understood the rules, witnessed the power of his opponent, and had deftly turned those rules and that power against Novagenica to his advantage.

The rules that had driven his anger and resentment were the same rules that allowed him to feel confident in his purpose. To have faith in his actions.

Manolo smoothed the gooseflesh on his arm with a weathered hand, checked the time again. 9:35.

He moved. Calmly entered the decontamination chamber. Blasted by a chemical cocktail and then through the doors. Slipped into the stairwell to the side, taking the steps two at a time to the landing at the bottom. Pushed open the door to the subterranean floor, checked the hallway.

Manolo had worked the timing out so that no one *should* be present. Hadn't really thought through what he would say if that wasn't the case, if he encountered another person roaming the halls. Especially a doc.

Just act like you didn't understand your schedule. Pretend to be stupid and they're more likely to believe you. Probably because they were dispositioned to believe it.

Head jutting out from the door frame, he looked left, then right, scanning the long corridor. Not a soul. He stepped out into the hallway, turned and pulled the door shut quietly behind him, then off to the left, five doors down.

He tried not to rush, but the adrenaline that flooded his system was making it hard. He forced himself to move deliberately. Measured steps, measured breaths. If he didn't maintain his calm, *Novagenica* would tag the measurements as suspicious. Alert the barracks.

For one terrible moment his ears were deafened by the shrieking wail of the alarm sirens, heard the stomping of heavy boots, visualized black-clad armored troops rushing in, weapons raised and aligned, muzzles leveled on him. He pushed the thought away along with the momentary panic that came with it, allowed the simple act of motion to still his mind.

When Manolo reached the room, he didn't bother to check the interior video. He pushed, and the door opened.

Onto an empty room.

His breath caught in his throat, something awful and anxiety-inducing that leapt up and got caught there. His mind refused to believe his eyes. A bed, a dresser, the ledge at the far end of the room. She was nowhere to be found.

Panic hit him, a lead rod to the gut. He didn't have time to go searching the commons, and the more areas he touched increased the likelihood of him coming into contact with people. People that would ask questions. *She should be here!*

Manolo had planned it meticulously, knew the rhythms of all the patients, or at least what was left of them, which was Kat. The patient counts had changed, but the routines had not. Breakfast was over, so she should've been in her room until 10:30.

And hell, it's not like there's anybody left in this place for her to talk to.

He stood there, left hand on the cold metal frame of the door, the other hand curled around the side of the door itself, struck numb with shock.

Move dammit!

He left the doorway. Got down on his hands and knees and looked under the bed. Nothing.

Shit. Where where where? He stood and his gaze settled on the bathroom door.

He ran around the bed to the door, flung it open. Another empty space. A narrow corridor that connected this room to the next one over, with a countertop and two sinks.

It isn't the only explanation, though, is it? he asked himself. He felt a terrible sadness, then, as the realization settled over him like fallout from the sky.

She's dead. The sickness took her too and you, you stupid fool, were too late. You couldn't even save this one child.

Something hard hit him in the shoulder.

"Hey," came a voice from his left, then "*what the...*" It was the voice of a girl, loaded with frustration.

A girl, the thought pierced the din in his brain.

Manolo stepped back and the door to the toilet swung open wide. To reveal the girl. Patient 165D.

Her name's Kat, Manolo reminded himself. *Not a number.*

She stood there, in shock, face haggard and pale with dark circles under her eyes.

God, she looks like she's been to hell and back. Please let her be strong enough for this.

As if in slow motion, her eyes widened into twin full moons, huge, bright, and frightful.

Manolo backed farther away so his back was against the other door, pressed a finger to his lips, urging her to remain silent. She yelped anyway. Jumped backward into the small room that held the toilet and shower. She reached out for the handle, pulled hard on the door. Manolo slid to the side, put his body in-between the door and the frame. He managed to get his whole body in, shut the door behind him in case someone checked the feed.

"You here to finish me off?" the girl asked, lips trembling, voice quavering.

Manolo held out his hand in a gesture of peace, once again pressed a finger to his lips.

Val, he mouthed.

A curious looked passed over her face. Manolo leaned in, whispered so low only she could hear.

"My name is Manolo. I'm taking you to Val." He stepped back, looked into her eyes, questing for approval. There was confusion in her face, yet she nodded, almost imperceptibly.

Manolo raised his hands in a *wait* gesture and hurriedly took off his scrubs to reveal the second set of clothing he wore beneath: the ill-fitting garments of his dead wife.

He took off the yellow blouse, Esme's favorite, and held it out to the girl.

She understood immediately, snatching the garment from his fingers. Turned around to face the opposite door that led into the next room. Yanked the Novagenica-issued flannel shirt off her torso, and pulled Esme's top down over her head.

She turned back around to see Manolo holding out a pair of blue jeans. She grabbed them and turned around again. Manolo pulled his scrubs back on in two swift motions.

When the girl, Kat, turned back around fully dressed in Esme's clothing, Manolo couldn't help but smile. The top was a little loose, but Esme hadn't been a big person, not much bigger than this girl. Esme would've approved of what he was doing.

For once, he thought with a little internal cackle. A private joke that, he thought, most married men would understand.

But there wasn't time for sentimentality. He leaned in again, whispered, "Follow me. Act natural, like you belong."

There was fear in her gray-blue eyes, like that of a lost child. But as he watched, a hardness took over. A resolve.

He nodded in affirmation, then turned. Exiting the bathroom, Manolo stepped around the bed to the door that led out. It had no handle on the inside, just a flat rectangle set in the wall. As Manolo approached, the door clicked softly open.

He began to extend his head out into the hallway when he heard it. The sound of a door closing.

Shit. He pulled back, swiftly, hoping he hadn't been spotted. He was about to shut the door when he decided against it. He angled his head so one ear was just at the edge of the door frame. Squeezed his eyes shut and listened intently. Nothing.

No rustling of fabric, no squeaking of shoes. The docs rarely entered the rooms, so it had to have been an orderly. In the room now.

If he was wrong...*Well, let's not think about that. Let's just pray I'm right.*

He checked the time. 9:39. No time left. The tour would be over in a minute, then another five minutes for sampling and selection. At 9:45 the group would split between those accepted and those leaving the Complex. Five minutes, roughly the time it took to get from Research to the Egg.

And still they had to get out of the building. They wouldn't get a better chance than now.

Manolo look to Kat, raised his eyebrows as if to say *Ready?*

She nodded once. He grabbed her wrist and they shot out into the corridor, pulling her along, down toward the stairwell. Manolo found himself holding his breath, forced himself to breathe.

A surge of adrenaline hit him as they approached the stairwell. *Almost there!* He reached out, felt the cool metal of the handle. He heard the click again, whispering softly from the hallway behind them. It felt like the cocking of a pistol's hammer.

Manolo slammed his shoulder into the door, yanked the girl in behind him. She skidded into the landing at the bottom of the stairwell. Gently, Manolo held the handle in the down position as he eased the door shut, then slowly released it so the mechanism didn't make any noise.

He allowed himself to breathe. He had to assume whoever was back there didn't see them. It was the only way forward. Something tickled at the back of his brain. A small detail, something out of place. Was it all going too smoothly? How long had Maria said she'd make him invisible to the AI? A part of him suspected the amnesty was up, yet the AI remained still, silent. No barking in his ears, no images overlaid on his vision showing him how he'd mis-stepped, which he'd half-expected and prepared for.

But what could he do?

Nothing. That was the answer. Not a damn thing.

"So...is Val meeting us here in a dark stairwell, or..."

He looked to Kat. Her face was composed, but she tapped a finger nervously on the hand-rail.

Manolo checked the time again. 9:40

"Up," he said, pointing a finger and climbing up the stairs.

She fell in line, taking the steps two at a time behind him. A few moments later they topped the last step to the ground level landing. She was winded, breathing heavily. Probably weakened from the treatment.

"We have five minutes to get where we're going," Manolo said, careful to avoid any words that might be flagged for alarm. "People will be out," he jabbed a thumb toward the outside. Just remember, it's..." he searched for the right word, "*natural* for us to be here. Understand?"

She was bent over, hands on knees. She gave him the thumbs up. "Yeah. Natural, got it."

"Stay by my side," Manolo said, then pushed the door open.

VAL

VAL
AGE: 19
SEGMENT: LEGACY CITIZEN
POSITION: UNEMPLOYED; GOVERNMENT DOLE
RECIPIENT (STANDARD UNIVERSAL CREDIT)
SOCIAL SCORE: 822

OH SHIT OH SHIT OH SHIT.
Terror, pure and stark, gripped her mind. Her lower mind was telling her to get up, to run. *He KNOWS!* it screamed at her, demanding she bolt from the room. Tear out of here and never look back. Yet a sliver of her higher mind remained intact. It knew what would happen if she ran. This wasn't a match that could be won physically. Her mind was her only asset.

Despite that knowledge, her mind was so at war with itself that the only thing she could utter was a simple "Uh..."

"Most people who come to the tour don't understand this process, this *place* as well as you seem to," Tori said.

"Oh?" Val said. Below the table, she gripped her own hands so tightly that the fingernails dug painfully into the flesh.

What do I do? She was on the verge of full-blown, mind-erasing panic. Somewhere out there, Kat was making her way to this foyer of this building. What would she do if Val wasn't there? Would she be able to get out? Would she even know to move on without Val?

"Of course," the man said matter-of-factly. He eyed her shrewdly, a hint of a smirk playing on his lips.

How can he be so calm?

Unless...unless she was reading the situation all wrong. He hadn't actually accused her of anything.

No choice, now. She sucked in a long, slow breath and exhaled over the course of five seconds. Somewhere outside this wall a man, a complete stranger, was breaking her sister out and bringing her to Val. She had to trust that he would do his part, now Val needed to see her own part though.

Alright, let's see where this takes us.

"But most aren't as smart as you, either," Tori said.

Val showed a mouthful of teeth. "Sorry," she said unapologetically, "the stupid girl act dies hard. But it works great most of the time."

"I knew it," Tori said, gloating and triumphant. He pounded the table lightly with a fist. "I knew you were different. I could tell you were looking for something, to get more out of this than just food, shelter and all the other banal crap that most of these subs are looking for."

The realization hit her like a bucket of cold water and she almost burst out laughing. *Holy shit, he thinks I'm* negotiating! *Okay,* she thought, pulling on her deep well of confidence, *this I can do.*

"You're right of course," she said bluntly, "I want management."

"Exactly! That's the spirit," Tori clapped his hands together, the sharp crack making Val flinch. "Of course, you know it's out of the question. You have no experience."

Val scoffed. "What better experience than surviving in the Arc? Every minute of every day is do or die. And if you're weak-willed then it's die. Obviously, that's not me."

"Mmm, you have a point. However, I have no allowable provisions for straight to management subscriptions. Best I can do is Supervisor Trainee, but with many of the lifetime perks of management."

Val stood up abruptly, the backs of her knees forcing the chair back violently. It clattered as it skidded then toppled backward.

"We're done then," she said with an air of finality, a tone that left no doubt as to her decision. Taken aback, the guide just watched her in shock. Val savored the feeling of power. "I'm leaving now. My offer stands for two days. Think about it."

She walked across the room regarded the handle for a moment before realizing she had no choice. She gripped it and pulled, half expecting it to remain locked. It swung open freely, throwing her a little off-balance. She gripped the edge of the door with her free hand to steady herself. Then she stepped through and didn't look back.

The door let out onto the mezzanine that overlooked the building's lofty central space. Below, in the foyer, a handful of tour-goers milled about, waiting for some obvious signal that it was time to leave.

She counted the heads. *Five out of thirty. Well, six including myself.* Those were not good numbers for Legacy. The Complexes of the world continued to siphon off the loyalty of the world's population.

Val followed the curve of the balcony railing to the stairs, resisting the urge to run her hand along its smooth metal. She tried to move quickly without looking like she was in a rush. There wasn't much time left. She *had* to be down in that foyer by 9:45.

Screw it, she thought, and jogged down the steps, jumping over the last two to land on the hard, glittering floor. She joined the group that was huddled together near the middle of the floor.

"Where'd everybody go?" she joked. Nobody laughed.

"Joined up," said a tall man who obviously didn't get the joke. Val looked at him , not surprised. He had crags like canyons working their way down his face, with eyes that dropped and jowls that sagged.

"Yeah, figured," Val responded.

"Just wanted to see what it was all about," he droned on, his voice deep and inherently mournful. "Always been curious."

Val nodded, looked away.

"I see now I ain't missin much," the man continued. "Looks pretty, but then again I never been too high on appearances."

Val turned her body away, rolled her eyes. *Goddamn dude, take a hint,* she thought, growing more perturbed with each passing second. It was 9:45. *Where the hell's Kat? She should be here by now.*

A chime rang through the foyer, then a voice sounded over an intercom.

The tour is now complete. If you have decided to decline our invitation to join, proceed to the exit. The moving walkway will escort you back outside the Complex. The next train departs in ten minutes. If you do not exit this building within the next two minutes, your choice to remain within the Complex will be deemed as implicit acceptance of the Novagenica, CPX Contract of Subscription. Thank you for visiting Novagenica, CPX.

Immediately, the group began to shuffle toward the exit. Val hung back at the rear of the pack, her head swiveling in every direction, looking around in a panic. There was the main entrance where they'd come in. Kat wouldn't be coming from that direction. Off to the sides were smaller, almost hidden doors that Val assumed led to the interior of the Complex. That was where Kat would be.

She looked around again, made sure nobody was watching. Nobody. That was the nice thing about automation. Sometimes humans were more effective in raising an immediate alarm.

Where are you, Kat? She wondered as she bolted for the nearest side door, the one to her right. *Get your ass out here because I'm* not *leaving without you.*

KAT

KAT
AGE: 13
SEGMENT: LEGACY CITIZEN
POSITION: UNDERAGE; GOVERNMENT DOLE
RECIPIENT (REDUCED BENEFIT)
SOCIAL SCORE: 300

S HE WAS BLINDED. It was the first time she'd been outside in a
month. Maybe more. She'd sort of lost track of time. Last thing she
knew it had been fall, but this definitely felt more like winter.

"Act natural," Manolo, her accomplice, said. "Walk with me." They
began to walk along a path that wound away from the building they'd just
exited. Her feet felt like they were someone else's; she barely had control
of them. Snow dusted the ground to either side of the path, green tips of
grass peeking through the blanket of white.

Manolo walked swiftly and Kat tried to keep up, the wind buffeting
her face. The air was cold, but tasted good. Clean. She took a huge drink
of it, then hacked it back out in a fit of coughing. Her throat felt red and
raw, and the frigid air made it burn.

"Great job, Kat. Way to not draw attention to yourself," she muttered.

Manolo shot her a sharp look. He walked to her right. He was older,
bald except for two short strips of stubby brown hair above his ears. Kat
suspected he was as anxious as she was, but when she studied his weathered, lined face she found not one trace of worry. Exactly the opposite, in
fact. He had a sort of beatific look, the kind of expression she'd see on the
faces of saints in the tapestries at *Deus Ex*. He saw her studying him and
looked over, said, "Doing great," in a kind, soft voice. The voice she imagined a grandfather would have when encouraging a child.

He seemed pre-occupied, looking off in the upper right corner of his
vision repeatedly.

Probably looking at his system.

He glanced upward again and this time he instantly hurried his pace, uttering the simple command, "*Come.*"

Kat hurried to catch up.

"Spry little thing, aren't ya?" she asked, huffing. Her body felt less like a and more like a sack filled with crap.

Useless, heavy crap. Like a stuffed animal filled with lead.

They were walking along a path that cut between two fields, headed toward a large building of steel and glass that resembled an egg. She jogged lightly, caught up to him. Looking around, she noted that they hadn't spotted anyone since they'd left the building. "Is it strange," she asked, "that nobody else is out here?"

"Count your blessings," he said.

"Huh," she grunted. "Not many before today."

He cursed under his breath.

"What?" she asked.

"Can you run?" he asked. He began to jog.

Kat tried to run, to catch up, but instead her body locked up. She saw a weird flash of green on the sidewalk that looked like the images of footprints.

*What the...*she had time to think before she clattered awkwardly to the ground.

Her mind went blank, black, swept clean of all thought except a vague recognition of pain in her face.

It works.

It was a thought, but not hers. She wasn't sure where it had come from and didn't have the capacity to pursue it.

Awareness returned to her in a sudden torrent, like she'd been granted permission to once again inhabit her body. She'd lost time, but it was also like she'd never been gone in the first place.

Head angled on the cold concrete, she saw a couple on the opposite path across the field staring their direction.

"Perfect fucking timing," she spat, then punctuated the comment with "*assholes*". She tried to push up but the muscles in her arms were emaciated, wasted away. Then Manolo was at her side, hauling her up.

"You okay? What happened?" he asked, brown eyes full of concern, compassion.

"Dunno. Blacked out I guess."

"Can you move?" he asked.

"Think so," she closed her eyes briefly, shook her head momentarily as though shaking something out. She couldn't explain what had just happened, but they didn't have time to sit around and pontificate all of life's little trivialities and nuances.

Maybe a seizure, she thought. That seemed likely.

"You sure?" Manolo asked again.

"Yeah, it's just I saw this weird thing and then I fell. Footprints on the sidewalk."

Manolo's face went slack, eyes wide. "Were they green? Looked kind of like a meter inside each footprint?"

She took a half step backward. "You're creepin me out now," she said.

The little old man looked like he was thinking very hard, straining at a giant weight internally. He grabbed her by the upper arm and pushed her along the path.

"We don't have time to worry about it," he hissed. "You have thirty seconds to get in that building. Now run!"

She did as she was told. It was liberating at first. An explosion of adrenaline fueled a renewed vigor in her legs. Her lungs sucked deeply at the cold, fresh air. The wind pushed her hair back and bit at her face. Warmth suffused her limbs, giving her body a buffer against the cold.

For the first time in a long time, she felt *good*. Free.

The constraints of her atrophied state came crashing back in. It began with her lungs. Halfway to the building her lungs began to burn as her once strong, measured breaths gave way to gasping, erratic gulps of air. Her legs were on fire.

She kept her eyes on the goal. The building. *A hundred yards. I can do this,* she thought.

The fire in her lungs and legs were spreading, attempting to converge on one another and consume her whole body. She ignored it, pressed on.

"Don't think, just run," Manolo said at her side. She cleared her mind, focused on the goal. Miraculously, it worked. She controlled her breathing, two in, two out. Her body picked up speed.

Get there get there get there

The building loomed as they entered its shadow. With one final burst of speed, she lunged at the door, yanked back on its handle, putting her body into it, felt something pop and strain in her shoulder.

It didn't budge.

A burgeoning horror dawned on Kat as she yanked again, ignoring the ache that flared from shoulder to back, this time with the manic energy of terror.

She dug her heels in, angled her body away.

Still nothing. She backed away, looked at the handle, unsure of what to do.

"No," she whimpered. She wanted to cry. Surely her thirty seconds were up. In the span of five minutes she'd gone from despairing, trapped in a prison, to the prospect of freedom. Almost believing it could happen, only to discover it was a cruel joke. The administering of hope, only to snatch it away again. Like dangling drugs in an addict's face.

I'm not *giving up now.* She ground her teeth, mustered what strength she could.

Two hands, fingers red and numb, on the cold steel. She mustered what strength she had and pulled.

The door gave. It flew open and a body slammed into her. Kat instinctively threw her arm around her assailant and rolled, landing on top of the person.

The air *oomphed* out of her, taking what remaining strength she had along with it. She tried to scramble away but the person held her tight.

"Let go," Kat hissed, struggling with her captor. She pushed and wriggled, but her arms were pinned. The person beneath her was a girl, not much older than Kat, but bigger. She wore a hat and a long, baggy shirt.

Kat went still and looked into the girl's face.

"Val?"

Before she could grasp what was happening, Val was all over Kat, arms wrapped around her, picking her up to her knees, hugging, supporting, gripping her tight.

"Oh my god oh my god," Val repeated. Tears streamed down her sister's cheeks, wet and messy.

Kat was dumbstruck. A firm hand gripped her shoulder. "*Get up.* Now." It was Manolo. His voice was sharp and high. "We need to leave."

Val looked up, over Kat's shoulder. "You Marco's dad?"

"Yes, but that's for later. Come, *now*." Manolo gently pried the two of them apart, pushed them through the open door.

Into the lobby of the crystalline, egg-shaped building. It was vast, airy and smelled of *new*.

"Here," Val said, taking her cap off. She handed it to Kat. "Put this on."

Manolo herded them in the direction of a group of people who were shuffling toward what Kat assumed was the front of the building.

"Exit, now," Manolo said. "No touching, no looking back. Walk quickly, don't run." With one last little shove, he pushed them toward the group.

They hurried side-by-side toward the exit. As they reached it, Kat did what Manolo had asked her not to do. She looked back.

Above, at the railing, a man waved to her. She ignored him; he wasn't her main concern.

Manolo.

The kindly old man stood in the foyer, watching her go with a sad little smile, eyes shining.

Why's he staying? And then the worse of the two thoughts. *What'll they do to him?*

But her worry was subsumed by the prospect of freedom. She faced forward again as they stepped through the wide opening at the front of the building, onto a moving walkway.

The sudden transition almost made her fall, but she grabbed at Val's arm, grasping the shirt to stabilize herself. They stood next to one another, silent, her hand maintaining its grip as though in need of constant confirmation that Val was indeed real, there, in the flesh and wouldn't simply disappear, a ghost of her imagination.

The walkway sped them through a forest of pines and to the white wall that pervaded the horizon beyond. The cold wind raised gooseflesh on her arms. The walkway passed through a tunnel in the wall, out the other side and deposited them in an open field.

Into freedom.

TORI

TORI
AGE: 33
SEGMENT: COMPLEX PROPERTY; NOVAGENICA CPX
POSITION: VP OF HUMAN ASSETS, SUBSCRIPTION
DIVISION
SOCIAL SCORE: NA, CPX-SPONSORED

T ORI STOOD AT THE BALCONY that overlooked the lobby. His
body tingled, hummed with energy, optimism. The tour was over,
which was likely the last one for the foreseeable future.

And good riddance. Except that last one—the girl. She'd been fun to toy
with.

Assuming everything went according to plan, which...how could it
not? He'd been assured by the all-knowing machine it was so. No, everything would go according to their very carefully thought-through plan.
All reports from the Arc suggested the situation was at the cusp of chaos,
which was just the right time for Novagenica to step in. All that remained
was the distribution of the medicine, tied to the new contract, of course,
which could begin as soon as he got back to NG's command center in the
Arc.

The new contract differed a little from the main subscription contract.
This one demanded the rights to a percentage of the sub's labor. For most
it wouldn't mean complete indentured servitude—although some would
certainly qualify. No, the majority of the population would just pledge to
be where Novagenica wanted them when called upon, whether to clean
out sewers or to test a new drug. It took a ton of manpower to build
a new society, and after today the resources that would be available to
Novagenica *we*re seemingly limitless.

Tori watched as the last of the stragglers hurriedly exited the building.
Don't want to get caught by the exit clause, do ya?

One of them was the girl who'd said she wanted management. She walked out with the rest of the stragglers, those few who had declined his offer of subscription.

Should've taken what you were offered, he thought with a smirk. Come tomorrow, maybe the day after, they'd all fall sick and then be in the same boat as everyone else in the Arc.

Begging. For help, for aid, for medicine. Anything to keep them alive.

The girl looked back, the bill of her cap covering her face in shadow, and Tori smiled, waved.

And we'll be happy to help, he thought, *for a price.*

With smug self-satisfaction he sauntered down the steps. He needed to get back to the Arc, oversee the distribution, but he had a little time to spare. Next train left in a little over an hour, which gave him enough time for one final errand.

He pushed his way out a door at the rear of the building, led to the Complex grounds. The air, cold and wet, hit him in the face. He pulled up the clothing's interface, turned the thermal element to seven.

Research wasn't far away, just beyond the next hedge. He walked quickly, step livened by the giddiness that always accompanied the end of one thing and the beginning of another. His warm breath puffed out clouds that crystallized around him in the winter air. Where the walkway cut through to the next sector, the hedge grew in a soft arc that swept up and over the path. Through it Tori could see the squat grayness of Research. A building made to be ignored, the gaze tended to slip around it, finding the fields and landscaping much more amenable to the eye.

He entered Research, made his way down to the bottom floor. Banged his way out of the stairwell, footfalls lightly *tap-tap-tapping* in the narrow space. Down the hall to the fifth door.

He laid a hand lightly on the door to Patient 165D's room. The screen came to life, showing a feed of the interior overlaid with a hub of the patient's bios. Just a bed, unmade. Nobody inside the room, which wasn't all that unusual. She could've been in the bathroom or down in the commons. What *was* unusual was the fact that none of her bios showed on the screen. Flatline heart rate. Water, muscle, fat percentages—no data. No temperature, no hormone levels.

Tori moved to the door, gently eased it open, and entered the room.

Reached the edge of the bed and peeked over the far side, half-expecting to see the girl's cold, blue body lying on the floor.

But of course, it was as he'd been assured. The room was empty.

What did you expect to find? the AI's voice spoke directly in his ear.

Tori's body jerked ever so slightly. He wasn't accustomed to the AI addressing him out of the blue.

He scratched at his forehead then sighed. "Oh, I don't know. You machines can be an untrustworthy lot," he responded to the air, to nothing, feeling slightly foolish, knowing the machine that ran Novagenica could hear him. "Just had to be sure. Not to say I understand why we let our Patient Zero go scot-free, but I suppose you're the all-knowing engineer of the plan to kill off Legacy. Is it rude of me to second-guess your decision-making process?"

It was necessary, its voice slithered in his ear. *We assure you.*

Tori froze. He was used to Novagenica's AI speaking for the soundness of its plan using statistics. But there wasn't even an attempt made here. Just a proclamation meant to force the issue of trust, but coming from a machine it had the effect of instilling in Tori something quite the opposite.

He left the room in a state of quiet dread, unsure of what to do next.

INTERLUDE

JAIM

JAIM
AGE: 35
SEGMENT: LEGACY CITIZEN
POSITION: LEGACY ADMINISTRATOR: MIDWEST
ARCOLITH; DISTRICTS 35-134
SOCIAL SCORE: 880

THE BENCH WAS FREEZING beneath Jaim's legs and back, the cold seeping in through his clothing. He rubbed his hands together, blew hot air into them. They were, like the rest of him, dull, weary, desiccated.

A patch of gray sky remained visible through the buildings. Jaim's body ached for the touch of sunlight, but it wasn't happening today. The gloom was persistent, the type that looked as though it'd last for weeks. The kids hadn't gotten the memo, didn't understand they were supposed to be quiet, subdued, comply with the weather. They ran around the park, chased one another in-between trees, darting in and out of the little circle of playground equipment, laughing and screaming, delightful peals filling the air. Completely oblivious to the cold. Oblivious to the circumstances two levels below their feet.

The transfer park was dotted with moms and their children, strollers, dirigibles, the myriad attendant paraphernalia of parenthood. The activity was, as always, pervasive, which was why Jaim considered it the perfect place for a meeting. Too many eyes for an assassination. It was foolish reasoning and he knew it, yet the thought gave him comfort.

The bench was flanked by bushes on either side. Jaim sat in the middle, arms splayed out to the sides along the backrest to discourage anyone else from joining him. His visitor should be along shortly, and then he'd have an explanation to the cryptic note he'd received from McAlister not long after the broadcast.

TRANSFER PARK BY YOUR OFFICE. 11 AM. NEED TO PLAN. FOR OUR MUTUAL BENEFIT.

McAlister. What was he up to? And had he seen through Jaim's plans and wanted in? Regardless of what his fellow Admin wanted, Jaim was on high alert. After the emergency broadcast, everyone would be gunning for him. Everyone except for Moon. Jaim had given *him* de facto control over 1 and 2. He'd be buried under busy work right up to the point the flames started licking at his doorstep. The thought gave Jaim a feeling of warm satisfaction. The smug, scheming asshole would get what was coming to him.

"This seat taken?"

Jaim looked up, saw the round, red face of McAlister looking down at him. He looked ridiculous in jogging pants and hooded sweatshirt. The hood was pulled up, limning his puffy, bloated face like a marshmallow puckered at the edges. There was something different about him, disturbing. It was in the eyes, an odd mixture of disdain, discomfort, anxiety. McAlister the sycophant was gone, replaced by a man who seemed both full of fear and scorn.

Jaim scooted over on the bench. "Sit," he said. McAlister did so hastily, rear-end barely touching the bench before the words came spilling out of him.

"I don't know what you think you're doing but this, this has got to stop. We discussed a meticulously planned takeover. *Whatever* it is we just saw on the EBS was *not* according to plan."

Jaim looked at McAlister, face contorted with disgust. "What did you think this looked like John? Did you think it was just gonna be nice and orderly? Everybody waiting patiently, slotting in neatly to their new surroundings? No chaos? No *death*? Man, I need to come live in *your* fairytale, because all of that sounds fantastic, simply fucking unreal."

McAlister looked as though he'd been slapped. "Okay, point taken," he said, "you don't have to be an asshole about it. I'm worried, alright? *We're* worried. Have you even *talked* to the other Admins? And what about Moon? Was he part of the plan all along? Did you plan on double-crossing Talksmall the whole time?" McAlister was getting worked up again, practically shouting. His face looked like a volcano ready to blow.

"Listen...John," Jaim took the man by the shoulders. "You—need—to—relax," he said slowly, methodically. "Just fucking breathe. This was all engineered from day one. All according to plan, my friend. All according to plan."

McAlister did as told, closing his eyes and meekly resting his hands in his own lap. His whole body language seemed to fold, as though the anger was a false front, crumpling inward to reveal his true self. *There we go*, Jaim thought. *There's the Welcome Mat we all know and love.* After a moment, McAlister opened his eyes again.

"You good?" Jaim asked.

"I...I think so."

"Good. Because now, buddy, we can talk about how we're gonna capitalize on this opportunity. Look around," Jaim invited with an expansive arm extended. "And see the little slice of heaven we have here on 3."

"Yeah, okay. It's nice," McAlister agreed blandly.

"Surely you can be more descriptive, John. *Nice* doesn't cut it. Have you *seen* what life's like below us?"

"Well sure, who hasn't, but..."

"No buts. It's hell. A vicious arena soaked in blood, deceit and graft. No productivity, no fraternity. Hell in every sense of the word. As compared to our heaven here on 3. But what if, *what if*, we could start over?"

"I'd say it's a pipe dream. Impossible," McAlister said.

"Is it? Or is that just your short-sightedness talking? Because I believe it *can* be done. And that we're already halfway there. Part one," he held up a finger, "Get rid of the people. This mysterious illness, whatever it is, is the perfect catalyst. It won't completely eradicate the population, but it gets the party started. Once the refugees discover Moon and Talksmall don't have any answers *or* solutions, the riots will start. They're already so close," Jaim made an indication with forefinger and thumb, "just right there, on the verge. The whole thing only needs the tiniest of sparks to blow up."

McAlister rubbed his forehead with one hand, a bead of sweat forming along his hairline despite the temperature. "What's, uh, part two of the plan?"

"Right. Two," Jaim held up two fingers, "once everyone's gone, we," he gestured to McAlister and himself, "you and I, we pass a bill to fund a massive reconstruction project. Remake 1 and 2 in the image of 3 and up. Redistrict so that the two levels are divided in half between the two of us. Maintain strict standards for who can reside and work there. I mean, for God's sake, we already have a ready-made system currently in place in PASS. We can use it to restrict population movement.

A shadow of thought passed over McAlister's face as he stared off into the park. His head started to bob minutely, then gradually more emphatically. "It could work, DP."

Jaim smirked. He had him. "Of course it'll work!" He slapped McAlister on the shoulder enthusiastically, smiling broadly. "And we all stand to gain from it. The Complexes think they can build their own societies and just quietly take over, like a fungus growing beneath our feet. This is our one and only chance to change that, Moon wasn't wrong about that."

A vestige of worry returned to McAlister's face. "You're okay with burning Moon?"

Jaim scoffed. "TS is the least of our worries. Plus," Jaim said, his eyes narrowing so he looked like a cat on the prowl, "that snake in the grass is better off dead. I can guarantee he would've done the same to us."

McAlister's eyes widened and it happened all at once. He scrambled backward on the bench, away from Jaim. A voice spoke from behind Jaim. "Well, you're right on that account."

Jaim started to turn when his head exploded in pain. His body pitched forward from the force of the blow. Hands clutched at the back of his shirt, threw him to the ground. He opened his eyes, vision a blur, unable to focus on anything except the playground equipment. The grass was cool, wet on his face and fingers. Unfamiliar voices continued to talk over him.

"Two for the price a one."

"Lucky day. Brought my rabbit's foot, 's why."

"Man, these two's weak. Just one shot. With the old five-fingered hammer, no less."

"Hard to smuggle bats through with all these ArcSec cops."

"Pssh, just manning their stations like good little robots. 'Sides, they don't know bout the Golden Road."

"Yeah, ya went ol' fashioned, Franky. Hey, that'd be a good nickname. Ol' fashioned Frankie."

At the top of his vision he saw a blurry, dark form scramble away, toward the playground. Another figure, this one standing, stepped over Jaim and calmly kicked the crawling form in the face. The body recoiled and began to twitch violently.

"Nice shot!"

"Oh shit, ya *offed* 'im!"

"Nah, that's just seizures."

"Either way, 's a win for the people."

"Mm-hmm," agreed a voice, this one close, harsh, guttural. Warm, moist air on Jaim's ear. The voice's owner bent over him, face inches from Jaim's, a visage mastered by a dark energy. It crowded in close, dominated by a leer that revealed crowded but perfectly white teeth, a predator inspecting its downed prey. "Crowd's gonna *feast* on these two," he said, hot breath filling Jaim's face.

The face disappeared abruptly, and darkness slammed down.

BOOK FOUR
THE HUNT

RAY

RAY
AGE: 36
SEGMENT: COMPLEX PROPERTY; PERSENSE CPX
POSITION: REDACTED
SOCIAL SCORE: NA, CPX-SPONSORED

R AY GOT THE CALL, and he couldn't have been happier. Despite the general excitement of the last few days, there had been little for him to do. Nothing localized, or emergent, that wasn't already being dealt with by the local thugs. Or by Ogunwe, the snake.

Ray didn't trust him. He'd sat through the meeting, watching Ogunwe. Watched how he held himself, sized him up as he would any adversary. And he *was* an adversary. Ray knew it the instant he saw the man. Couldn't explain it, he just knew.

After the meeting, he'd returned to his hole on Level 2. He'd begun to feel jittery, antsy. He'd started to bump *Calm,* dosing himself once every few hours, to cope with the mental strain the waiting took on him.

Then he'd received his orders, and was instantly swept clean of the restlessness that had beset him. This time, however, the call came directly. No encryption, no written message. Just a simple statement in his ear.

Ready your gear, the AI commanded, and Ray complied.

It's finally time, he thought. *No more waiting, no more hiding in the shadows.*

He bumped *Elate* to celebrate. The flood of chemicals flooded through him and he threw his head back, let it loll there, and exhaled. It had been weeks since he'd had a proper shower, and days since he'd even bothered giving himself the 'ol whore's bath with a wet rag and some soap. Despite that, the *Elate* left him feeling bright and clean inside, washing out the jitters into a sort of looseness of limb. A relaxed readiness. He set out to prep and oil his gear, then charge and pack fresh batteries for the drone.

He sat cross-legged, naked except the black briefs he favored, on the scratchy plastic turf they put in these coffins. Black and indestructible, but uncomfortable as hell. A dim light flickered overhead. He'd switched the bulb that came with the place when he'd first rented it—a blue-light bulb that simulated daylight—for one that glowed a dull red. He preferred the dark. There was something *romantic* about it. Not romantic in the sense of some mushy love story that only existed on feeds. Romantic in the sense of this moment in history having import. A lone man readying his gear in the low light of revolution. A scene that meant something, that would have an impact on the world.

VAL

VAL
AGE: 19
SEGMENT: LEGACY CITIZEN
POSITION: UNEMPLOYED; GOVERNMENT DOLE
RECIPIENT (STANDARD UNIVERSAL CREDIT)
SOCIAL SCORE: 822

SHE'D HELD HER BREATH all the way through the thick walls of Novagenica. At least that was how it felt. She held her silence even longer. She choked back the questions, the professions of love—and shame—to the point of bursting.

She feared it was too good to be true and if she opened her mouth it would end up being the first leak in a cataract of emotion she'd be incapable of holding back.

But when they entered the train and sat in those same deeply cushioned chairs, she looked over at her sister and found Kat there, smiling. Very real, very present, not a figment of her imagination.

Val put her hand out, palm up on the armrest between their seats. Kat took it. The hand was cold, clammy, which seemed to match Kat's sallow complexion. Val squeezed and layered her other hand atop Kat's, rubbing heat and life back into it. Then the doors *whooshed* shut and they were moving.

Val exhaled like she'd just surfaced from a deep-sea dive, and everything came out at once.

"Oh my god I can't believe..."

"You came..."

"It's really you..."

Val pulled Kat in tight, fingers clawing at her sister's shoulders as though fearful she'd lose her again. When she finally opened her eyes, tears fell in a wash down her face. She ignored them. She held Kat's cheeks in both hands, turned her face in inspection.

"You okay? What happened? What…" And the next question, the one she didn't want to ask but knew she had to. "What did they do to you?"

Kat, strangely, wasn't a blubbering mess, not like Val. She seemed like she'd aged more than the couple months she'd been gone.

*Not just older. She's…*she searched for the right word. *More mature, yeah, but it's more than that. Stronger.* That was it. She could see it in her sister's jaw, especially in those eyes that seemed calm and wise at the same time.

"I'm okay, Val," Kat said. "I'm just so…*so* happy to see you."

Then Val remembered their last conversation and all the guilt came bubbling up to the surface.

"You sure about that?" Val asked. Kat looked confused. "I mean, are you sure you're *really* happy to see me? After all the shitty things I said…and did."

"Uh-uh, no," Kat shook a finger at Val. "You do *not* apologize to me. I should be apologizing to you."

"Kat," Val said forcefully. "It's my fault. I'm the one responsible for you getting kidnapped. Me," Val jabbed her own thumb into her sternum. "Remember when we went and got food from Deus Ex?"

"Listen, it's not important…"

"Remember how I didn't have enough SUC and you offered to pay?" Val asked.

"Yeah, sure, so what?" Kat said, irritated.

"That was when it happened. Some shitty little marketing company traded your genetic information to Novagenica. You must've matched some sort of profile, so they tagged you and *took* you and goddammit Kat it *is* my fault. Do you see?"

When Val finished, she was breathless and yet, somehow, she felt lighter. Relieved. The guilt had pressed outward and she hadn't planned the confession, it had just come out of its own volition.

Kat looked at her with something like pity. "It's funny you say that. I had plenty of time to sit and stew about this, and it's me who should be apologizing to you. You didn't ask for any of this. It's not your fault Dad died. But you stepped up and stayed and tried to take care of me. You tried to do the right thing. And…" She put a hand on Val's arm. The coldness of her fingers made Val want to flinch but she held her arm there. Kat began to tear up again, the first evidence of cracks in her stoic composure, the words coming out hitched and staggered as she sobbed. "And holy shit, here you are. You rescued me from a goddamn *Complex* and I'm just so so so happy you're here and not a dream. I mean, Jesus, you're some sort of superwoman or some shit."

Val couldn't resist, she pulled her sister in again, and they rested their heads on one another's shoulders. Kat's hair was greasy and smelled like anti-septic, but Val inhaled deeply, holding the scent in.

Thank you God, or whatever's out there.

Val's attention wandered across the aisle, out the window. Fields transitioned to barren suburbs, alternating between walled stronghold neighborhoods and blighted, burned-out sections. Soon, they'd be over the city and then pulling into the Arc.

"You okay to walk?" Val asked, pulling back.

"Well, I've got this bum knee," Kat said, slapping a palm at a thigh.

"Ha ha, get kidnapped and it turns you into a comedian." Val's eyebrows drew downward. "The train station's on three. We'll have to go down to level two at least. Maybe to one. I'm not

exactly sure, I'll have to ask Trevor."

Kat's attention perked up, looking for all the world like a dog begging for a treat. "And who's Trevor?" she asked.

Val bent her head to the side and scratched at the crown. "Oh, he's just...I met this guy and he helped me, sorta, find you and..."

"And is he cute?" Kat poked Val in the ribs and she shied away. Kat continued to prod at her.

"Well yeah, but hey, stop that," she pushed Kat's finger away, which was poking around Val's swatting hand.

"Is he your *boyfriend*?"

"No," Val protested, but then she thought about it. *Is he?* She didn't know. There was something between them, it was undeniable, and now that she had Kat back her excuse for maintaining somewhat of a distance was gone. *Poof.* But was that a good thing? She thought maybe it was.

"Ew gross, that weird smile on your face means you're thinking about him," Kat said. "God knows what."

"No," Val said, looking away.

"Uh-huh."

This is a distraction, she thought. *She never really answered my question. She's avoiding telling me something.*

"Will you really be able to walk? I need to know, for real," Val said.

"I'm *good,*" Kat said, a little too nonchalantly.

Val eyed her dubiously. "You sure?"

"Seriously. I mean, yeah, I'm still a little weak from whatever death-potion they tried to kill me with, but I'm getting better. You should've seen me yesterday."

"Yeah?" The thought of her sister being forced to stay in that place against her will, poked and prodded like some test animal, enraged Val all over again.

"Uh-huh. Trust me, I'm g-"

The last word stuck in Kat's throat, like the word was a physical thing and she was choking on it. Her eyelids fluttered and her eyes rolled back in her head. Only half-orbs of green peeked out from beneath the twitching lids.

At first, Val thought it was a joke. "Very funny, little sister. Ha ha."

"*Ggggguuuuhhh,*" the formless sound continued to seep from her Kat's throat, a thick, almost liquid utterance, and now her hands were trembling. A bead of saliva welled up at the corner of her mouth.

Val stared, emptied of everything except terror, bright and alarming.

A rivulet of blood trickled out from the black hole of Kat's nostril, tracking its way over the curve of her upper lip, *drip-drip-dripping* into her lap, staining the jeans there.

Val took her by the shoulders, shook once.

"Kat?"

She shook again, harder. "*Kat!*" The lady across the aisle, a white woman with puffy, altered lips and a severe ponytail of blond hair looked at Val like she was intruding on her personal space.

"Fuck off," Val said, then focused on her sister. Questions raged through Val, too many to answer. *Is this a seizure? Is this what Novagenica did to her? Is she going to die?*

The last thought was too much; it put Val on emotional overload. *No no no, don't do this, Kat. Don't go away again.* She shook Kat, harder than she ever had before, and the girl's body went entirely rigid, locking up. Her eyelids popped fully open, the eyes staring straight forward, unfocused.

The trembling ceased but her body remained stiff, an automaton waiting for a command, until finally she seemed to relax and sink back into the seat.

"Ow, stop Val, that hurts."

Kat's fingers pried at Val's, pulling them back from her arms one by one.

"Kat?" Val asked, smoothing her sister's hair. "Are you back?"

"Am I back? The hell does that mean?" her sister asked. "Why're you acting so weird?"

"Are you feeling...okay?" Val asked.

"Didn't we just go over this?" Kat said.

Val reached out to her sister, but was surprised when Kat swatted the hand away. "I'm not a baby," she said. "Trust me. I'm okay."

Val looked out the window again, the miniscule pull of slowing inertia tugging at her gut. She felt like she was being pulled, slowly but surely, into an alternate reality and was questioning whether or not what she thought she'd witnessed had in fact happened.

Until she looked back to Kat.

The focus had returned to her eyes, but they were wide with doubt, uncertainty.

She'd wiped her nose with her palm, smearing the trail of blood across her cheek. She looked at the red stain on her palm like it was something foreign, alien.

"What is this?" she asked. "What is this? What happened?"

Val's heart sank.

It's not over, she thought. She was a fool to believe it could be.

She shook her head and said, "I don't know."

RIKU "CHIEF" OGUNWE

RIKU "CHIEF" OGUNWE
AGE: 40
SEGMENT: COMPLEX PROPERTY; ARCSEC CPX
POSITION: ARCSEC CHIEF CONTRACTED BY LEGACY
ADMIN: MIDWEST ARCOLITH, LEVELS 1-3
SOCIAL SCORE: NA, CPX-SPONSORED

THE PLAN WAS SIMPLE. At least from Ogunwe's perspective. Go in, talk to the lead engineer. Convince him or her that he had authorization to begin distribution of the medicine, *without* the contract a Complex like Novagenica typically required. That last part was the tricky bit. Not only from a confidence perspective, since Complexes generally didn't give shit away for free, but also because it relied on Moon.

A shiver ran through him. He detested the idea of being at the mercy of that...thug wasn't the right word, but below the smooth-talking, shiny façade, Ogunwe could see the brutalist at the heart of the man.

The leader of Talksmall had assured Ogunwe he had somehow acquired high level access to Novagenica, which he'd then use to intercept the communications of the lead engineer and give all the distribution point orders as though they were coming directly from the operation's man in charge—Tori Fanning.

He didn't like it, but what could he do? The medicine was the only thing that mattered right now. The quicker he assumed control, the more lives he'd save. And then finally he'd be free to hunt for Olivia in earnest.

And I won't give up until I find you.

He lurked in the dark of the alleyway, one hand rested against the cool metal of the door, the other on his pharma gun. He wasn't sure what to expect and he certainly didn't like going in blind. While he didn't remember seeing any of Novagenica's troops the first time he'd been in the building, over a day had passed since he'd last set foot there. Who knew what had happened in the interim?

He pushed and the door opened. No resistance, no clicking of locks, nothing. It simply swung open soundlessly. *Moon came through, somehow.*

The space was as he remembered it. Open, with concrete floors, pallets of wrapped boxes packing the main level. The bustle of white lab-coated workers and bright lights illuminating the space from high above, giving the room a pale glare. Ogunwe walked to the center of the area to where the small army of NARKs was parked in a line. Black, boxy. Silent yet somehow menacing.

"Listen up!" Ogunwe shouted, his voice booming throughout the open space. He thrust one arm high in the air, fingers spread wide like the mitt of a glove. "I need to talk to whoever's in charge!"

The scurrying whitecoats stopped, looked up from their tasks. All eyes were on him. A woman with short auburn hair that curved in a bell to her neck approached from a corner where she'd been directing two male whitecoats. Her face was marked by pinched, diminutive features set in a stern expression. Her skin was waxy, splotched, belying some internal deficiency.

"Well that's a hell of a way to get attention," she said. "My people are already spooked enough having to work in these conditions."

"Sorry...I'm sorry, what was your name?"

"Agatha," she said.

Ogunwe snapped an image of the woman and sent it to Moon, tagging the image with her name.

As fast as you can, Moon, as fast as you can.

"Agatha, you said?"

"Doctor Agatha," she corrected. "Agatha is my last name."

"My apologies, doctor. It's nice to meet you. I'm Chief Ogunwe, Arc-Sec CPX."

"Well?" She stood there staring at him.

"Sorry to interrupt, but we're out of time. I have orders to get these machines rolling." He gestured to the inert NARKs.

"Orders from who?" she demanded.

"From my boss, Commander Briggs, ArcSec."

She put her hands on her hips. "First of all, wrong Complex. I don't belong to ArcSec." She gestured to the NARKS. "And neither do these. I'm not authorizing the movement of the product."

C'mon Moon, he thought urgently, *give me the sign*. But nothing came through. He'd have to stall.

Ogunwe leaned in, spoke softly in Doctor Agatha's ear. "You may not have noticed, I know you've been very busy here, but outside these walls," he leveled an index finger at the front of the building, "the world is literally breaking." He stepped back, watched her. "This isn't a sub issue. Not a *your* CPX or *my* CPX anymore. It's life or death," he said, touching two fingers to his sternum, "for all of us."

He watched closely, looking for a sign of the effect his words had on her. He could see the façade begin to crumble, a tremor at the edge of her mouth, hands clasping one another, fidgeting.

Still nothing from Moon.

"We'll be protected here," she offered weakly.

"You think so?" he asked, a little incredulous.

"The building has defensive measures," she countered.

"And what if all of Level 1 burns?" Ogunwe asked. "What if that mob outside your doors sets fire to everything they see? Do you really believe this building will still be standing after the entire Level burns out?"

Ogunwe noted the slightest hint of tension in the eyes, a contraction of skin at the outer corners. One hand moved from her hip to scratch at her back. He almost had her.

"Listen," he shifted into a warm, reasonable tone. "I understand. You're under a mountain of pressure. From what I've seen you've done an admirable job. No one'll blame you for delivering this medicine early."

And there was his gambit. If Moon didn't come through soon he'd have to resort to force to get what he wanted. Not ideal, and he wasn't sure it'd work. More than anything, he needed the information in her head. Needed the knowledge on how to get the NARKs moving and, more importantly, how to distribute the medicine without a contract, Novagenica's Terms of Service, tied to it.

The doctor scoffed. "Do you really believe that? No one will blame me?"

She had a point. They both knew it was a lie. Of course the Complex would blame her. They'd string her up and use her as an example of what happened when you didn't follow orders.

Come on Moon!

"I—" he stammered. His system pinged.

Got it. Sorry for the delay.

Ogunwe had to stop himself from breathing a sigh of relief. *Thank God.* He sent the confirmation back to Moon immediately.

Telling her now. Expect a message soon.

"I do," he said smoothly, recovering his composure. "But we're just wasting time. Just do me a favor and confirm with your superior, whoever's in charge."

She looked at him, head back, eyes appraising. Undoubtedly she was busy composing a message to her boss, the man he'd met with the day before, Tori Fanning. But it wouldn't be Tori responding to her, it'd be Moon.

Ogunwe watched as shock bloomed on her face, widening the naturally contracted features. What little color she had seemed to bleed away, making the freckles stand out. Her eyes, staring in the distance, stayed that way for several moments. Reading and re-reading the message. Eventually her eyes adjusted as they found Ogunwe's. They were still wide with shock.

"You win," she said.

Ogunwe maintained a grave countenance. "Good. Now, about your other warehouses. We need to get them moving as well. Is ten minutes too soon?"

"No, no," she said, shaking her head. Her short red hair swung about her jaw line. "It just takes a simple word, assuming your people have cordoned the distribution points off already."

"My people are ready," Ogunwe confirmed. He realized that one hand still rested on his pharmagun. He removed it, clasped both hands in front of him.

"Alright," she said, her shellshock beginning to wear off, resolve slowly taking hold again. "Alright. In exactly ten minutes from...now. Have your escorts ready at each location." She shook her head again once, a confused little smile showing on her face, whispered, "Incredible."

Ogunwe had begun to turn away but stopped. "What?" he asked, curious.

"Nothing...well, not nothing really. It's just..." she looked him square in the eyes. "I've belonged to Novagenica for twenty-three years. In all that time, you know how many times I've given away product without a sub mandate?"

"No idea," Ogunwe lied. He had a pretty good idea.

She formed a circle with the thumb and index finger on her right hand, indicating a zero. "Until now." She shook her head again. "We're just gonna give it away. Must really be the end of it all out there," she mused.

"I wasn't blowing smoke. It really *is* life or death."

Ogunwe turned to exit. *Excellent work, Moon,* he thought, feeling a spreading sense of hope in his chest and a strength, a purpose in his step. He didn't want to jinx it, but it looked something had finally gone his way.

The AWOLs stood clear and rigid, ten feet tall, forming an empty column running straight through the middle of the packed street. On the outside, pressing up against the walls, milling about like cattle, the stupefied crowd Ogunwe's men had gassed. 'Treated' was the proper word, but Ogunwe preferred blunt accuracy over political correctness.

Inside the AWOLs, an obscene abundance of space. Empty volume barely filled by the NARKs and Ogunwe's men. Outside the clear walls the people mostly sat in the streets, heads lolling, pacified by *yoop,* the crowd-controlling chemical. Some stood, turning in circles as there was no room to pace. While the chemical euphoric and the AWOLs managed to keep the crowds at bay, they did nothing to ameliorate the stench emanating from the refugees. Piss and shit and a sourness that smelled vaguely of sickness, vomit. Ogunwe wrinkled his nose under the mask. Damn thing wasn't helping.

"Vander, status?" Ogunwe asked. His black-booted toes hung over the lip of the sidewalk outside the building. Fifty yards in the dim distance, at the head of the column of soldiers, Ogunwe saw Vander give the thumbs up.

"Itchin' to get started, boss." The rest was implied. *Ready to get it over with it and get the hell out. Me too, Vander. Not sure I'll be that lucky though.*

Ogunwe looked to Doctor Agatha, who stood by his side. She still wore her white lab coat but had buttoned it, pulling the collar all the way up to beneath her jaw. Her blue-gray eyes flitted nervously from one side to the other.

"Ready?" Ogunwe asked. She nodded, one curt jerk of the head. Her skin looked almost green in the dead, evil lighting that rained down from the Arc's superstructure. "Go," Ogunwe ordered Vander.

"Copy," Vander replied, the words filtering directly to Ogunwe's ear canal.

Agatha controlled the NARKs, starting them forward. They idled smoothly on their wheels down the center of the open space. Ogunwe's men marched along both sides of the machines, splitting into two columns. They didn't have weapons out, not yet. No need for provocation.

Ogunwe walked at the rear, his long, lean frame making him feel like a pharaoh marching through the masses of poor. The matte black machines glided steadily along the cracked asphalt, boxy lines connoting aggression. The soldiers alongside them, hard eyes watching. Ogunwe in his vantablack ops gear at the rear, the clothing blending into his skin, making the whole of him little more than a shadow, save the white mask. Agatha, cold, pale, imperious, at his side. All in all, a display of might to contrast with the utter wretchedness pressing in on the clear walls.

Ogunwe held his head high, took in the details of the crowd as they made their way steadily to the distribution point. A man in a tattered orange plastic jacket watched the procession with manic delight, a few days of stubble on his dirt-smeared face, a shock of yellow hair that matched yellowed teeth as he pressed up against the AWOL. Snot dribbled out his nose and down over his lip, dripping steadily into his mouth and down his beard. On the opposite side, two girls were enmeshed, little more than teenagers, kissing and petting one another, oblivious to the writhing mass around them, black-caked fingers tangled in one another's rat's nest hair.

The procession continued and the stupor of the crowd held. But the gas would only work for so long.

It took them five interminable minutes to reach the distribution point. Each passing second ratcheted up the tension within Ogunwe, like a cord wound tighter and tighter. He felt it not just in himself, but in the mass of teeming people. As they neared the open area in the intersection of 12th and Baltimore, a restlessness bled into the crowd. Ogunwe watched with a critical eye as the bystanders pushed in, more forcefully now, red and raw palms pressed up against the clear surface. There was a lucidity in their eyes, an alertness not normally present in those treated with *yoop*.

A woman pounded with a fist on the AWOL. "*Help us!*"

She continued to pound, beating out a rhythm with her fist. "*Help us! Help us!*" Another woman sat cross-legged beside her; a girl, maybe five years old, lay with her head in the woman's lap. Eyes closed, skin colorless. Ogunwe couldn't tell if she was breathing. A man stood now, also began to pound away at the AWOL. His dirty palm slapped at the clear surface, leaving a muddy palmprint. His face was half-covered in a tangled, patchy beard and he sneered at Ogunwe, looking directly at him, eyes blazing like fiery coals, as he slapped at the AWOL. He said nothing, which Ogunwe found more disturbing than the woman pleading for help.

Ogunwe felt the urgency infect him then. He looked to Agatha. *Can't those NARKs move any faster?* he thought. His instincts were telling him time was a luxury they no longer had.

Ogunwe opened the command comms channel. "Let's get this show on the road, Vander," he said.

"Yeah, Chief. I'm with ya."

In the middle of the intersection, the troops spread around the circumference, forming a phalanx with the NARKs positioned at the center.

Ogunwe joined Doctor Agatha in the middle, next to the machines. Her head was bent to some unseen task on her system. A soft hum and warmth emanated from the NARKs. Instinctively, Ogunwe put his hands out as though warming them over a campfire.

"Any way you can boost the signal here, Doctor?"

"What do you mean?" she asked.

"Never mind," he said, "just hurry please."

She held out a hand, palm out, pink and slender fingers splayed. Ogunwe couldn't tell if she was indicating for him to simply wait or whether it would take five minutes. He hoped it was the former. By the look of the crowd they didn't have five minutes.

"Vander." Ogunwe motioned for his second in command to join him by the machines.

"What's up boss?" Vander said, approaching in his bow-legged cowboy saunter.

"You seeing what I am?" Ogunwe nodded to the people crowding around the outside of the clear box.

"Think so." Vander coughed behind the mask, pulled it up and turned his head to the left, spit on the ground. *Should treat our crew first*, Ogunwe thought, watching the man hack up another wad of phlegm.

"What percent you figure are immune to the yoop?" Ogunwe asked.

"Permanent? Pssh, not much. I'd say it's other drugs interacting, blocking our stuff."

"Grab Green," Ogunwe ordered.

Vander let the mask fall back into place over his stubbled chin, said something unintelligible behind it. Green trotted over. She stood at attention, back pathologically straight, head held high and proud.

"Sergeant," Ogunwe addressed her.

"Sir," Green said.

"Here's the scenario. Forty percent of a riot population has yoop immunity conferred by some unknown substance. What's the next best option?"

"Put 'em to sleep, sir," she said without hesitation.

"Not an option. We need them awake."

"What for, sir?"

"To save their own damn lives, sergeant."

Green looked around, her expression darkening beneath the mask.

"I have a solution, sir, but it's not without risk," she offered.

"Every day's a risk, let's have it."

"Pressure drop."

She said it like it was supposed to mean something to him. Something *ominous*. Ogunwe gestured with his hands, in a circular motion, for her to go on.

"Involuntary blood pressure drop. Invokes a more peaceful, relaxed state brought on by lower heart rate. Not like the euphoric, it actually makes the physiology do its work. Puts the horse before the cart, so to speak. Decent success rate, in the seventieth percentile."

"Risks?"

She hesitated. "It can be a little *too* successful. Anyone with existing heart issues can experience loss of consciousness or even cardiac arrest."

"Rate?"

"Can't say exactly, sir, but there are higher risk factors with this population. These people were already at the lower end of the health spectrum, now we're adding drug on top of drug..."

"And," Vander interjected, "the crowd sees us gassin—er, treatin' em with gas—and then people start droppin' like flies right after? People are bound to overreact."

Green nodded in agreement. "And that'll probably happen *before* we can get them the medicine they *really* need."

Ogunwe mulled it over. "Gonna have to risk it. Looks like it could be Kopa to me, and if it is then we can't let it take root. Our window is tiny, like pea sized. Reichmann, you there?"

"Yes, lover," Reichmann's voice hissed softly with a feminine sibilance in his ear. Vander snickered beneath his mask. All of Ogunwe's command shared the same channel.

"Task the video and GPS feeds for locations of known Kopa dealers up to a 3-block radius of all the distribution points. Find any and you send the drones up and drop 'em."

"Like drop 'em drop 'em? For good?" Reichmann asked.

"Drop 'em," Ogunwe confirmed. "For good."

He turned to Green. They'd have to get the *Pressure Drop* up and over the AWOLs, and they needed to know what people out in the crowd were using. If it *was* Kopa, as he suspected, this could quickly become less about saving the lives of those outside the AWOLs and more about protecting the lives of his people inside them.

"Run extensions up the walls. Ready the *Pressure Drop*, dose and duration at your discretion, but wait for my command. I want to make sure our people get treated before we take the risk. Also, run sniffers up to taste the air, figure out what drugs they're eating out there."

"Copy," Green said, then trotted away to round up her engineers.

"Vander?" They locked eyes. "Get those damn slots ready ASAP. We need to be ready to distribute to the crowd soon as the doctor gives us the happy thumbs up, but I'm going to try to get her to dose us first. Medicine won't be any good if everyone this side of the AWOLs is dead."

"Right on boss. And thanks."

He felt a tap on his shoulder. Ogunwe turned to see the pale mask of Agatha, framed by that bright red hair. Concern pulled everything on her face inward, like the middle of her face was a screw that had been tightened as far as it could go.

"We have a problem," she said.

He stifled a groan and said, "Hit me."

"The virus. We sequenced it yesterday, when we first showed up. I just ran the sequence again, from the same sample, just to, you know, double-check."

She was wringing her hands now and Ogunwe reached out, clasped them both in one of his huge hands. "Slow down and tell me. We'll figure it out together," he said.

She breathed heavily, deeply. "Okay. The sequence...has changed." When the words had no impact on Ogunwe, she continued. "It was from the same sample but, somehow, it's different now." She looked to him, questing for some semblance of understanding.

"Spell it out for me, doc," Ogunwe said. "I can see the water but might need a nudge or two to drink."

"Okay, well, it's normal for viruses to evolve, to mutate over time. But this one evolved in a matter of *one day*."

And then he understood.

"What happens if you print the medicine, but the virus evolves again?" Ogunwe asked.

"It'll probably be useless. A virus, if it's deadly, that can evolve this quickly," she shook her head, eyes wide and looking up, "it could wipe out this entire population. All of us."

"As in, 'everyone in the Arc', all of us?" Ogunwe asked.

The look she gave him was grim. "Think bigger."

"Fuck," Ogunwe said.

"Yes," she agreed.

"We have to tell someone," Ogunwe said.

"Who?" she asked. It was a simple question, and it felt like he should've been able to answer it. But the way she'd said it, combined with the cynical tilt of her eyes, made him realize it was rhetorical. And he realized she was right.

Who, other than Novagenica, could tackle a problem like this? There was no federal government to speak of, all the local governments were in decay and disarray, and the most powerful and advanced entities—the Complexes of the world—had no incentive to address the issue. It was better for them and their subscribers to simply retreat behind their tall walls, institute a lockdown and recede from the world. Blast to pieces anyone that came knocking on the door.

Everyone else caught outside, their lives were all forfeit. His own. His men and women. Everyone he knew and loved. Including Olivia and Stef.

Wherever they are. A black pit widened inside of him; he could feel himself teetering at the brink of it, weirdly close to collapsing inside of himself, down a recursive hole where he'd continue to fall into smaller versions of himself, tumbling forever and ever...

"Hey." The doctor waved a hand in front of his face, snapped her fingers twice, a crisp, dry *click click.* "Now is *not* the time to check out."

Ogunwe blinked twice.

"You back? Good. We'll do what we can, and for now it has to be good enough." The doctor turned her back on Ogunwe, bent over one of the NARKs, returning to work.

They were surrounded on all sides by a clear cage, there in the middle of the intersection. Ogunwe pivoted on a heel, taking it all in. Green's team of engineers had slung poles over the top of the AWOLs and was assessing the chemical composition of the air.

Ogunwe was first taken by horror, then by rage.

Every single one of them was as good as dead. Ogunwe fumed at the injustice of it. He wanted to hit something, *someone,* whoever was responsible. And that, he knew, were the people at Novagenica and PerSense.

And PerSense owns ArcSec, which means we're in league with the whole stinking mess.

He should've seen it before, his own role in the coming atrocity. First the PASS mandate handed down by Briggs. Then the Exodus. The raising of the floodgates. The mysterious illness.

And lo and behold, let's have a meet and greet with Novagenica, who says they can solve the problem. But don't forget the fee! There's a price to save your life. There's always a price.

It all pointed to a very specific end goal. The consolidation of power and the ultimate demise of Legacy governments.

He wanted to stomp his boots on their bodies until they cried for mercy, to kick their heads in until the brains ran out their ears. Briggs. Tori Fanning from Novagenica. The mercenary from PerSense, that Ray character. *Even that fucking prick of an Admin, Jaim-whatsisname.*

But mostly Briggs. Ogunwe's fury toward Briggs burned white hot. For his own part in dismantling everything in life Ogunwe loved and held dear. His work. His purpose. The lives of his colleagues. The lives of his loved ones.

But none of that can happen if you get trapped in here.

The thought was like smelling salts for his anger; it snuffed out the heat of the fire and brought the cool clarity of logic back to him. In order to enact whatever revenge he could, he first had to have a plan. To get out of this stinking, raging hellhole. That was step one.

He made the call. Moon picked up immediately.

"You've changed your mind?" Moon asked. Not gloating, simply a factual evenness.

"Yeah, but not in the way you think," Ogunwe said.

"Talk to me."

And Ogunwe did. He told him about the mutation, about how it had to have been engineered, and tasked him with finding a cure. When he was done, he had only one demand, one he relayed with a firmness that couldn't be brooked.

"You tell them, Moon. The media, the world, whoever will listen. Tell them who did this. Tell them it was Novagenica, PerSense, and ArcSec. Tell them it was the power and hubris of Complexes, that they're destroying lives for their own gain. Promise me you'll tell them."

"I will," Moon said gravely. "Good luck. The world needs more people like you. Do your best to survive."

Ogunwe grunted. "Survival is the furthest thing from my mind," he said, ending the call. And it was true. He was beyond survival now, beyond the confused, muddled hash of what it meant to protect and save lives. There was no gray area where he was headed. The way he saw it, the best way to protect lives, now, was to take them.

For you, Olivia. And for all the other victims out there.

TREVOR

TREVOR
AGE: 18
SEGMENT: LEGACY CITIZEN
POSITION: UNKNOWN
SOCIAL SCORE: 821; FLAGGED FOR VERACITY

TREVOR EMERGED FROM THE DARK of the stairwell, onto the tight, byzantine passages that marked the southwest residential district on 3 known as Watership.

He found people going about their business, a typical workday.

No crowding. No screaming. No crying, pleading, breaking down in the middle of the street. No pissing on the walls, no shitting in the darkest corners and wiping up with a sleeve. No druggies wailing, clawing at their own chests as though infested by worms just below the skin, and if they could just, just make an opening the worms would come pouring out. No sprays of vomit, flung against the walls and windows like toddler art. No bloodless cheeks, no hollow eyes, no corpses lying stiff in the street, and none of it capped with the ever-present fine mist of effluence that *drip-drip-dripped* from above.

Instead, what he witnessed was mundane, banal, and it sent his mind reeling.

A breeze, light and temperate, ruffling his hair. The smell of grills being started up. Not the sugary burning of bars being infused with scent molecules and pressed together. This was the smell of real food, meat, sizzling overheat.

Orderly movement of people, not dressed in rags and plastic ponchos, but casual, nice, pressed, put together people, going to and fro, in and out of glass-walled offices, homes, cafés.

A man, jabbering on his system, arguing about where to meet for lunch and what horrible things he was going to do to his lover after (and how much she'd enjoy it).

A sweeping machine, boxy rectangular form painted dark green, flashing lights, crawling down its dedicated lane, moving inexorably to the next trash bin and then the one after that.

Moms jogging, dressed singularly in bras and tight leggings that hugged their sculpted, paid-for asses, towing floating babies in dirigibles, who giggled with glee as they swayed in side-to-side arcs, fleshy balloons in the air.

Flowers in pots, hanging from eaves. Trees spaced just so along the pedestrian way. Birds—*birds* – flitting from branch to branch, tree to tree, a rustle of wings, the occasional chirp.

It was like an inverted assault on his sense. Knowing, knowing for certain, that the world was cruel, hard, cold to the touch, he'd entered into a fairytale.

An alternate reality, he thought. It was the only way he could process what he saw without totally losing his shit.

He made his way through Sunlight district, where he and Val had first met Ian in his posh, sun-filled condo. Today the light that filtered through the break in the buildings was gray, muted. But the smells remained the same—cooked meat, ethnic spices and an undercurrent of baked goods and pumpkin that lent the area a cozy warmth.

Then from Sunlight to the grid of thoroughfares, populated with offices, cafes and restaurants, that comprised the central region of 3. Past a transfer gate and its park, replete with children running, screaming, playing. He noted the clear wall at the mouth of the stairs, and how no one was queued up to head down the stairs.

He took the pedestrian lane west, then north, walking mechanically through the mixed-use low-rise developments toward the westside train platform, feeling like an interloper among wolves. Those wolves occasionally smiled at him, but mostly they scowled. They could sniff out someone who didn't belong. And Trevor, with his long hair, tattered sneakers and fake leather jacket, *reeked* of outsider. He knew it, and they knew it.

Normally, he wouldn't have given two shits. He played the game of life loose and fast. But from the moment earlier that morning when he'd been turned away at the lightbox on 1, watching Val enter the elevator alone, lost amid a hundred strangers as the teeth of the vault-like doors clamped shut, a low-level anxiety had steadily gnawed at him, eating away at his calm.

Alright dude, get it together, he thought, hands fluttering, smoothing at the wrinkles on his jacket, pulling down on the hem, snapping it tight. *Mind-fucked is not the ideal way to greet the love of your life and her newly freed-from-the-jaws-of-death kid sister.*

He arrived at the platform, stepping up onto its broad, elevated expanse in time to see the long, sleek train come slashing through the uprights of the Arc outer wall. The train, a white blur, slowed quickly and resolved as it stopped, then lowered onto the tracks like a cat settling itself for a nap. The doors opened and people flowed out.

Like a miracle, unexpected yet hoped for, Val was there, and Trevor permitted himself to breathe again.

She emerged from the boxy black opening, a brightness to her that screamed strength and something else, something better inside of himself. It was corny, and trite, and overdone, yet the truth remained that she made him better. Made him *want* to be better. And no matter how corny, that in itself was important.

Behind Val exited a younger girl that looked similar, almost familiar, dressed in a bright yellow shirt, jeans and ballcap. It was Kat, Trevor was certain.

He grinned, thrust both arms over his head and waved them back and forth. Val spotted him. Recognition in her eyes, but no smiles on the face. Something else. *Relief, but not exactly relieved.* If that made sense.

Something isn't right. Even from fifty yards away he could feel it, kick-starting the worry-engine in him all over again.

She rushed over, pulling Kat behind her, and threw herself on him. Not a reunion borne of joy, but desperation. Her arms clasped his middle, vice-like, her head to his chest. He hugged her fiercely, eyes squeezed shut, head buried in her hair, which smelled like it needed a wash, but it smelled of her and that was what mattered, that she was *here.*

A cough from over Val's shoulder.

"Ahem. Hello?" Trevor opened his eyes to a finger tapping Val's shoulder. "Hate to break up the reunion, but maybe we, uh..."

Val pulled back. "Yeah, definitely. Let's move."

Something weird was going on. Val's attention was scattered, incoherent.

"But an introduction *would* be nice," Kat said, arms folded over her chest, shoulders pulled up and back. While she tried to project confidence, humor, she looked two missed meals away from death. Her skin was sallow, translucent to the point he could see the blood vessels snaking their way beneath. Huge dark rings encircled her eyes and her flesh was pulled tight against the bones, most notably on her arms where they were all jutting angles and delicate, bird-like wrists. She wore a baggy yellow shirt that seemed to accentuate this fact, hung about her body like a sack with holes cut in it. The shirt had a dark, brown stain where it billowed over her hips.

"This is Trevor," Val put a hand behind Trevor's back. "He's my, uh..."

"Boyfriend," Trevor blurted, offering a little half-wave. "The word she's searching for is boyfriend." Val said nothing, so Trevor steamrolled on, filling what would've been awkward silence with noise. "It's your typical modern romance. A relationship rooted in fear, violence, all that."

Kat cracked a grin. There was an exhaustion to her, made more evident by the effort it took for her to smile. "I like this guy, sister," she said to Val, jabbing a thumb in Trevor's direction.

"Nice to meet you," Trevor said, "and don't take this as me trying to rush the intro, but we gotta scoot. Like now."

"Hey, I'm just the broken teenage lab rat in this scenario," Kat said, and Trevor grunted a laugh. "Just tell me what to do."

She puts on a tough face, have to admire that, he thought. *But she's hurting. Maybe that's why Val's so quiet.*

"We'll go someplace to lay low for a bit. Sound okay?" Trevor asked, stepping down off the platform and into the pedestrian traffic on the street.

"You lead, I follow." She panted a quick breath, taking the steps slowly, carefully, hands at the ready like she was walking a tightrope.

Val draped an arm across Kat's back. "Where you thinking?" she asked Trevor, positioning herself opposite so Kat was between the two of them as they headed south, back the way he'd come.

Weird. What is up with her? Was it what I said? The boyfriend thing? But he didn't mention any of these things, instead allowed the talk of necessities take precedence and wash over, mask the deeper emotional currents between them.

"Figured we'd at least stop by my coffin. Get some supplies, maybe rest for a couple hours, then move on..."

"I don't know," she said, looking away. "We need to go someplace quiet, with enough space to rest." Her eyes flickered toward Kat, then away.

A flash of clarity struck him, allowed him to see outside himself, to realize everything wasn't always about him. It was Kat. *That's why she's acting weird. Something's wrong with her sister.* He glanced at Kat, who quietly walked between them, her eyes on her feet, looking shrunken, how a sick child will look younger than their actual years.

"You seen the situation on 1?" Trevor asked.

"Not since we were separated," Val replied.

"Hard as it is to believe, it's gotten worse. But..." he paused. *No better way to say this other than just get it over with.* "We need to head back down."

Val's eyes narrowed. "Well that's dumb."

"While I'm fond of being wrong when we're together," Trevor said with a fair dose of sarcasm, "this time it's actually the sensible thing to do. I mean sure, it's beautiful and nice and shiny up here, but we're sitting ducks. You know for a fact NG is," he looked at Kat, softened his words, "they're gonna wonder where Kat went, and when they find out they probably won't be too happy. Now that Ian's down on 1 we've lost our only friend on 3. On 1 we have help, we can get lost easier. I dunno, maybe hide out at Talksmall for a while?"

As he talked, he noticed Kat rubbing her arms, the goosepimples making the tiny blond hairs stand on edge. He looked at his own jacket and instantly felt like an asshole.

"Here," he said, shrugged out of the jacket and draped it over the girl's bony shoulders. "Wear this. It has some minor environmental controls. I'll turn the heat on."

"Thanks," she smiled gratefully, pushing her arms through the sleeves, tugged the wide, bronze zipper up. "*Oh,*" her eyelids fluttered, "oh my god that feels *so* much better." The jacket was about four sizes too big, but she rolled the ends of the sleeves back so her hands poked through the cuffs. Val regarded him with some beatific look of appreciation and admiration. He waved it away.

"C'mon. I'd have to be some sort of Grade-A premium asshole not to give it to her." The girls said nothing, simply smiled at him as though they were in on a joke. "Okay, whatever," he said, embarrassed for a reason he couldn't quite say.

"So," Val said, "where…" But her words were cut short.

A message flashed across Trevor's system.

GLAD TO SEE YOU'RE OUT. SITUATION ON 1 AND 2 DETERIORATING.

Trevor looked to Val. "You getting this too?"

"Ian," she said, nodding.

Kat looked between the both of them. "Who's Ian? Getting what?" she asked, but the next message came through and both Trevor and Val were focused entirely on the words.

REFUGEES GETTING SICK FROM A PREVIOUSLY UNCATALOGUED VIRUS. SOME ARE DYING FROM IT. ON THE VERGE OF TOTAL BREAKDOWN. TALKSMALL JUST RECEIVED INFORMATION FROM AN INSIDE SOURCE WHO SUSPECTS THE VIRUS ORIGINATED FROM NOVAGENICA.

Val's hand flew to the open *O* of her mouth, a gasp of air sucked inward.

So that's what Novagenica was doing. Fuckers were testing a virus on her…and the fact that she's alive means…

WE HAVE MEDICINE, HIJACKED FROM NOVAGENICA BY THE INFORMANT. WE'RE DISTRIBUTING RIGHT NOW. THIS MAY WORK FOR A WHILE, BUT THERE'S ANOTHER PROBLEM.

WHAT? Trevor sent. *STOP DICKING AROUND AND GIVE IT TO US.*

THE VIRUS IS MUTATING. QUICKLY. WE CAN DISTRIBUTE MEDICINE, BUT IF THE VIRUS MUTATES

THE MEDICINE WON'T HELP, Val sent, finishing the thought.

CORRECT. AT THE VERY LEAST, HUNDREDS OF THOUSANDS WILL DIE LOCALLY. AT WORST, A GLOBAL PANDEMIC ENSUES, KILLING MILLIONS. PERHAPS A MAJOR DIE OFF.

NO, NO WAY, Trevor sent. YOU'RE TELLING ME THIS COULD BE THE END OF HUMANITY??? I DON'T FUCKING BELIEVE IT.

PREDICTIVE MODELS DON'T LIE. BUT THE WAY IT MUTATES SEEMS TO FOLLOW A SCHEDULE, WHICH MEANS IT MIGHT BE PROGRAMMED THAT WAY. AND IF IT'S PROGRAMMED THEN WE CAN DECONSTRUCT ITS CODE, LINK THE MEDICINE TO THE MUTATING CODE.

Val looked to Kat. She seemed to sense where this was going. Kat turned both palms up, asking a wordless *What?*

VAL, YOUR SISTER MAY HAVE THE KEY TO THE MUTATIONS OF THE VIRUS. THIS IS THE MOST LIKELY EXPLANATION OF HER ABDUCTION. I NEED YOU TO BRING HER TO ME. QUICKLY. BEFORE WE RUN OUT OF TIME.

Trevor stopped, stood in the middle of the ped lane, unsure of where to go or what to do. The revelation that everyone and everything he cared about was on the verge of extinction dawned slowly, then shone with the full force of a noonday sun. The image of Nicolette swam up from the depths in his mind, the picture he'd captured from the CCTV system. Her brown hair curling at the tips, the sad look in her eyes an almost prescient recognition of the inevitable disaster to come.

"Fuck," he said. "Fuck fuck fuck fuck *FUCK!*"

The other walkers in the ped lane gave him a wide berth, looked at him like he was a curiosity. *What is this outrageous behavior?* he could hear them thinking. He flipped them off with both hands.

"Fuck you, you fuck. Go die in a dark hole," he snarled, voice laden with derision. A man, fleshy white face indignant, mouth open about to say something, index finger pointing. "You too, dickpiece. Fuck right the fuck off before I break that pasty face of yours." The dark look Trevor shot his way was enough to scare the man into scurrying on, casting the occasional glance backward.

"Come on," Val said. "We have to move." She grabbed Trevor and spun him back around, then took Kat by the arm and began to pull.

"Nope," Kat said, ripped her arm out of her sister's grasp. "We're not doing this." She stood, feet planted wide, facing down Val. She put her face up close to her sister's, spoke with authority. "Tell me what's going on. I deserve it."

In that moment, she looked more a mother than a child. Demanding, authoritative, calm. It was enough to make Trevor realize how petulant his own reaction had been. *Here's this wisp of a teenager, acting more like an adult than you are. If she's not throwing tantrums, you sure as shit shouldn't be. So snap out of it and do something useful.*

He stood beside Kat, facing Val. "She's right," Trevor said.

He watched tremors of indecision ripple through Val's face until finally everything stilled, a placid lake at dawn, and one tear trickled out from a red-rimmed eye.

"I'm sorry," she said, and the look of sheer misery on her face communicated everything Kat needed to know.

Kat's shoulders slumped, rolled forward. Resignation suffused her, a sponge soaked in the sadness and duty of her fate.

"I get it," Kat said, attempting to look bright but failing. The corners of her mouth twitched, trying for a smile but unable to hold it. She blinked, smoothed the corners of her eyes with the tips of her fingers then, noticing what she was doing, folded them over her chest. She looked to Trevor, and when she spoke her voice was high, tight, full of the strain of control. "You'd figure I'd be used to it by now. More bad news, right?"

Trevor's heart broke at the sight and sound of her, at the little girl who had been torn from her home, taken as a test-subject and prisoner, gotten sick and seen who knew how many horrors, then broken free only to be dragged into the midst of a city-wide conflict and that, no, as it turns out you're not free, not exactly, because now it's not just about *your* life, it's *everyone's*.

You're the key. To the survival of humanity. How, exactly, did you tell someone this? Trevor wondered. *The only way you can, dipshit. Like she said, she deserves the truth.*

He nodded once, curtly, and told her what they knew.

After all, she wasn't wrong. As bad news went, it was of the highest order.

RAY

RAY
AGE: 36
SEGMENT: COMPLEX PROPERTY; PERSENSE CPX
POSITION: REDACTED
SOCIAL SCORE: NA, CPX-SPONSORED

T *HE GIRL KNOWS.*

"Knows what?" Ray asked. The voice in his ear had become so commonplace, so familiar that he no longer startled at what he'd once considered its intrusions on his thoughts.

He sat on bench at the transfer park on 3, enjoying the mixed urban-idyll panorama before him. People socializing, children running, playing, the occasional shaft of sunlight streaming in. A breeze that held the scent of drying leaves.

She knows our plans, the AI said.

Ray scratched at the side of his nose. "Won't stop the inevitable," he said with something like self-satisfaction. "But she's further on down the list."

Silence, and along with it the sense of displeasure. One of Ray's eyes twitched.

"Unless, of course, you want me to bump her up to the top?" Ray suggested. "Take her out first?"

No. Continue your investigations. I will track her and her companions.

"I can have someone else, y'know, take her out. It'd be easy."

The silence that ensued was a version Ray interpreted to mean the AI was considering taking this particular tack.

Yes. But only an attempt. Make it look convincing but keep her alive.

"Until when?" Ray asked. "When can I kill her?" he asked, and couldn't keep the need out of his voice. Another time his professionalism would've bristled at the suggestion he could go so far as to take a job personally.

But it was all personal now.

Not until I tell you, the AI commanded.

Ray felt the slightest let-down and bumped *Calm* to erase the disappointment. The chemicals eased into him, and he seemed to notice the impact less and less.

It's okay, he told himself. *PerSense—no, the AI—the AI knows what's best.*

The thought was confusing to him, so he bumped another dose and finally felt the rush.

Truth be told, he was having difficulty discerning the difference between the two—AI and PerSense—but the *Calm* flooded over him and told him the distinction wasn't worth worrying over.

You're finally on the path, that's what matters. Prove your worth and they—PerSense, the AI, whoever—they'll see it. They have to.

RIKU "CHIEF" OGUNWE

RIKU "CHIEF" OGUNWE
AGE: 40
SEGMENT: COMPLEX PROPERTY; ARCSEC CPX
POSITION: ARCSEC CHIEF CONTRACTED BY LEGACY
ADMIN: MIDWEST ARCOLITH, LEVELS 1-3
SOCIAL SCORE: NA, CPX-SPONSORED

THE AIR WAS COLD, SHARP there in the middle of the intersection. The winds that blew through the outermost corridors of the Arc whipped viciously around the buildings. The NARKs hummed softly, barely audible over the noise of the wind and the crowd.

Dr. Agatha stood in the very center of the intersection, flanked by the double-row created by the NARKS. Her white coat flapped in the wind, but she ignored it. She was all business, jitters gone as she focused intently on her work. A port on the rear of the machines extruded centimeter cubes, black and yellow-striped to indicate they were medicine. The machines popped them out in fifteen-second intervals. She bent to inspect the cubes as they slid out of the machines and dropped into collection trays.

Ogunwe's men shifted, surreptitiously, one by one, each taking the man's spot before him. Once they reached the far end of the box in the intersection, the dead-end so to speak, the lead man in each of the two lines presented themselves to Agatha whereupon she dropped a cube into their open palms. They took the medicine then walked to the rear of the line that snaked its way all the way back to the building that served as NG's forward operations base. The building that somehow, miraculously, Ogunwe had *not* been required to take by force.

The process was achingly slow, killing Ogunwe one second at a time. The men continued to shift in their line. After ten agonizing minutes, the inoculation was finally complete.

Ogunwe's eyes went to the tubes that hooked over the AWOLs. Each one had a horn at its end, with a curved neck so the engineers could simply telescope them out and hang them across the top of the walls.

"Reichmann, what news?" Ogunwe said.

"Well, I've already pooed twice today," came Reichmann's disembodied voice, followed by silence. "No? Not the news you were looking for?"

Jesus, talk about the king of tone-deaf. Not exactly the time for jokes.

"What are the sniffers telling you?" Ogunwe persisted.

"Alright fine, hold on, lemme pull it up."

"Bringing Green in," Ogunwe said.

"Here, Chief," Green affirmed. She stopped conferring with one of her engineers on the wall and joined Ogunwe, standing at his shoulder. Her body seemed to bounce in minute increments, energy, muscles, everything about her wound tightly.

"Yeah, okay, just a sec," Reichmann stalled. "Looks like nothing out of the ordinary. Mostly Tango and Buena. Just a cun– " he stopped himself, "just a hair of Kopa, but sniffers have a hard time detecting that stuff anyway. That's about it. Well, that *and* the ageless scent of good ol' fashioned speed."

Ogunwe looked to Green. "We a go?"

She shifted from one foot to the other. "The speed's probably compounded with the Kopa. Tango and Buena aren't the problem. If there's only a tiny bit of Kopa, it *should* be okay."

"What do you need to be a hundred percent?" Ogunwe asked.

She snorted. "No such thing, Chief."

"Okay, stupid question. What'll make you feel more comfortable?" he asked.

"Getting this whole thing over with," she said, despair creeping in.

Ogunwe felt the same but wasn't about to reveal it. He stared, waited her out.

She relented. "I just need to know the Kopa really *is* just in trace amounts. One in a hundred we can handle. Twenty-five or thirty percent could cause a stampede.

"I'll do some digging," Reichmann assented.

"Do it *quickly*," Ogunwe ordered. He looked to Green. "And Green, be ready the instant we know." She nodded, moved off, leaving Ogunwe to himself. The window was narrowing and he felt it, like he was in a giant man-sized press and the walls were steadily moving in. How much longer before found out they'd been hacked? How much time before the crowd, separated by a thin clear wall—strong, certainly, but not indestructible—began demanding blood instead of answers?

"Excuse me, Mr. Ogunwe?"

Ogunwe's head snapped in the direction of the voice. Dr. Agatha jumped back a half a step, startled.

"Sorry. On high alert."

She took a deep breath, chest visibly expanding. "I can sympathize."

"What is it?" Ogunwe asked.

She held a hand out to him, palm up. A black and yellow-striped cube lay there.

"Were you planning on being a martyr today, or...?"

Ogunwe pulled down on the mask that covered the lower part of his face, snapped it off, let it flutter to the ground. "Damn thing's useless anyway." He offered the doctor a weary smile. "Thanks. Apparently I need a little looking after myself."

"Don't we all," she said kindly. She smiled back at Ogunwe, more warmly than the situation warranted. There was something in it, in the way her eyes searched his, that spoke of a want for intimacy.

*Maybe another time, doctor. My heart's already taken. Perhaps in another life. Another...life...*A thought occurred to him.

"Doctor, would you be able to spare another one?" he asked, holding up the cube pinched between forefinger and thumb.

She smirked, ran a hand through her red hair. "Well, hell froze over, pigs are flying and NG is giving out product for free. It's the devil's hat trick, so sure, what the hell." She took another cube from the tray at the rear of the closest machine, pressed it into his other palm, letting her fingertips linger there a second too long. The thrill of human contact made something leap inside of him.

"Here," she said, "now take that. Doctor Agatha's orders."

"Thank you, doctor. Sincerely."

"Don't mention it. Just don't go dying today."

"Do my best. You ready to distribute to the refugees?"

"Whenever you are."

Ogunwe waved Vander over.

"What's up, Chief?" Vander said.

"The doctor's ready. Do whatever she asks."

"Aye aye, captain."

The doctor turned, letting her gaze linger on Ogunwe, and mouthed *thank you*.

"Reichmann," Ogunwe said.

"Yeah boss."

"What about chemical analysis of the crowd?" Ogunwe asked.

Reichmann's voice fell. "Yeah, not a great success. It isn't much more refined than what I initially got. On a positive note, I've deployed drones and snuffed out a few bad guys. Cowards were mostly sitting on their asses in their heated apartments, watching feeds of the fallout."

"Well that's something positive," Ogunwe said. "Good work."

"Aw shucks, Chiefy, you say the sweetest things."

Ogunwe shook his head in chagrin. Reichmann was a good man, but incorrigible.

Despite his cohort's attempts to lighten the mood, Ogunwe felt that implacable suffocating force working on him, pressing in, taking his breath away. The invisible window continued to close. He leaned on one of the NARKs, tapped a foot nervously. He needed this whole operation to be open and shut, to conclude without event. To be able to get out and away quickly. How much longer did he have before Olivia and Stef were completely lost to him? How much longer did they all have? He looked at the crowd in all their diversity. Noted the hollow eyes of some, the drained pallor of others and, more disturbingly, the animated pacing of a few seemingly lone wolves. If it was Kopa, though, those lone wolves could become a pack in a heartbeat. All the data in the world, and they still didn't have all the answers. *Reduced to prayers and the magic of positive thinking*, Ogunwe thought. Well, that was if he'd gone in for that sort of thing, and he most certainly didn't. *No more*, he thought, *it's do or die time.*

"Vander, you have the panopticon up?"

"Yessir. All identities tagged, now just associating drug usage and chemical composition. It's taking longer than I'd like," he said in his slow southern drawl, the irony lost on him.

"Get your men ready for more than just tagging."

"Copy that."

"Green," Ogunwe said.

"Sir?"

"You're a go."

"Sir."

The change was near-instantaneous. Ogunwe let his eyes unfocus, taking in the whole of the crowd. There was a slow undulation to it, a rhythm. As soon as Green turned on the jets of gas, the rhythm changed. There were pockets that slowed, punctuated with currents that began to agitate into swirling eddies. A tiny sliver of doubt wedged itself in Ogunwe's mind. Had they done the right thing?

"What's going on, Vander?" Ogunwe asked.

"Nothing of major concern, Chief. A few reacting adversely to the gas."

"Adversely how?" Green demanded.

"Kinda weird, but they're, uh, biting at the people around them."

"Shit," Green cursed over the line.

"What –" Vander started, but Green interrupted.

"Target them, *NOW!*" she shouted.

"I can't just take 'em out," Vander protested, "they're too hidden. I'll hit everyone around 'em. There's kids out there, for chrissakes."

"Enough yelling. Explain yourself Green," Ogunwe said.

"We *really* don't have time. Kopa junkies start a stampede by infecting the others around them with the drug. It's normally done through sexual activity but can be done just as easily through transfer of saliva, or saliva directly to the bloodstream."

But they had taken too long. Those eddies, in just under a minute, had become swirling vortices of humanity. Bodies lunged up against the AWOLs, hands slapping, crazed eyes darting to and fro. The crowd was caught up in the stampede.

"Chief?" It was Agatha. The doctor looked at him, terror chiseled into her features. He had no reassuring words; the battle was on them and there was nothing left to say but to fight for their lives. He wanted to apologize for putting her in that position, but the words caught in his throat.

Something peculiar happened then. It was the hum. The machines had gone silent. An instant later, a tremor passed through the doctor's face, erasing the fear with one of curiosity.

"That's weird," she said, cocking her head to the side. Then her face went slack and her body went inert, fell forward like so much dead wood. Her head smacked the pavement dully.

"*Doctor!*" Ogunwe shouted, springing forward. She'd come to rest face down on the pavement, wedged in-between the NARKs. The machines, now dead, allowed the cold to move in. Ogunwe gripped the woman by the shoulders, turned her over. Lifeless eyes stared at nothing. She was gone.

Ogunwe looked up, expecting help, expecting at least recognition of what had just occurred.

Instead, his soldiers were fixated on the crowd. On the manic activity just outside the AWOLs. The horde surged up against the barriers, scrabbling, pounding at it. Climbing atop one another to slap and punch and ram it with whatever they could—their legs, feet, hands, heads. The Arc-Sec soldiers flinched, took a step back as the clear, thin walls swayed violently, displacing the vibrations into the ground.

At this rate, the AWOLs will crack within minutes. As soon as that happens, we're toast.

Alarmed, Ogunwe stood swiftly, spared one last glance at the dead doctor. Novagenica knew. They'd flipped the switch on the NARKs *and* the woman, executed the killswitch for her implicit guilt in distributing free product. Which meant Moon's access was undoubtedly revoked, and there'd be no retreat into the Novagenica building.

He'd feel the guilt later. No room for that kind of thought now. Just planning, action, instinct.

Do or die.

Circuits of self-preservation warred with his responsibility for his people. He looked around at the soldiers around him. They were just foot. They'd do whatever command wanted of them, whatever Briggs wanted, and that included the enforcement of PASS. No, Ogunwe's true responsibility was to those closest to him. To Olivia, his friend and secret love. To Stef, the son of Thiago. To the tightly-knit group of officers that had sat around a gnarled wooden table taking shots of whiskey. They had pledged their trust in him, put their lives in his hands. He wasn't about to let them down.

"Reichmann," Ogunwe said, shifting the channel to private.

"Yeah boss?"

"It's about to get ugly. You need to move, now. Novagenica knows their security was penetrated by Talksmall..."

"So right about now they're about to connect the dots, huh?" Reichmann said, strangely jovial. It left Ogunwe with a sinking feeling that he knew how this conversation would progress.

"You can't stay there any longer. If I'm compromised, so are you. Get to Riverside..."

"Sorry, Chief, just saw a huge spike universally. Crowds across the board everywhere are getting restless again. But it's worst at distribution points," Reichmann said, pushing aside Ogunwe's concern.

Ogunwe moved away from the soldiers who stood in formation, cupped a hand around his mouth. "Listen to me, you bullheaded idiot," he hissed. "Get. Out. If any of us get caught, the Golden Road was a waste of time. Pull the cabal to Riverside. Place the riot heatmap on everyone's HUD. Hopefully we can avoid the worst."

"And what happens to you?" Reichmann asked.

"I have other business to take care of," Ogunwe said with an air of finality.

"And you're gonna need help gettin' it done. Think I'll stay a minute, Chief. See this thing through."

"Goddammit," Ogunwe muttered.

The line went silent, except for the sound of Reichmann's breathing. A thundercrack pierced the air like a sonic boom, Ogunwe's head snapping to the source. A giant fissure, punctured by innumerable perpendicular striations, ran diagonally, from the ground to the top, through the AWOL.

Ogunwe pulled his weapon from its holster with his right hand, telescoping baton gripped tightly in his left. He could feel the tautness of his entire body, the heightened awareness of his surroundings. The wall wouldn't last another thirty seconds. It was time. *Do or die.*

RAY

RAY
AGE: 36
SEGMENT: COMPLEX PROPERTY; PERSENSE CPX
POSITION: REDACTED
SOCIAL SCORE: NA, CPX-SPONSORED

T HE REGIONAL HQ OF ARCSEC ON 3 was a five-story building near the level's central column. Near the transfer gate, close enough to major thoroughfares for its officers to move quickly at a moment's notice. Not that the office on 3 had much in the way of barracks. Maybe enough space for a few hundred soldiers. Nothing like their fortress on 5, Briggs' fiefdom, replete with his own private army. That had been impressive. This was just a bland, squat office building with a logo on its front.

As such, Ray had little trouble bypassing security. It was all automated locks, solid doors meant to bar access. No threat of violence to deter. No manned checkpoints. Being that Ray's Complex, PerSense, was now a parent entity to ArcSec, Ray's credentials were grandfathered in to ArcSec's system. Entry was simply a matter of walking in.

He entered on the main level, the transparent door revealing no foot traffic inside. The main level was dedicated to the booking, tagging and temporary storage of offenders. It was a cramped space with one small window that typically had an attendant working the desk behind it. Ray stepped over, peeked in. Like the lobby, the narrow room behind the window was deserted.

All resources committed, Ray thought. It was good. It would make this easier.

The schematics of the building showed a stairwell in the northeast corner. Ray found the door, pushed it open and entered the stairwell.

Third floor, the AI spoke in his ear. It wasn't too long ago he'd have found the intrusion irritating, but as his relationship with it grew, he'd come to view it as almost a 3RD party extension of his self. A part of him, yet a thing that could reach out and leave him, extend out into the ether.

As he climbed the steps, taking the time to plod upward methodically, he considered the evidence. Ogunwe's meeting with the Talksmall people, its very nature concealed by measures of extreme caution so not even his own Complex knew the contents of the conversation. An escape that coincided with the hacking of not only 's tour logs but all of their distribution printers. Permissions freely given for medicine, no contract required, an operation overseen by Ogunwe himself. Perhaps even engineered by the man.

Not much to go on, really, but Ray wasn't a public prosecutor constrained by legalities. And the mandate given to him had been satisfying and simple:

Find everyone who knows, and find out what they know, the AI had whispered in his ear. *Then kill them.*

Ray reached the landing at the top of the stairwell, eased the door open. He stepped out into an unoccupied foyer, low lights casting a warm glow on flooring that was some form of plastic designed to look like wood. To his right, the elevator bank, all three dormant, stainless fronts gleaming, blue lights glowing in the dim.

Beyond the foyer, a wide frame that opened onto a darkened room. Ray padded softly on his toes to the edge of the opening. The room was cavernous and gridded by pillar workstations. Dark except for one illuminated pillar in the far corner, a spotlight casting a focused glow amid the room as though the operator were on-stage performing a soliloquy.

Ray readied himself, pulled the flechette pistol from its holster, cycled through to a five-minute dose of sedative. With his left hand, he slipped the long knife from its sheath on his thigh. He inspected the blade in the dull light. Not even a drop of blood showed as evidence from its previous victims.

Luckily for Ray, the lone operator had his back turned to the entrance. Ray crept into the room, crossed it rapidly, suppressing the urge to weave through the pillars, instead hewing close to the outer wall. The closer Ray got, the more cautious he became, his movements exaggerated. The suit would insulate and wall off the sound of his cracking and popping joints, but the fabric would rustle if he moved too quickly.

At fifty yards he could discern the frame of a man, upper body wide and rippling with muscle, looking uncomfortable and cramped perched on the tiny chair that jutted off the pillar.

At 10 yards Ray holstered the flechette gun. The focused pool of light that washed over the stranger was almost blinding. The man started humming softly to himself.

Ray rushed forward on silent, perfectly balanced feet, a black wraith amid the shadows. He came up behind his victim, brought the knife around so the man could see it, simultaneously pushed two fingers into the man's lower back and expelled warm breath into the man's ear.

"Shh," he purred softly.

Ray inched the long knife closer to the man's neck, careful not to touch the skin.

The man inhaled sharply, body bucking, head instinctively snapping back and away from the knife. His head connected with the pillar, made a dull *clunk*.

"Ah ah ah," Ray chided, ramming his fingers deep into the node of nerves. "I mean it. Not a word, transmission, nothing," Ray whispered in his ear. "Just breathe. Now, snap your fingers if you understand." The snap came immediately, the sound of dry fingers compressed together, briefly punctuating the silence in the hollow space.

"Good," Ray said brightly. "The first thing I want you to understand is that you're mine. It's important you know that. Now if you decide at any point you're not mine, that's fine. All well and good. It simply means your death. Slow or quick are irrelevant. I won't bore you with vivid images of your torture. I've found it's the anxiety about *when* that gets people the most. Makes them do stupid things. Now, I'd like you to acknowledge that this is *your* decision to make. Snap once if you agree, two if not."

One snap. The man wasn't a complete idiot then. Ray might be able to make use of him.

"Sometimes death is the easier decision to make, given no alternative. But that's not me. Not how I operate. I find people are much more useful if they feel they can lead long, productive lives. I know it sounds like a silly question, but would you like to lead a long, productive life?"

One snap.

"Great. This is just great. We'll have a fine partnership, but before we discuss terms I want to lay out a few ground rules. Number one, I'm an asset of PerSense, covert recruitment ops. Any comms you send out to your buddies will eventually get parsed by our *joint* AI. PerSense and Arc-Sec, working together. It's a beautiful thing, right?" Ray paused, waited for the response. Another snap, this one delayed.

"But I'm rambling. Point is, there's nowhere to hide, so don't bother hitting send on that SOS. Just delete it. Whether it's me today, another PerSense operative, or even one of your buddies here at ArcSec, plunging a knife deep in your back on some sunny day, your dues *will* be paid. Understand?"

Another snap, this one methodical. Ray felt more comfortable, could sense the man's panic begin to ebb. Ray eased the knife away from the prisoner's neck, slipped back and around to face him, keeping his own body out of arm's reach. Ray leveled the flechette gun at the man's torso. The shock that had undoubtedly gripped the man's face was now replaced by a simulacrum of calm. He had a broad, flat face with a thin, hatchet nose and eyes that were far too close to one another. It was a brutally ugly face, one undoubtedly assigned to hardened criminals in fictional feeds, but the signs of cowardice were there. Eyes that showed a touch too much white, lips compressed in a thin line, quavering, jaw clenched. Fat, short-fingered hands trembling ever-so-slightly at his side. He was trying so hard. Big man, undoubtedly, but little steel inside of him. This one would be easy. Ray slid the knife back into its sheath, gripped the seat on the next pillar five feet from the man, spun it around so he could sit and face his captive.

"Listen, whatever you want, I'll..."

"Hush," Ray interrupted, lifting the knife to his lips. "Choke that back, friend. There's nothing you can offer me other than what I want, so here it is. I want what's in your head. I can get access to your assistant any-time I want, but what I *need* more than anything is your knowledge. And your ongoing cooperation. Now I want you to be totally honest with me, okay?"

The man nodded his head. "Yeah. Okay."

Ask him what Chief Ogunwe's plans are, the AI said, and Ray felt a shiver of simpatico go through him. Or maybe that was the *Elate,* a friendly chemical passenger in his bloodstream, infusing his cells. He snorted and the man looked at Ray like he was crazy.

Well maybe he was, just a little.

"Good, good," Ray said. He stood and circled the pillar in a slow walk, a tiger prowling. "Now tell me—what's your Chief, Mr. Ogunwe, up to?"

Ray purposefully framed the question to be open-ended. He had an inkling of what was going on, something more than the above-board partnership with Talksmall, but if he led with what he knew then he'd only get confirmation instead of all the little details and treasons he didn't know. And what he needed to know was everything.

"Listen, I'll cooperate. Whatever you want," the man said, hands out, "but first I need to know what to do if my boss, Ogunwe, if he contacts me. We're all on an open channel and all hell just broke loose. They need my help."

"Wonderful. Don't do anything differently. Help away. Pretend I'm not here, just a fly on the wall. For now, though, tell me everything you know."

The man sucked in a giant breath, chest heaving. He rolled his neck, closed his eyes.

His eyes snapped open, held Ray's. "We call it the Golden Road."

KAT

KAT
AGE: 13
SEGMENT: LEGACY CITIZEN
POSITION: UNDERAGE; GOVERNMENT DOLE
RECIPIENT (REDUCED BENEFIT)
SOCIAL SCORE: 300

SHE COULD FEEL IT PROBING. Poking, prodding about the crevices in her mind, discovering what parts did what. An alien entity, searching her memories, testing the controls. She tried to ignore it, but its fingers were cold, foreign, tainting her emotion and thought. She shivered at each minute exploration.

Soon enough, her sister would notice how quiet she'd gotten and ask what was up. For now, Kat could pass the constant shivering off as the residual effects of whatever illness Novagenica had given her. But as soon as Val started asking questions, Kat wasn't sure she could simply shrug it off, casually lie about how she was feeling. Wasn't sure she wanted to lie about anything anymore.

Besides, they'd both seen the blood on the train.

And then there was the small matter of her (questionable) status as savior of the world.

Turns out there's a reason I was the only one to survive.

Which she still hadn't had time to fully process. Her friends, Zara, Antonio, Robbie—and of course the countless others trapped in that aseptic hellhole—had all died in service of finding the cure.

And aside from the general fucked-upedness of a Complex stealing people to test cures for a disease they created, the fact was none of those other kids would've had to give their lives if Kat had simply been taken first.

But is that accurate? How much do you really know of what's going on?

You're alive, yeah, but now you've got some creepy roommate in your head, trying to take over the joint. So the question remains, what do you really know?

Not much. If that wasn't the goddamn truth she didn't know what was.

These thoughts continued to career around her skull, fed by the perpetual engine of anxiety, and still she maintained her silence, folding inward, shivering constantly despite the warmth of Trevor's jacket.

He led and they followed, winding through the narrow, fringe streets of 3, trying to keep up with the quick pace he set. Not quite a panic-walk, but close. Eventually they came to a dead-end where the rust-colored beams of 4 were visible high above, and the flat, giant face of the arc's outer wall stretched up to eternity. Before them, a stocky, stuccoed building hemmed in by others that were stacked like a vertical village of adobe huts. Trevor punched some digits on a keypad and pulled the door back. They moved into a low and wide, unfinished office space. Trevor entered last, pulling the door shut firmly until it clicked.

"Just straight down here." He pointed and started forward, walking down a hall that extended off the receiving area. The office smelled of dry, weeks old dust mixed with the vaguest hint of printed construction materials. Glue, alcohol, burned sugar. "There's an exit that leads to a stairwell. And then..." he stopped at a door, turned, hesitating, the tension of worry in his eyes and the set of his jaw.

"And then what?" Val asked, eyes going to Kat, mirroring his concern. "What is it?"

"Stairs."

"How many?" Val asked.

"I'll be fine," Kat interjected with a weak sort of bravado. One look at their faces told her nobody was buying it.

"Well a lot, but I have something that might help," Trevor said. "Look in there," he pointed to one of the zippered pockets on the leather jacket Kat wore. She unzipped it, took out two hard centimeter-wide cubes that had the spongy texture of fried tofu. One was red, the other green.

"Food?" Kat asked, regarding them with a cocked eyebrow. "How long have these been sitting in this crusty jacket of yours?"

"Dunno, a week? Maybe two? They'll give you strength, that's what matters."

"Just take 'em, Kat," Val urged, the tone of a big sister creeping into her voice. Kat felt the old urge rise within her, the defensive, bitter comments wanting to find a way out.

But she was past that. Had seen too much to worry about petty nuisances.

"Okay," Kat said, then popped the cubes in her mouth, chewing. "Hey...actually pretty good," she said brightly around the food.

"Glad you approve. The red one's marinated chicken and the other is all your minerals and vitamins, with a long-bleed speed bump," Trevor said. He turned back to the door, punched in another code and pulled it open.

Kat finished chewing and swallowed. The change was near instantaneous. She felt stronger, stood taller, color flushing her pale cheeks, casting out the hollowness beneath her eyes. She could almost forget that she'd been sick in the first place.

"Better yeah?" Trevor said. There was a charm to him, a constant twinkle in his eye, that Kat found calming. She understood why Val liked him.

"Actually, yeah," she said, and went through the door into a stairwell. The first thing she noticed was the faint smell of spice, like the burning of incense. She pressed her abdomen up against the stairwell railing, looking up, then down into the narrow cleft between the columns of stairs. The stairwell extended out of sight, into darkness in both directions.

"How far this thing go?" she asked.

"Your guess is as good as mine," Trevor said. "Far as I know it goes all the way to seven. Down, though, it stops at two."

"Man," she said with wonder, letting the word hang in the air.

"You guys smell that?" Val said, wrinkling her nose. "What is it?"

"Blind Tiger. Opium den at the bottom of the stairwell," Trevor said.

"Oh, right," Val said half-heartedly.

"I know what you're thinking, but it's the only way to get around the PASS checkpoints at the transfer gates. It's how I got up here and how we'll get down to Riverside," Trevor said. His eyes shifted to Kat. "Where we'll get you help." He took the first step down, left hand on the railing, body half-turned to make sure Kat followed. He extended a hand. "Ready?"

They descended the well and the smell of incense grew stronger, commingling with another scent, this one more toxic, something associated with the creation and composition of polymers.

They reached a landing and Trevor stopped, said, "This is..." when a message flashed across Kat's system, interrupting Trevor.

MOVE FASTER. YOU'RE BEING TRACKED.

Kat froze, breath caught in her chest. "What the fuck is tha..."

Val was at her side instantly. "It's Ian. He's a friend," she said.

Kat allowed herself to breathe. For a brief moment she'd been reminded of the invader in her mind, thought it had moved from passive cohabitation to active communication.

"Okay," she sighed, trying to blink away the message, but closing her eyes only served to amplify the message inscribed there on the surface of her vision. *MOVE FASTER.* "But dammit, can you tell your buddy to give some notice before intruding on my privacy? I mean, c'mon. Literally almost scared the shit right out of me."

But Trevor and Val ignored her, seemed intent on the message. The conversation scrolled across Kat's vision.

WHO? Trevor sent.

NOVAGENICA AI. COMPOSITE VIDEO, AUDIO, NETWORK POSITIONING. PREDICTIVE.

CAN YOU FIGURE OUT WHO'S BEHIND IT? Trevor asked.

ALREADY DONE. TABS ON HIM NOW BUT IT'S DIFFICULT TO MAINTAIN. THE VIDEO FEEDS ARE ARCSEC, OWNED BY PERSENSE. THEY'RE PROGRAMMED TO GHOST HIM. TALKSMALL HAS SOME ROGUE COMPONENTS, SPREAD AROUND THE ARC, WHICH ALLOWS US TO CATCH THE OCCASIONAL GLIMPSE. WE HAVE A BETTER NETWORK ON 1. YOU'LL BE SAFER IF YOU CAN MANAGE TO MAKE IT DOWN HERE.

Well shit that didn't exactly inspire confidence, did it? Kat thought. *Who is this asshole?*

IF??? Trevor punched out, mirroring Kat's thoughts. *WHERE IS THIS DUDE?*

THREE.

The response, one word, came without sound, but Kat could've sworn she heard a tolling, like an accompanying death-knell. They were just on 3. For all they knew he could be right behind them, creeping silently down the stairwell. Reflexively, Kat looked back up the staircase. Nothing but gray emptiness.

WHERE ON 3? Val sent.

There was a pause of a few seconds.

ARCSEC. BUT HEADED YOUR WAY. AT LEAST THAT'S WHERE VECTORS POINT. MAYBE THIRTY MINUTES BEHIND YOU.

Trevor breathed an audible sigh of relief, which didn't seem right. Thirty minutes head-start could be erased pretty quickly, especially the way Kat was moving. And Kat was suspicious of that word 'maybe'. There was too much wiggle in it.

HE COULD STILL EMPLOY A PROXY TO DO THE DIRTY WORK. BLIND TIGER AND FREE DRUG ZONE ARE GOOD PLACES TO ENLIST DESPERATE PEOPLE. A JUNKIE WILL DO ANYTHING FOR A FIX. WATCH YOUR BACK.

Thanks for that, dude, Kat thought. *Way to really drive the point home.*

Val turned, exchanged glances with Kat. Kat noted the stiffness in her sister's back and neck, the tension in her face, carving a stark line between cheekbone and jaw, her bloodless fingers gripping the railing.

"PerSense," she said. "Why are they a part of this now?"

Trevor went to the door at the landing, fingers probing a drawing below the doors handle. Kat looked closer, saw a minute rendition of a tiger's face there, like a spray-painted stencil. Despite the size, the detail was phenomenal. Striped black and white fur, arctic-blue eyes fogged over with cataracts.

"They own ArcSec, and ArcSec put up the PASS barriers. They're all in it together," he said, putting one hand on the door's handle. "Do me a favor and reach in there," he pointed to the leather jacket's right pocket. Kat put her hand in and pulled out a bundle of white masks.

"Hats?" she asked.

"Masks. For protection." He took two of the masks from Kat, handed one to Val. "Nasty shit in here. People using the old-fashioned way, which means smoke. Potential contact high. I'd rather not be high as an astronaut running for my life."

He pulled the mask down over his head. It was one piece, like a winter cap, that fitted snugly to his skin, covering him all the way to his neck. "Go ahead," he urged the girls through the mask. It had no holes for his mouth or eyes, but the fabric was sheer so he could see through it as though viewing through a veil.

The girls pulled the masks over their heads. "It has a scrubber," he explained. "Always on. Cleans everything that comes through it." He turned the door, but before punching in the code he turned back to Val and Kat. "Walk calmly, stay right beside me, heads straight forward. After we get through Blind Tiger, we'll be in the FDZ. Same rules apply—stay beside me, especially don't stare. People won't be able to see your eyes, but they *will* be able to see you turn your head. Do not take these masks off. Whatever it was infected you with," he pointed to Kat, "they've set it loose on everyone down here. We'll get through here, through the FDZ, then head to the transfer gates by Riverside. Should be the fastest way. Got it?" he asked, giving the thumbs up.

"Yeah, dad," Kat said. "Enough with the lecture."

Trevor chuckled. "Ha ha, daughter. Very funny."

Val looked nervously behind her, up the staircase. "Let's get on with it," she said.

"Exactly, less talking, more doing," Kat said.

Trevor pulled open the door and a blanket of smoke rolled over them. He entered first, plunged into the cloud, down a wood-paneled hallway. Kat rushed in behind him. The wood was mahogany, worn, dull from exposure to the toxic chemicals in the smoke. At the end of the hallway, a warm lighting illuminated the fog. They exited the hallway onto a wide balcony that ringed four stories of open space. On each level, the same balcony, ringed with low, partitioned platforms that served as beds, covered in pillows, draped in veils that were sometimes pulled shut, sometime wide open. The platform to their left had its veil pulled shut, but the next one was drawn back and through it Kat could see two men and two women tangled in a lazy sort of coupling. She felt like she should avert her eyes but found she couldn't. White robes that bore the blind tiger mark carelessly left open, flesh exposed, a mass of bodies slowly writhing. As they walked by, one of the men stopped what he was doing, stretched his body over the other three and pulled a hookah from a recess in the wall, proffered it to Kat. The man's eyes were dull and red, glassy, the barest hint of a smile sitting stupidly on his face. Kat shook her head vehemently, rushed ahead to Trevor's side.

They came to the corner of the square, where a bulky man with bare head barred the way to a staircase.

No words were exchanged. He simply let them through.

As they descended to the first story, moving past the bouncers at each set of stairs, the private partitions became smaller and smaller until finally they no longer held beds, but became tiny hollows that could house no more than two people, with no curtains and a red wooden bench nailed to the rear wall. The benches were old, rough with peeling paint and dotted with burn marks. The customers on the first floor were different as well. There were no acts of congress like they'd witnessed on the top floors, all actions taken were done so alone. The open area in the middle of the main floor was a dark place, littered with stools that held a host of people in the throes of various stages of usage, lighted only by the small, intermittent orange flames that burst to life in people hands just before they touched them to the ends of pipes, lighting up gaunt faces, wooden with a desperate sort of concentration. Mouths agape, eyes vacant. Some had overturned on their stools, and their inhabitants hadn't bothered to set them upright, instead laying on the floor, bodies curled, jaws working back and forth. Others lay prone or on their sides, unmoving.

"Jesus," Kat whispered.

"Shh," Trevor turned slightly, admonishing her, but his eyes slid past her to behind them. Kat followed his gaze, noticing the bouncer had moved from his position at the stairwell. Following them.

Ian's warning blared in her mind. *A JUNKIE WILL DO ANY-THING FOR A FIX. WATCH YOUR BACK.*

They walked quickly, Kat barely keeping her panic in check, stepping over and around the staggered collection of stools and bodies, winding a path through their middle. The smoke was thicker here and without inherent lighting it made the navigation difficult. More than once she stepped on a foot, sometimes the outstretched hand of a semi-conscious user. They were only 30 yards away from the exit, a hallway bathed in red light. Kat glanced back again. The bouncer, a black-haired giant of a man who looked like he was of mixed Middle Eastern and Latino descent, had gained on them.

Oh my god there's no way we'll make it, she thought.

Ahead, she saw Trevor patting his pockets. Looking for something. He put his hand where his jacket pocket should've been. It slipped into empty air. *Shit! I'm wearing his jacket!*

Ten yards to the red-lit hallway that exited the Blind Tiger and she knew instinctively that their time was up. The bouncer would be on them in another second. *Do something. Anything. But a* hollow sort of dread kept her moving onward, despite the fact she knew they should stop and face their pursuer.

That was when Trevor stopped and spun. Kat bumped into him. He caught her, and Kat felt him fumbling around in the jacket's left pocket. His fingers found what they sought and removed a smooth, thin rod.

"GET DOWN!" he roared, pushing downward forcefully on Kat's shoulder. Kat dropped, feeling Val hugging her from behind, her sister's weight covering her back. Trevor flicked his wrist, the rod telescoping out with an audible *snick.*

Kat toppled to her side and the bouncer was there, tight vantablack jumpsuit, implacable face chiseled from rock. He paused, for the slightest of moments. Trevor swung the rod at his knee. It bounced off.

Ah shit. Not good, she thought.

The bouncer swept a tree trunk of an arm in Trevor's direction. The man's open palm connected with the left side of Trevor's head, knocking him to the side, almost like an older brother affectionately swatting away the childish attempts of a younger sibling. It was more of a push than a slap, and Trevor stumbled, tripping over a wisp of a woman who was huddled into herself on a stool, bringing her down with him, crashing to the ground.

The woman uttered feeble cries of protest but Kat was still watching the bouncer and was confused by what she saw.

Another man stepped toward Val and Kat, but the bouncer stepped over them, cutting the man off. From somewhere, the man produced a long knife and jabbed it in the bouncer's direction. He looked like a junkie but moved with preternatural speed, augmented by *something*. Possibly tech, but more likely just drugs. The knife flashed, slashing at the bouncer's abdomen. It connected and Kat sucked her breath in, expecting to see a red crease of flesh split open and insides come spilling out. Instead the knife slid away, deflected by the material of the jumpsuit. The bouncer clamped one bear-like paw over the man's wrist, then smiled. Inexplicably the junkie smiled back, showing a rotten mouth full of black and broken teeth like a logged mountain forest, trees jagged and snapped at their trunks. The bouncer's right hand shot straight out with artificial speed. *Enhanced* speed. The fist connected with the junkie's head, snapping it backward with incredible violence. His body went limp, fell to the stained and scarred floorboards.

Trevor scrambled to his feet, rod still clenched in his grip. The bouncer looked at it, then Trevor.

"Need more practice," he said, voice low and thick.

"Uh, okay, you got it boss," Trevor replied. The bouncer offered his hand to Val while Trevor put a hand beneath Kat's arm, helped her up.

"What the hell was that about?" Val asked the bouncer. Kat could see the fury, the outrage building inside her. Kat looked down at the body of the attacker, shock wearing away, replaced with anger.

"Blind Tiger sees all comms traffic, in and out. This one," he gestured to the junkie on the floor, "was just bought by someone outside. Accepted a contract to take your lives," the bouncer replied.

"Bought? By who?" Val pressed. Trevor looked at her as though to say she already knew the answer.

Ian. He was right. He said they'd try to get us, here or in the Free Drug Zone.

The bouncer didn't reply, his face unreadable.

"Thanks anyway," Trevor said, nodding. The bouncer nodded back, then bent to the task of picking the limp body up. He heaved it over one massive shoulder and as he turned Kat noticed the white tiger tattoo behind his left ear. Identical to the mark in the stairwell. Then, without so much as a goodbye, the bouncer disappeared into the fog of acrid drug smoke, body swaying limply on his shoulder. *Probably to recycling.* Kat shuddered at the thought.

"C'mon," Trevor urged. "We gotta hustle. Sooner we're on 1, the better I'll feel."

They turned through the roiling smoke, away from the darkened host of users and junkies that lay hunched in the shadows, calcified and depleted versions of themselves, then away quickly as though escaping a nightmare.

Down the cramped hallway bathed in red light, out the door and into the FDZ.

They descended cracked stairs to wet cobblestones and Kat felt those cold, alien fingers slide into the spaces between her memories and motor function. Her body continued downward, smacking the pavement below, but she felt none of it, her consciousness retreating inward.

It was like being thrust beneath a freezing river at midnight. Immersed, dragged down to its middle depths, her breath was gone and the heat of her existence swept away in the current. Her entire being was reduced to a dull throb encased in frigid darkness.

"I'm okay," she heard herself say, and thought, *I sound weird. Wooden. Not like myself.*

"No really, guys," her voice again. "I'm okay. Just slipped, stupid fucking feet." *There,* she thought with weird satisfaction. *That's better, more like me,* and lay back in her dark prison, strangely contented.

RIKU "CHIEF" OGUNWE

RIKU "CHIEF" OGUNWE
AGE: 40
SEGMENT: COMPLEX PROPERTY; ARCSEC CPX
POSITION: ARCSEC CHIEF CONTRACTED BY LEGACY
ADMIN: MIDWEST ARCOLITH, LEVELS 1-3
SOCIAL SCORE: NA, CPX-SPONSORED

THE AWOL SHATTERED WHERE the fissure met the ground, leaving an opening several feet wide. The force of innumerable bodies pressed up against it spilled forward, the levee broken. Ogunwe knew, instinctively, there was only one solution to this problem. Out.

"Siege protocol!" he screamed across all channels.

The horde flooded through the opening, the first numbers of the group shoved through and onto their faces, the first drops in a flood. Those behind trampled those in front. The kopa infection had reached critical mass. The stampede was mindless and wouldn't abate until the drug left the system of its inhabitants. Which could mean hours. Ogunwe didn't have hours, much less minutes.

His soldiers were well-trained. As one, they split up into their respective groups, forming several diamond-shaped ranks. Weapons at the ready, each soldier held a pistol in one hand, set to wide shot, and a magic wand in the other. The wand was set to deliver a jolt of electricity on contact, the soldiers protected from the shock by their uniforms. The instant Ogunwe ordered the siege protocol, the soldiers' tactical gloves automatically formed a bond with their weapons. If an assailant wanted to take the weapon, they'd have to take the limb first.

Refugees scrambled over the bodies in front of them, hands and arms bloodied as they squeezed in at the edges of the broken, jagged AWOL. Ragged hands tore at the soldiers' faces, arms, anything they could get their hands on. The soldiers beat them back, refugees' bodies convulsing as the wands connected with heads, shoulders, torsos.

Each diamond held a group of engineers at its center. Ogunwe pushed his way to the center of the foremost formation, joining Green and another of her subordinates.

"Green," Ogunwe addressed the engineer across the command channel. Green stood next to him, body rigid, immobile. Her face was a ghost of itself, transfixed by the sudden violence, in shock.

Ogunwe slapped her, hard, across the face.

She cried out, eyes holding the glare of betrayal at first. Quickly, she understood.

"Birds ready, Green?" Ogunwe asked.

"*Shit*," she said softly, then uttered commands to her team. Ogunwe looked back, watched drones shoot up from the center of each diamond.

"Reichmann?"

A pause of a couple seconds. *Where was he?*

"Sorry. Here boss."

"Get us a path to the closest transfer gate. Fewest people. Send it to our drones."

"Ten four."

Ogunwe tapped Green on the shoulder. "Soften the middle. Use whatever you have to."

The drones shot forward down the middle of Baltimore, black streaks that moved in unison. A heavy gray gas, sprayed from internal tanks, fell softly in their wake.

"Time the release of the AWOLS. Take down the sections in-between our formations first. Ten seconds after, release the sections at the point of each formation. Got it?" She nodded. "Good. Do it. Now."

"Wands only!" Ogunwe commanded. An instant later the first set of AWOLs came down and drug-crazed refugees flooded in around the sides of the formations. Ogunwe's visibility went to zero in a heartbeat. A maddened horde moved in, teeth bared, eyes wild. Wands flashed through the cold air. Bodies piled up. Clouds of steam, exhalations of the soldiers' exertions, puffed out from the formations. Ogunwe could barely hear past the roar of the stampede.

The second set of AWOLs collapsed.

"Fire at will!" Ogunwe commanded. "Retreat to the gates! *Now now NOW!*"

The soldiers marched and fired simultaneously. Those who'd received the magic wand treatment were lucky. The first fusillade tore through the crowd. Bodies collapses in their wake, a fine red mist spraying into the cold air.

The front point of Ogunwe's formation plunged into the crowd first, crawling north through the intersection, down Baltimore. Ogunwe looked back, over the heads of the soldiers that swung their sticks back and forth. His plan to release the pressure from the points of the formations seemed to be working. All four diamonds, Alpha through Delta, moved at a steady march, mowing down assailants before they could hinder their path.

Ogunwe's foot caught on something and he reeled backward. A firm hand gripped his arm, steadied him. He looked over at Green, thankful. Then he looked down, knowing what he'd see, but not wanting to witness it all the same.

A body. A teenager, longish dirty blond hair caught in the throes of a drug he probably hadn't willfully taken, trampled underfoot, clothing shredded, riddle with tiny holes that seeped red.

Keep marching. Ignore it and just keep going.

"DO. NOT. STOP!" Ogunwe shouted, giving release to the thought inside of him. No room for reservations, not now. That was for later, if they survived.

"Got your map, boss." Reichmann's voice was subdued. Ogunwe couldn't blame him. "Shortest way is also the least crowded. Seems like a straight shot. What else you need?"

Ogunwe made sure the channel was private.

"How are the others? The Golden Road?"

A pause. "Better'n you, but not by much. AWOLs are holding for Carpenter in the Southeast and Sophin in the East. Riverside is contained, but that's also where most of the Talksmall apparatus is. Somehow they're still getting medicine to people."

"Any word from Moon on the mutation?" Ogunwe asked.

"I'll check." A pause. Ogunwe took the moment to look ahead. They'd managed to get a hundred yards up Baltimore. Signs that hung in windows of shops were dead, no lights beyond. Windows had been shattered, doors smashed in, dark shapes in the distance flitting in and out of the buildings. Maybe a quarter mile beyond, in the pervasive darkness that was level 1, the transfer station waited. Ogunwe couldn't see it but he knew it was there. As he peered down the length of the street, an alarm bell rang in Ogunwe's mind. Something was off, but the thought was interrupted by Reichmann's voice in his ear.

"They think they have a lead," Reichmann said. "They've got some whiz there who's modified the TS black boxes to spit the stuff out. Says he took one of the cubes and modeled it in, like, two minutes. Also, they busted some girl out of Novagenica—don't ask me how they did that— who might hold the key to the mutation. Patient Zero I guess. If Talksmall can get hold of the code they might be able to model the treatment after it."

Ogunwe closed his eyes briefly. Another variable. His sense of duty warred with what his heart wanted. Find Olivia, or find the girl, Patient Zero? He knew the answer, and his frustration mounted at being continually stymied.

"Good. That's, that's good," he said, collecting himself. "Can they start shunting the medicine out to the other distribution points?"

"Already ahead of ya there, Chief. Morales says TS has *all* of their eggs in one basket. Every single black box is at Riverside. Morales apparently tried to sweet talk your buddy Moon into releasing a roving contingent, but he wasn't having it. Said he needed every man. Morales thinks he's up to something, but she can't guess what."

Ogunwe grunted. *Not surprising. And really, it's peripheral. Most important thing now is to protect that girl. Patient Zero, he called her.* "Reichmann? Can you track the girl?"

"She's probably already been tagged by Novagenica, so it shouldn't be hard," Reichmann said.

"Do it then. She'll need help getting to Riverside. Find her, then get back to me."

"Ten four."

The connection went silent and the soldiers continued their march, reaching the point Ogunwe had noted earlier when Reichmann had interrupted. Then it struck him what was wrong. It was so obvious, but he'd been distracted by thoughts of the bigger picture, of his conversation with Reichmann. The crowd had thinned, their path to the transfer gate deserted. On both sides of the road furniture had been piled up, starting at the buildings and then working their way into the streets. Down the middle, a narrow space through which the formations would have to march. Beyond it, the giant steps of the transfer station, darkly illuminated bronze barely visible, an empty staircase of giants. The whole setup looked like a gauntlet. Someone had set a trap. They'd pushed through the kopa stampede but were now confronted with a far more lucid, far more dangerous opponent.

"HALT!" Ogunwe screamed. The soldiers stopped. Some of them slumped, breathing hard. Most swiveled their heads to and fro. A man, a boy really, standing next to him, was wild-eyed, panicky. Ogunwe put a hand on his shoulder, gripped it firmly. "You're doing fine, son," he said, feeling the muscles beneath his hand relax. Not much, but it was something. Ogunwe looked up, trying to find the drones. Nowhere in sight, undoubtedly taken out. He needed a moment to think.

"Chief?" Vander's voice, across the command channel. He'd taken up position at the rear, didn't wait for Ogunwe's response. "Do kopa junkies usually destroy property? Like, throw shit into the street?"

No, they don't have the capacity to plan. The thought came like a warning shot. Something was off.

It was Green's voice that answered. "No. The drug makes them fixate on people, faces. Objects hold no meaning to someone in the grips of the stampede."

"Suspicion confirmed then. We're gonna need to address this," Vander said. "I've got people throwing furniture into the street back here. And they're not showin' their faces, just piling shit up, tossing it from the windows on either side."

Ogunwe peered over the heads of black-clad soldiers, down the road, past where Vander stood. He could barely see through the haze, but thought he saw the gray outline of a bulk growing, stretching from each side, tapering to a point in the middle of the road. Silhouettes of large items falling from windows. It was the rear of the trap. Whoever was responsible was going to try to force them through the gauntlet.

"Take your formation and clear that debris, Vander. Now."

"Sir," Vander said. No sooner had the command been accepted than the rear diamond moved, at a trot, toward the wall that now spanned the width of Baltimore. Vander's men reached the barricade, spreading out along it, began to drag items out of the pile.

"Retreat, back through that hole! Follow Delta group." The formations reversed course, trotted toward Vander's group, Delta.

They were halfway down the street when an orange light blossomed. Thunder erupted.

The explosion backlit the black figures of Delta group in their various poses, searing their shapes onto Ogunwe's retinas. He raised his arms to block the light, but it was a meaningless afterthought. A wave of heat swept over him. The formations broke, men and women threw themselves to the ground. Some ran back up the street to where they'd come from, others bolted for the buildings on either side of the street, climbing in through shattered storefronts.

"Vander!" Ogunwe called out. Screaming filled his ears. On the ground, through the orange flare that had seared his vision, there were no signs of movement. The black shapes of Delta group, some torn to shreds, lay inert.

Ogunwe spun. The barricade in-between them and the transfer station had been filled in while Ogunwe had retreated. It, like the rear barricade, created a solid line across the street. They'd been baited to wait, thinking the gauntlet was a trap. It had bought their enemy enough time to fill in the barricade to the transfer station. There was no doubt in Ogunwe's mind what would happen if they were to approach it.

The same that happened to Vander.

As Ogunwe stood and stared, dumbfounded, the incongruous pile of material burst into flames. The sudden eruption shocked him back into movement.

"Reichmann! Get me a path through these buildings!"

Nothing.

"REICHMANN!"

Ogunwe spun, taking in his surroundings. *There has to be a way another way out of here.*

"Reichmann!" The silence beat, throbbed in his head to the time of his own heartbeat. Still no response.

Dammit, he cursed inwardly. *There goes my eyes, my ears,* and *my path to Patient Zero.*

Another soldier bolted for the buildings, disappeared into their black depths. From right and left, Ogunwe saw flashes of light, accompanied by the roar of gunfire, indication of the enemy sitting, waiting, picking them off one by one. Some of the flashes appeared from high above, in the darkened windows four and five stories above them. Ogunwe's soldiers returned fire, shooting blindly at phantoms.

"Reichmann you son-of-a-bitch, you better be dead." Ogunwe said, then issued a command over all channels. "All groups to me!" he boomed.

He grabbed Green by the shoulder, shoved her in the direction of a blank-faced brick building on their left, no windows, but a door that looked like solid steel. "Get me through that door and drop gas. We're gonna find a way out of this damn maze!" he shouted, his voice barely audible over the roar of the blaze and the staccato pop of weapons. Firelight played over Green's face, creating flickering shadows, highlighting the strain in her eyes.

She sprinted for the door, practically dragging one of her engineers along beside her. The remnants of Alpha, Bravo and Charlie groups clustered around Ogunwe. A cursory count told him only forty of his hundred and fifty soldiers remained.

"Stay on my heels," he ordered. "We're going to find a path through these buildings, make one if we have to."

Ogunwe ran to the door where Green was busy working on cutting the lock out. The soldiers moved with him, pressing themselves up against the side of the building, watching for muzzle flashes from across the street.

Green's method was crude but effective. The white-hot torch cut through a small u-shaped section of steel around the doorknob. She stepped back, kicking at the knob, black boot lashing out once, twice before it caved inward. The door banged open. Before it could swing shut, another engineer lobbed two small cylinders in the open space. They *clinked* down the hall, hissing as they released their contents.

No sounds of chaos, of people running. No screaming or hastily yelled commands. No firing of weapons.

Ogunwe pushed at the door, his weapons up. It creaked open to reveal a rectangle of complete blackness.

The doorway opened onto a central hallway that fed into individual rooms. He couldn't see anything, but what he could smell hinted at what purpose the building served. The hallway smelled of cinnamon, sage, spruce. It was at odds with the stress of the situation, as though they'd entered another world. Ogunwe moved straight down the hall, Green falling in directly behind him, the few remaining soldiers piling in quick as they could.

He moved without thought. HUD switched to IR, the white heat of the fires in his periphery receding as he stepped through the opening, holding his body sideways to present a smaller target. There was a certain speed that was ideal in cases like this; a swiftness that took the enemy by surprise that had to be tempered with caution at every turn, at every point of potential ambush. Ogunwe moved at that speed, looking for a heat signature that either moved or resembled a human form.

The sound of boots made hollow *thunks* as they pounded down the concrete corridor. A hundred feet down, the hallway ended in a *T*. Ogunwe paused near the intersection, shoulder to the wall. Soldiers piled in behind him, Green first among them. Ogunwe turned to her. Her shoulders heaved; she struggled to catch her breath. "Got schematics yet?"

"Think I've got it. Yeah," she said tersely in-between breaths. "Take a right here. At the end of the hallway should be a door. Beyond it, there's a shared wall with the next building over. It'll get us past that barricade."

"Exit?"

Green shook her head, a hint of a smile curved at the corner of her mouth. "Doesn't mean we can't make one."

Ogunwe glanced one last time down both extensions of the corridor. No telltale heat signatures from either side. He spun to his right and sprinted down the dark hallway, his pistol pointed straight ahead. He could hear the remainder of the soldiers behind him. Shouts, thundering echoes of shots from behind, the booming sounds amplified throughout the narrow hall.

So much for the element of surprise, he thought with chagrin. Ogunwe sprinted at top speed, not slowing down as he approached the door. He threw his shoulder into it, uttering one last silent prayer that the door wasn't reinforced as he did so.

The door crashed inward, slamming open. Agony flared in his right shoulder as his body catapulted forward, thrown to the ground. He rolled with the momentum, turned it into a somersault, landed in a crouch.

He was surrounded by the yellow, orange and ultimately white cores of heat signatures. Everywhere.

And the screams of women. No, not women. Children. Huddled all around him. They filled the room. Forty or fifty of them cramped into a space the size of a large living room.

What the...

The ArcSec soldiers flew into the room behind him, abruptly stopping and bumping into one another.

"Hold your fire!" Ogunwe screamed.

He pushed his troops to line the walls as they entered, and sized up the space. They were in a large rectangular room. Corridors stretched into blackness at each far corner. Both led north, presumably past where the barricade stood outside the walls. As his men filled the room, no adults stepped up or came forward. The children had been abandoned.

Why would there be so many kids, by themselves, together? The scent of spice, cinnamon and perfume, was heaviest here and triggered a vague suspicion within him.

They're not here by choice. They were kidnapped. A vile hatred arose within him then as his suspicions were confirmed. *It's a sex-trafficking ring, but the sicko coward leaders got scared and bugged out to someplace safer.* If he got his hands on whoever was responsible, he'd personally crack their necks for them, one by one.

"Green!" Ogunwe shouted.

"Here, Chief." She approached carefully but quickly, stepping around the huddled shapes.

"Get that door sealed and get us an exit out the far wall, down the corridor." He looked around at the children. The screams had died off, but a muted mewling—some sniffling—persisted. "We're taking these with us."

A pause. Ogunwe could almost see the objections running through her head. He knew she'd come to the only conclusion possible. Whatever losses would justify the gains. If they didn't do it, didn't at least try to bring them to freedom, then ultimately the responsibility for the children's fate was theirs as well, which included consigning them to a lifetime—whatever remained of it—of abuse.

Maybe it would make up for the men and women already on his conscience. *Delta group. Vander.* Ogunwe's throat tightened. *Leave it for later*, he thought.

Green moved, assigning her lead engineer to the sealing of the doorway. The existing door held nothing in the way of structural integrity. He pulled an AWOL from his pack, set it lengthwise across the entryway. There wasn't much time. Whoever had followed them into the building from the street would be on the door in minutes. *Maybe seconds.*

Ogunwe slid a flare out from his thigh-pocket and cracked it open, his shoulder aching with the effort. Haunting green light spilled over the room, illuminating Green's shadow as she disappeared down the northwest corridor. Tiny ovals of faces looked up to Ogunwe. Knees pulled to chest, arms hugging knees, some with heads buried in their legs. Tears and snot streaming down faces. Not a single adult among them.

He moved to the center of the room, raised his voice enough to be audible but not loud enough to be considered intimidating. "My name is Mister Riku Ogunwe," he said slowly, letting his voice dip and rise in a sort of musical wave with each syllable. "I'm with ArcSec security. We're here to rescue you kids." The mewling gradually died off. "Are there any big kids here?"

Hesitantly, three hands rose into the air, one female, two male. Each of them an indistinguishable mezcla. "You three," Ogunwe said, "please stand up. I need your help." The three rose tentatively, the girl the tallest of them, frizzy brown hair bound behind her head which just barely reached Ogunwe's chest. All three were frail and underfed, innocent. Bile rose in Ogunwe's throat. Whoever was in control of this operation deserved to die a thousand horrific deaths. But the pissants had already fled. It sickened him, but he had no time for emotions and no time to lament what he couldn't control. He had to focus on the present, on what he *could* control.

Ogunwe gathered the three children, gave them instructions to separate the rest of the younger ones into three equal groups. Told them when he called their group then it was their turn to get their group down the hall and out the exit.

"But for now, I want you to go to each of your groups. Get them standing around you. Huddle up and hold hands, okay? Then wait for my signal."

The kids nodded. Ogunwe noted their muted expressions, reserved, solemn eyes and closed mouths.

"Okay, go!" he shooed them away and they moved quickly.

No sooner had they gotten their groups to their feet than the sound of the door crashing open incited another round of screams. Fortunately, the first AWOL was already up. They'd closed the door that entered onto the large living area where the children were seated and placed the AWOL in its frame. Their assailants wouldn't be able to open the door or see in.

The door slammed into the AWOL again and again. Ogunwe shook his head. Where was Reichmann? He needed to know where that girl was, the patient with the key to the viral mutations. But he couldn't wait any longer. There wasn't any time left.

He contacted Moon.

As soon as the man answered, Ogunwe spoke, bulldozing over the niceties. "I need your help," he panted.

"What?" Moon asked.

"The girl. Patient Zero."

"What about her?" Moon asked.

"I need you to track her. Can you do it?"

"Sure, but ArcSec's surveillance network is superior..."

"Just do it!" Ogunwe shouted. "Send her location to me."

"Got it," Moon said.

"And Moon, by chance you have any pull on vigilantes on 1, tell 'em to stop shooting at us," Ogunwe said, then disconnected.

The pieces of his plan were falling into place. *Assuming Moon can get you what you need.* But there was plenty that needed to be done in the meantime. *Like getting these kids out safely.*

"Green, status," Ogunwe commanded.

An explosion sounded from the corridor. Ogunwe whipped around to view the source of the sound. A flume of dust and debris shot out of the hallway where Green had disappeared.

"That status enough, Chief?" came Green's voice.

"Think so," Ogunwe said. "Anything beyond that wall we need to worry about?"

"I don't know. Might wanna get some of our people through this hole first, see what's waiting. Do it quick though. They've probably got eyes on the outside of the building."

"Who's the ranking officer?" Ogunwe called out to his men. He left off the implied. *Beneath Vander.* The soldiers stood at the ready to either side of the entrance, weapons pointed at the AWOL. A stocky man, not much taller than the tallest of the children, peeled off from the group, stepped up. He had a proud bearing—arched back, haughty chin, firm gaze.

"Vazquez, sir. Whaddya need?"

"Take half these men, form a forward guard. Scout the way to the transfer gate. I've added you to the command channel with Green. Take her with you. I'll command the rear-guard. Kids in-between."

Vazquez nodded. "Sir."

"Above all, protect the children."

"Goes without sayin' sir. My life before theirs."

"Hop to it, soldier. And good luck."

Vazquez was efficient in gathering up his men. They filed down the corridor and out of sight swiftly. Ogunwe called on the leader of each group of children, ushered them to where the corridor began. Told them to hold hands of their buddy that was next in line and wait. Meanwhile, he gathered his own group of soldiers.

They were a mixture of nerves and adrenaline. Black jumpsuits covered in dust, blood. Some bore wounds. Punctures to the torso, lacerations to the face and arms, legs. Their jumpsuits had done the job of stabilizing the wounds, feeding the sites with localized coagulants and painkillers.

The banging on the door reached a pitch and frequency that was almost unbearable. Then it stopped, the silence eerie, unsettling. Pulses pounded and sweat dripped from temples. Fingers twitched outside trigger guards.

Uh-oh, Ogunwe thought, knowing whoever was out there wouldn't just give up. They'd gone to too much trouble to try and trap him and his soldiers. *This* can't *be good.*

Then the firing began. Thunder filled the air. Rounds cracked and splintered the door, pinged off the AWOL. Ogunwe winced as the screams of children rose once again, adding to the cacophony.

The thousands of rounds their unseen enemy had unleashed shredded the door in a matter of seconds, leaving nothing but smoke and a pile of splintered debris at the foot of the AWOL. The clear surface was pocked and scarred yet remained intact. Beyond it, two darkly illuminated men, green wraiths in roiling smoke, holding large weapons, black barrels fixed on Ogunwe. Ugly smiles fixed on cruel faces. Caricatures of evil. Over their shoulders, in the shadows, more adversaries.

Ogunwe pulled his face back around the corner as the firing began again. This time at the walls of the entryway that surrounded the AWOL.

"Green, Vazquez, can we get this party started or what?!" Ogunwe shouted.

"We're at the base of the steps to 2 but haven't had time to scout the way to 3. Need more t –"

Ogunwe cut Vazquez off. "No more time! We're coming now, hot on your heels. Make a path or die trying!"

Ogunwe sprinted to the head of the corridor. He kneeled down, looked into the eyes of the girl who'd volunteered earlier. Her lip quivered but she managed to keep from breaking into hysterics. *Strong girl*, he thought with a touch of sadness. The Arc bred them a little too strong, sometimes to the point of breaking.

"Okay, it's time," he said. "Move quickly but stay behind the men in front of you. They're here to protect you, understand?"

She nodded.

"Good. Now go!"

The girl held the hand of another girl, right behind her, who held the hand of a boy, no more than three, behind her. And so on. The leader started down the hall, tugging all the children in a line.

"Don't let go!" Ogunwe urged as they passed by him, fear and bewilderment in their eyes.

The second group followed, then the third.

"Alright men, to me." Ogunwe said as calmly as he could. "Follow those kids. Make sure they get to safety." The soldiers filed past Ogunwe, who stood at the mouth of the hallway. Green's lead engineer was last. Ogunwe put a hand to the man's chest as the rest of the soldiers disappeared down the hall, out the barely illuminated hole that led outside.

"Got any of those AWOLs left?"

"Course," he affirmed, unslinging his bag from his shoulder. He began to reach in when Ogunwe stopped him.

"Just leave me the whole thing," he said.

A wariness overtook the man's face but was coupled with something else. Urgency. Anxiety. Fear. His eyes flickered up, above Ogunwe's shoulders, to the exit. To freedom.

"It's okay, soldier. Leave it to me. Just get the hell out of here, help save those kids."

The man dropped the bag at Ogunwe's feet, and fled.

Ogunwe didn't watch him go. There was only one way his plan would work, and there wasn't any time to spare. He pulled an AWOL from the bag, began to set it up at the head of the corridor when a voice shouted in his head.

"TALKSMALL Enforcers at the stairs to 3!" It was Vazquez, the stutter of automatic weapons coming through over the channel.

Shit, Ogunwe thought, but then was reminded he had more immediate problems to deal with. The rebels who were trying to gain entry to the room had gone dark. The deafening noise of weapons fire died out; Ogunwe looked up to see a fist explode through the final layer of drywall. Another hand tore at the wall surrounding the hole they'd made. They were circumventing the AWOL. A booted foot lashed out at a mangled piece of framing, bending it kick by kick until there was a hole big enough to pass through.

Ogunwe watched with curiosity, strangely mesmerized. A man crawled through head first, breaking Ogunwe's trance. He prodded at the AWOL, activating it just as the man leveled his weapon and pulled the trigger. Ogunwe flung himself to the ground as the AWOL ballooned upward. Bullets sang inches from his head, tore through the wall behind him before the expanding AWOL caught and deflected them. He flinched involuntarily as the bullets pinged the shield. The man opposite him lowered his weapon and in that instant Ogunwe knew his intent. It was only the slightest of movements, but the vector was there in his body language.

The man knew the AWOL's bond wasn't set.

He let his weapon dangle on its strap and propelled his body into the clear shield. Ogunwe braced it with hands and a shoulder, wedged the rubber of his boots into the concrete. The impact came, shoving Ogunwe back. The assailant managed to knock Ogunwe off, pushing the wall a few inches on one side.

Ogunwe pushed at the edge, trying to level it up, when he felt lightning strike down his left ear, across his cheek. He felt a jet of hot liquid course down his face, followed by a white-hot pain that turned the vision in his left eye to white. He turned to view the knife coming at him again. Ogunwe flicked his head to the side, thrust both hands up, and grabbed the man's forearm and pushed it high, away from his face, before yanking it back down over his shoulder. Something cracked and the man wailed in agony. *Fuck you, motherfucker.*

The bond was forming again. It gave him an idea, a way to buy some time. The man pulled on his broken arm, trying to get it back, but Ogunwe held on like a dog to a bone and forced it to where the AWOL met the ground. The bond formed around the man's limb, sealing him in with the autowall. As the man pushed and kicked at the base, trying to get some sort of traction, Ogunwe held fast, using the opportunity to make a call. *Pick up pick up pick up*, he thought urgently as the line rang.

The man beyond screamed in frustration. He stopped struggling, raised his weapon.

Ogunwe let go, turned and ran as the shots rang out, hoping like hell the bond had formed successfully. He dove through Green's makeshift exit, which glowed the sickly yellow of sodium vapor, a ragged hole blasted through layers of drywall, framing and brick.

His system continued to ring. No answer.

"C'mon Moon, answer dammit!" Ogunwe shouted to nobody. He scrambled to his feet outside the building, sent a hasty message:

*MOON, CALL OFF YOUR THUGS AT CENTRAL
TRANSFER GATE, 2 TO 3. MY PEOPLE ARE
ESCORTING CHILDREN TO 3. WE THINK THEY'RE
SEX SLAVES. CALL OFF YOUR MEN NOW!*

He stood there at the verge of the wide-open area. To his right, the wide expanse of stairs. Just ahead, the area surrounding the elevator, cordoned off by a thicker version of AWOLs. To his left, the west, the unknown. The whole area populated by a less dense crowd of refugees. He could still see the swath his people had made as they'd cut straight to the stairs. None of his soldiers were visible; they'd already pressed onward, upward to 3.

It was done. He'd achieved what he could. All that was left now was to hope that Moon would honor the request, that his soldiers and those children would make their way to safety. His duty now wasn't just to the cabal, to the Golden Road, but to the greater good.

Which meant Olivia would have to wait. Again. He was nothing if not practical, and a part of him suspected he wouldn't survive beyond the day. Knew it deep in his core.

Before him, the stairs to 2 stood. Beckoning. Taunting.

"Boss, we're in!" Green's voice in his head was breathless, jubilant. He could almost see her normally stern, angular face light up. "TS guards backed off and let us through!"

Ogunwe smiled. He was weary, exhausted, but triumphant. He'd managed to do something right. His eyes again went to the wide stairway, bronze and black antiques that led upward into the continued gloom of 2. A little farther, beyond 2, lay salvation.

"Chief? You read me?" came Green again, the note of doubt hanging there.

A message came through on his system, from Moon. A pulsating pin on a Map.

The girl's location. Sonofabitch came through.

Ogunwe knew what he had to do. And maybe, just maybe he'd be able to exact some of that revenge he was seeking in the process.

"I read you, Green," he said. "You're in charge now. Take care of those kids."

"Chief?" came Green's voice, limned with fear. Ogunwe closed the comms channel.

The deafening roar of automatic fire continued behind him, rounds shredding the walls surrounding the corridor. Ogunwe glimpsed tracers in the darkened hole.

I must be insane, he thought, and turned to the left, shunning the steps and certain escape, and took off at top speed into the crowd.

RAY

RAY
AGE: 36
SEGMENT: COMPLEX PROPERTY; PERSENSE CPX
POSITION: REDACTED
SOCIAL SCORE: NA, CPX-SPONSORED

RAY LEFT JONAS REICHMANN where he'd found him, just a little less alive.

The man had spilled every little secret he knew before spilling his blood. The bit about the Golden Road was interesting, but not necessarily novel. Just confirmation of his suspicions of Ogunwe, the traitor he was. The part Ray and the AI found most interesting was *how* Ogunwe and his little band of confederates had managed to steal access to Novagenica. Turns out it had come from Talksmall. But how Talksmall had acquired it was the real riddle. They'd stolen it from a thief who'd stolen it in turn from Tori Fanning, the handsome but obviously feckless lead of the Exodus project.

With this information, the AI did some digging and found the petty thief's name was Saul.

Ray promptly added his name to the list. *You're next, buddy,* he thought, a familiar excitement stirring within him.

As soon as this little nugget had been relayed, Ray dispatched Reichmann to hell with a slash of his blade across his thick, stupid neck. He propped the corpse up, balanced it so the man's bulk slumped against the pillar, torso covered in red, a sheet of blood that poured in a viscous waterfall down his neck. The body was bathed in a spill of light that looked, ironically to Ray, like a halo.

Just a little gift for your buddies. A warning of what's coming to meet them too.

Ray exited ArcSec's command post on 3 with renewed vigor, a spring in his step.

He took the shortcut to 1, through the NG-owned building on Vine. The entrance was neither publicized nor accessible to those without credentials. A scuffed, gray door in a non-descript maintenance shed a few blocks south of the transfer park on 3.

For Ray, the door had simply clicked open, exposing a gasket-lined frame that bordered a rectangle of black. He entered to complete darkness. The air was stale, tasted of dust. As his eyes adjusted, he noted the outlines of a railing and steps leading downward. He took them carefully, down a flight. They terminated at a landing, where another door awaited. This one, like the first, clicked open as he neared. Around the frame of the door was a thick foam seal, same as the outer door.

He could appreciate the security, intuiting the purpose of the stairwell. It was a killbox. For those who'd somehow managed to get past that first, public-facing door. Once confined to the space, the building's defenses would activate, neutralizing all intruders.

Probably gas, he mused. *Or the lack thereof. It's the easiest method.* All you had to do was pump the oxygen out of the room, create a vacuum.

Fortunately, Ray was recognized as a friend to Novagenica, and thus a friend to the building. He passed into the main column of the building itself, the door clicking softly against its seals behind him. The lobby wasn't furnished, an empty space with open framing and unfinished concrete floors. Following the theme, it was dark, the only light coming from glowing control screens at the elevator stack in the center of the space.

Ray crossed the floor, kicking up dust, jabbed a thumb at the down icon on the screen. A vague whirring, the rushing of air. The elevator whisked him down to the main level. In a matter of seconds, he began on 3, passed through 2 and terminated his journey to 1.

The doors slid open to a brightly lighted floor littered with pallets and packing materials. He stepped out, poked around the stacks of product. The fleet of NARKs that had once stood arrayed in a regiment were no longer present, leaving a blank space just outside the elevator doors. No people either. If they were smart, the NG workers had either found a secure space somewhere high in the building, or had fled altogether, exiting out the top of the building onto 3.

Regardless, it was none of Ray's concern. What *did* concern him was a lonely shack of a building not far from this one. The place the AI had tracked the criminal Saul to, where Tori's permissions had first been accessed.

Ray stepped out the side door into a darkened alley. To the sour, over-powering smell of rancid meat. He could almost visualize the cloud of vapor emanating from whatever rotted there in the alley. It served as a reminder. He pulled his hood forward over his head, toggling controls to complete the seal between hood and jumpsuit.

He wasn't just hunting soft urbanites willing to accept a little graft for inserting a piece of code. Whoever had the means and skills to infiltrate a CPX like would also be prepared for the eventuality of an assassin. Ray was betting the perpetrator was still present, but he needed to proceed with caution.

At the mouth of the alley, just inside the building's protective perime-ter, he reconnoitered the street. The clear, jagged remnants of broken AWOLs were scattered along the street. Now only a smattering of people populated the street, setting up lean-tos that had been shredded in the tumult that had just recently passed through.

He proceeded out into the open, made his way to Saul's hideout. A brisk five minute walk brought him to a brick-building with a darkened, dirty face that seemed, like so many other places on 1, to have been forgot-ten. Boarded windows and a crenellated roofline that had been subjected to the erosive depredations of time and neglect.

He's in there, the AI said. *We have all the information we need now.*

The rest was implied. No need for questions, he'd go in fast. It was eas-ier when he didn't need information. Just move in like a wrecking ball; a violent, massive attack with little warning.

Ray unslung his pack, removed the slim bullet of a drone, took it up to thirty feet. Trained his sights on the window that was criss-crossed with boards. Sights still trained, he started across the street, at first warily, weaving through the squatters. He pumped a unit of speed from his inter-nal tanks, picked up the pace. Halfway there and he loped along at a full sprint. Heads turned in his wake, watching. He ignored them. He jumped up to the curb, simultaneously letting loose two pencil missiles from the drone. They *psshed* in from above, making their telltale fizzing sound, and hit the window a second before he did, exploding the frame and half the surrounding wall inward. He threw himself through the empty space that was now just a cloud of dust, landed on splintered wood and shards of glass. Nimble feet pitched him forward into a roll. His shoulder crashed into something. A wall maybe. He switched to IR.

Everything in his vision was blanketed with red and orange. Whoever was here had been smart enough to heat the building until it felt like a jungle. Not quite a hundred degrees, but close. Enough to cause confusion. He switched off IR.

The air was choked with dust, but Ray could make out dim outlines, grayed-out features. He was crouched against a knee-wall. Above him, a framed opening that served as a pass-through window from one area to the next. He was in what looked like a kitchen nook.

He brought the drone in from the street, took it in low so it was just outside the blasted-out hole in the wall. Instead of popping his head above the frame and presenting a target, he viewed the space through the drone's camera: the dust settled and the space beyond the nook materialized into view. A wide living area. Couch in the middle. On the far side, a workbench backed by a wall held weapons on hooks. Pistols, knives and larger, automatic weapons on the far right.

Two of the hooks stood empty.

Always good to know, he thought with wry humor. Whoever was waiting for him did so with a fully automatic weapon for sure, what kind of rounds though remained a mystery. Only a few weapons manufacturers had the skill and know-how to make something that would cut through his suit. The DIYers though, they made him anxious. They were generally more creative and invented things that could kill in previously unimagined ways.

Ray eased the flechette airgun from its holster and interfaced with his system, setting it to autokill. He stood abruptly, muscles tensed. The speed coursed throughout him, trying to hijack his nerves. It made him want to run, to move, to smash. He controlled the urge, clenching his one empty fist, digging gloved fingers into the palm. Through the window, no movement.

He moved around the edge of the wall, moving as quietly as possible. It was no good. Debris littered the floor, tiny bits of wood and brick crunching beneath his boots.

He pointed the pistol at the rear of the couch, rounding it rapidly.

Nothing. He looked beyond where the work bench terminated at the corner of the wall. A darkened hallway led to the rear of the building.

Aside from the sound of his own breathing, the building was cocooned in silence.

Is he gone? he wondered, frustration mounting. He knew he shouldn't allow the chemicals in his system to influence his decisions, that caution was always called for, but he couldn't help it. He moved more out of impatience than anything else.

"Should've waited to bump that speed," he whispered to himself.

No, your instincts are right, the AI said, *he's here. Use the gift of speed. Smash our enemies.*

Ray obeyed, and it felt good.

He moved rapidly down the darkened hall. Kicked open a door on his right. A bathroom. Yellow and brown-stained toilet. He fixed his sights on the end of the hallway. Another closed door. Despite the urgency he felt, he schooled himself to patience. He began to walk in place, gradually stomping his feet ever louder. When his stomping reached a crescendo, Ray stopped abruptly, quietly pressed his back up against the wall. Still a good fifteen feet from the end of the hallway, he kept his eyes trained on the door, hoping they'd fall for the ruse.

Come out come out, wherever you are.

The door burst apart. Splinters flew down the length of the hallway toward Ray, ricocheting off his suit. He fell to the floor, flechette extended forward, aimed at the gray rectangle. Smoke roiled throughout the hallway, above his head. Beyond the doorway, he saw a dark shadow of movement through the cloud of smoke.

Ray let the speed take over, his body thanking him for finally utilizing it as he pushed up and into a sprint with inhuman quickness. He exploded through the frame, threw himself into the dark shape. Gunfire erupted but it was too late. The rounds flew errant, pocking the ceiling. Ray's whole body was a projectile; it flew into his attacker, slamming the body into the wall at the far end of the room.

He heard the breath go out of the man in a loud *oof.* More gunfire, the man's finger clutching involuntarily at the trigger. Ray stood, stomping on the arm that held the weapon. The man shrieked in agony as the enhanced force from Ray's suit aided in the snapping of both bones in his forearm.

The man went limp, must have passed out in shock from the pain. Ray put one flechette into his neck. The gun making *thwump* as he pulled the trigger. The dart, loaded with poison, pierced the flesh, burrowed below the surface. He wouldn't wake up.

Movement, in his periphery to his left. Ray turned to see a man somehow standing and cowering all at once.

He was a filthy weasel of a man. Skinny and bent, his pinched, grease-slicked face shook back and forth in denial of the situation. He held nothing of danger, his outstretched palms his only weapon. A plea, perhaps, to Ray's better nature? *Fat chance, fucker.*

Ray approached the man, lowered his weapon. "Saul?" he asked.

The man nodded fervently.

Ray studied the man. "So you're the guy that hacked NG?" Ray said, allowing the smallest note of praise. Amazing that someone who could perform such a feat could be so outwardly cowardly.

"I—I din't know," the man stammered. Ray looked in his eyes. There was something there that didn't match. The body screamed fear, but the eyes held nothing. No fear, no remorse, nothing. It was like looking in a mirror.

It's an act.

The thought flashed through his mind. Saul's eyes flickered around Ray.

Ray brought his backup in swiftly, silently. It was strange, vertiginous, watching himself from behind, from the viewpoint of the drone, standing there facing the bait while a huge hulk of a man approached him stealthily from behind.

Ray accessed the fire controls on his system, was about to aim the minicannon when he was cut off.

Let me do it, the AI said, and before he could protest or even relinquish control, the reticules were already locking on the back of the man's head.

Two quick, successive *pop pop*s and the man collapsed at Ray's feet, the back of his head a yawning bloody maw. He didn't bother looking back. He could see everything through the drone's video feed.

"Nice shot," Ray said aloud.

Thank you. I enjoyed it immensely, the AI said.

The confidence bled from the coward's eyes.

There it is, Ray thought. There's *the fear.* He gripped Saul's throat, shoved the barrel of the flechette in the man's broken mouth, and fired.

He left the bodies where they lay, exiting the smoky confines of the building to a street full of people who pointedly did not look in his direction.

He consulted his list. One by one the names of the guilty dwindled.

Just a few left now, and as he checked the map he noted with a muted sort of delight that the remainder were all grouped together. Riverside. They were making this easy on him.

It's almost time now. They're joining forces. Follow them to Riverside. I will tell you when it's time.

"What about Ogunwe? Can I at least take the sonofabitch out?" he asked, holstering his weapons.

As long as the girl remains unharmed.

Yes! Yes. "I can do that. I can. Easily," he repeated with a sort of manic glee.

Here we come, he thought. *Time to put an end to this little rebellion.*

VAL

VAL
AGE: 19
SEGMENT: LEGACY CITIZEN
POSITION: UNEMPLOYED; GOVERNMENT DOLE
RECIPIENT (STANDARD UNIVERSAL CREDIT)
SOCIAL SCORE: 822

THE BLACK HOLE WAS A DEAD REMNANT of its former self. At least to Val's eyes.

It had only been two weeks since Trevor had first taken her to the Black Hole, introduced her to his Group Involuntary Ejaculation concept. She had remembered feeling overwhelmed, incredulous, but also a growing admiration. In Trevor, she saw a spirit that was willing to take risks to create, to find a new way. That's what the Black Hole had come to represent in her mind. Possibility. Hope.

Where these values had once thrived alongside a sort of peer to peer commerce, the Black Hole now was simply another place where the destitute came to wait—for alms, rescue, an ignominious death. Packed in shoulder to shoulder, the stench of waste and the unbathed began to overpower the dampening effects of the cold.

The three of them wound their way through the middle of the crowded area. Val looked up. The catwalks were cleared of the usual urchins and in their place stood guards, peering over the railings, dressed in dark blue uniforms, old-fashioned assault rifles cradled lazily in the crooks of elbows. The white *TS* emblazoned on their chests glowed so that it was more prominent than their features. A disembodied reminder of the threat of force.

Val shuddered. *Never a good feeling to have a gun trained on your back.*

She tapped on Kat's shoulder, who was crammed in-between her and Trevor.

"You doin' okay?" Val asked.

"I'm fine," her sister said, and Val was more disconcerted than ever. After debarking the train, Kat had seemed to grow weaker by the moment, her body wracked by constant shivers. Val knew she was sick, but didn't want to do what she would've done in a former life—badger Kat until finally the girl exploded in anger, then shut down completely.

So she'd remained quiet about her suspicions.

And then, through the terrifying ordeal in the Blind Tiger, Kat had seemed to come out of it stronger. She'd fallen outside the Blind Tiger, a spill that made Val's guts jump and quiver, but Kat had popped right up like it had never happened. Blamed it on her clumsy feet.

And from there forward, nary a shiver. Conversation that seemed like it fit Kat's personality, but not the moment, if that made sense.

"You sure?" Val asked again. Worry was beginning to gnaw at her. She wanted to grant Kat her space, but she if there was something wrong and she could help...

"I told you I'm fine. Quitcher worryin'. Sheesh," Kat said, rolling her eyes.

Val, behind her, shook her head. *Something's not right.*

They shuffled through at a crawl, snaking their way along, uttering apologies as they went. People grumbled and complained, sniffled, coughed in their faces. Simply crossing the hundred or so yards of cobblestone took a full twenty minutes. Finally they broke through to the far north-western side of the market, to the narrow passage between buildings that led to the park where Talksmall had built its new ghetto. Only to find the entrance to the Riverside project barred.

A wall of Talksmall Enforcers barricaded the path. Here, the crowd was more unruly, shouting epithets at the guards, occasionally lobbing the odd stone. Still, the refugees left a field of space ten feet deep. They knew the reputation of Talksmall Enforcers. Everyone did. Val made her way to the edge of the crowd, stood on tiptoes, eyes focused on the black tenements beyond the guards, jumbled shapes shrouded in the natural darkness of Level 1. In a matter of days, they'd grown the project into a community, like a geometric monster that had crystallized, piece by piece.

"Gimme a sec," Trevor said to the girls, then went to approach the Enforcers.

"Mm-hmm," Val muttered, making a point of not looking at him, silently relieved as he stepped away.

She should've been overjoyed to see him. She knew that. And she had been the instant she'd exited the train and laid eyes on him. But as she'd hugged him fiercely, that tapping on her shoulder from Kat had brought her swiftly plummeting to reality.

Inside, in that instant, something flipped.

It was a weird feeling, something she wasn't sure she could explain. Like being a single mother and uncertain of how to behave, or even if it was okay to show the way she felt about Trevor. She was caught somewhere in-between her love for Kat and her love for Trevor. Like one had to take precedence over the other.

And if I had to choose...

It was probably stupid. And if she could just bring herself to open her dumb mouth and say what she was feeling to Trevor, and to Kat for that matter, the feeling would probably go away.

He'll say something equally dumb and totally adorable and this, this awful sick feeling inside of me will just slide into nothing.

But she couldn't do it. Couldn't bring herself to say the things that needed to be said.

So she ignored it, focused on her surroundings, became transfixed by the buildings.

Vaguely at first, then more forcefully, she heard Trevor's voice reach a crescendo.

"Just fucking *CALL* her!" he shouted at one of the TS guards, shaking a fist in the direction of the park.

"Get the fuck outta here," Val heard the Enforcer say before shoving Trevor. He fell back on his ass, hard, hands planted in the mud.

Val saw an expression on Trevor's face she hadn't seen before. Something like murder in his eyes.

Shit. She rushed forward but Trevor was already up and on his feet, stalking toward the line of Enforcers.

She grabbed his wrist with both hands, dug her feet in, but her heels slid through the mud. "Stop!" she pleaded, but he continued forward, dragging her behind him.

He was a few feet from the line when the entire line of Enforcers raised their bats. An evil little grin on the face of the one who'd shoved Trevor. The man beckoned him forward with the curl of his fingers.

"Bring it pussyboy," the man said. He lunged forward, swinging his bat. From behind him, another man was suddenly there. Tall, imposing in tactical gear. He grabbed the Enforcer by the collar and yanked back, hard. The Enforcer's black bat *swished* through empty air and his arms flailed as he was flung to the ground.

The Enforcer grunted and his face flushed violet with rage. "Who the fuck..."

The imposter planted a boot on the man's chest, pushed him back down into the mud. He wore a debris-littered ArcSec uniform and his face, dark brown skin covered in a sheen of a week's worth of sweat and effort, was impassive, carved of stone. A wide swath of blood glistened, ran from his left temple, down his neck and into his uniform.

"He's one of us," the man said. As one the other Enforcers relaxed, lowering their bats, turning their attention back toward the refugees.

Strange, Val thought. *He belongs to ArcSec and Talksmall? How is that even possible?*

The man focused his gaze on Trevor. "You have the girl?"

"What girl? Who the hell are you?" Trevor spat.

From behind, Kat stepped forward, around Val. "He's talking about me." She turned toward the man. "Aren't you?" she asked.

"Patient Zero?" the man asked.

Kat coughed a mirthless laugh. "Yeah, but you can call me Jesus. Or Almighty Savior. Or sometimes just Kat, when I'm feeling humble. But I have to say, I *am* preferential to Your Lordship. Has a nice ring to it."

The features on the man's face softened into something like appreciation.

"Name's Riku, but you can call me Chief if you want. People used to, but now..." he hesitated, a brief sadness flitting across his face before hardening once more. He offered a hand. "Let's get you guys somewhere safe."

Kat took his hand, her tiny paw swallowed up in his palm.

The four moved forward swiftly, the guards shifting wordlessly, shuffling just enough to allow them to pass through in single file. The crowd behind them burst in a renewed explosion of jeers, exclamations of unfairness. Val looked back to see, however, that none had dared to try and follow in their footsteps.

The warmth of satisfaction eroded, bled away to a cold shame, guilt. They were people, just like her, searching for a solution. For something as simple as shelter, a thing taken for granted, unappreciated until it was lost. She looked forward again as they wended through the hoard of refugees, scaling the gentle rise of earth that preceded the park.

At her side, Kat breathed heavily at the exertion. It was the first sign since they'd exited the Blind Tiger that she was still sick. Her sister glanced over furtively, offered a small smile and took Val's hand. Rubbed it as if Val was the one in need of comfort. Val's gut jumped at the touch and, as she looked into Kat's eyes, saw a horrifying blankness that caused her to look away.

What's happened to you? Where did you go? she thought, then felt shame at her own presumption. Just because Kat was feeling better, stronger, was no excuse for her to assume something was wrong. She could just hear Kat's response... *You just don't* want *me to be strong.*

And maybe she was right.

It was something Val hadn't considered before now, but now that she had the thought she couldn't simply discard it. Had she been holding Kat down instead of doing what she should have? Lifting her little sister up, guiding her to become a strong woman?

Maybe she'd been operating for too long under the low-level assumption that there was no future for people like them. Just a seamless flow of days that offered little variety and even less opportunity.

She didn't know the answer, but she *did* know that it didn't have to be this way. That every passing moment was one in which she could change the course of the future. That, it seemed as she took in the tumult of her surroundings, the opportunity was ripe for just such a change.

This thought in mind, Val looked back to Kat, gave her a reassuring look and squeezed her sister's hand twice.

"Ready to save the world?" Val joked.

"It's what I was born to do," Kat responded, but her face turned down into concern. She gestured at the people they passed.

Val paid closer attention, noticed a fair half of them afflicted with a yellow crust that accreted at the corners of their eyes and the rims of their noses.

"I just worry about what happens if it mutates, like they said," Kat said.

"Me too," Val said. "But hey, I've met the guy who's gonna decode the virus. If anyone can figure it out, he can."

"What's his name?" Kat asked, and her

eyes held something like avarice, but before Val could process it she was blurting out his name as if compelled.

"Ian," she said. "He's with Talksmall."

Kat nodded and went silent.

Val sank into silence as well, feeling unsettled. She couldn't ignore it. Something still wasn't right. She couldn't pin it down, but it was there, a subtle energy, like unseen currents disturbing, kicking up silt on the ocean floor.

Ahead, the man in the ArcSec uniform, Riku, led their small group to the top of the hill, just past the buildings that created the border between market and park. He stopped at the crest and they pressed in-between the people to join him, standing in a horizontal line, shoulder's touching.

Val gasped at the sight before them. A new civilization was being erected before her eyes.

The ghetto, if it could be called that, was an agglomeration of staggered box-like structures piled atop one another, built up and out into a cubist work of art. It now filled up the entire park, a district unto itself. Down the slope that was dotted with the multitudes of refugees, the path extended then broadened into a diamond-shaped open area bordered on all sides by the compilation of connected boxes stretching hundreds of feet up to the underside of 2.

"Whoa," Trevor breathed.

"How did they do this? In what, like two days?" Val said.

"Really is something, isn't it?" Riku said. His voice had a resounding timbre that made her think of the warmth and safety of home.

What home? Val thought, and teetered on the edge of despair. After all this was over, one way or another, they'd still be without a home, aimless, wandering people, scrounging just to survive another day.

"Let's not waste time," Riku said, interrupting her thoughts.

He picked his way down the slope and t

hey followed him, stepping lightly over and around the people that had been allowed to enter the park, some standing, many sitting or even sleeping. At the bowl of the hill, a path was sandwiched between the black Talksmall high-rises, like a river running through a canyon. Crushed rock crunched beneath their feet as they moved through another line of Talks-mall guards to the center of an open area that resembled a town square. The crowd beyond the second cordon had thinned, seemed restricted to Talksmall personnel only.

The ArcSec man, Riku, led them to a stage, thirty feet in diameter in the shape of a hexagon, that stood in the middle of the open space. Like the structures that surrounded it, it was also a matte black.

A tall pole thrust up from its center. Val eyed the pole with unease. It gave the stage the look of a gallows.

They stood there, dwarfed by the giant structures around them. Val rested her arms on the stage, leaning her chest up against it. The housing boxes that ringed the square all had balconies attached. She'd once seen an image of caves that had been dug into cliffs. This reminded her a little of that, save for the fact that in this case the dwellings took up the entire cliff face *and* extended, like the warren of a rabbit, deep into the structure.

"So," Val started, "what now? You try to get a hold of Estelle?"

"She's not responding." Trevor said.

"Now, we wait," Riku said, folding his arms and looking southward, up the hill.

Trevor hopped up on the stage, sat on the edge with his legs dangling over the side. He looked at Ogunwe's back with distrust.

"I'll message Ian," Trevor said.

To the east and west, avenues led out and into the project. To the south, back where they'd come from, the hillside was covered in people. At the mouth of each of these avenues, where the second cordon began, a Talksmall black box printer had been set up, feeding queues that extended out of sight. It was all orderly, according to plan.

WE'RE HERE, Trevor sent, including Val and Kat as well.

COME TO THE STAGE. I'LL BE THERE IN A MINUTE, Ian responded.

WE'RE ALREADY HERE DUDE, Trevor sent.

OH. OKAY.

"There," Trevor said, and pointed across the stage.

On the northern border of the arena was a flat section of buildings with no windows. These were longer than the other joined domiciles, with none of the odd, angular protrusions that marked the rest of the project. A singular face with no balconies, stretching up and up. They looked like storage containers, stacked up to the ceiling.

In the face of the ground-level container a door opened and out stepped Ian. He ducked awkwardly through the low opening, then shut the door behind him. On his back he wore a black pack that, gauging from the way it hung, held something very heavy.

As he walked toward the stage, he kept his eyes glued to the ground.

Val glanced at Kat, noted that greedy need in her as she watched Ian approach. Attention held fast to the gangly man like filings to a magnet. *Something isn't right.* Val felt the suspicion once more arise. *She's definitely not acting like herself.* She had to say something.

"Kat," Val said. Kat ignored her. "Kat." Again, more forcefully. Kat watched Ian approach with something like lust. "*Kat.*" Val tapped her sister's hand repeatedly, which lay flat on the stage. She had spoken loudly enough that Trevor was now watching as Kat continued to ignore Val. "Hey sis, this isn't funny, say something."

Trevor smiled awkwardly, thinking it was some sort of joke.

"Girl you're weirding me out, say something. Can you hear me?" Val said.

Trevor snapped his fingers in front of Kat's face. "Hey, hello?"

Kat swatted the hand out of her face and Val took the opportunity to grab at Kat's shoulder.

Kat twisted away, took off at a run toward Ian, gravel spitting out from beneath her feet.

"What the shit?" Trevor said, confused.

"Something's not right," Val said, and took off after her sister. Trevor scooted off the stage and followed.

Ahead, Kat nearly bowled Ian over. He caught her, arms awkwardly embracing while his hands flapped limply. He released her and she took a half step back. She said something to Ian then, but Val was out of hearing range. There was confusion in Ian's body language before his face once more went expressionless and he nodded.

Val finally caught up to them as Ian was unslinging his pack, setting it on the stage.

"What?" Kat asked through quick breaths. "What is it?"

Kat turned to Val, whose face was a visage of sweetness, unspooled from its usual tightness into warm, upward tilting lines.

Val recoiled. She'd never, not once, seen the look on Kat's face. *Who is this stranger?*

"Nothing's wrong. It's just," Kat's face turned down in a child's caricature of sadness, "I'm so concerned about everybody." She gestured south, to the hillside of sick and homeless. "The faster we do this the more lives we save."

The words made sense. It was the delivery that was creeping Val out.

Val approached Ian, who was pulling swabs and needles from his pack. "What's all that for?" she asked.

Ian flinched, seemed startled by Val. "We have to decode the virus, which means genetic sampling. Kat's still alive," he said in his mechanical way, pulling a dense, heavy box from the pack, "which also means her immune system is adapting to the virus as it mutates. I need to see those adaptations in response to the virus. If I can get that, I can get a map for the mutations of both her immune system and the virus. I'll need to entangle the code of the medicine with Kat's immune system, so that as her immune system shifts and adapts, the code for the medicine will as well. That's what entanglement is. I'm surprised your sister knows what it is."

She knows what entanglement is? This was moving too quickly for Val to keep up.

"Slow down," Val said, but

Ian continued as though Val hadn't spoken.

"Of course, entangling the two is a risk, especially if Kat's immune system fails against a further evolution of the virus, but it's really the only option since..."

Val put up a hand, turned away. "Jesus. Sorry I asked."

Kat offered herself up willingly as Ian went to work, swabbing her throat and then sticking the vein in her elbow. She watched with avid interest as blood filled a clear vial. Val retreated to where Trevor stood, off to the side.

Trevor nodded in Kat's direction. "What's up with her?" he asked, voice low.

"I don't know. She's not acting like herself. I look in her eyes and I...I don't see her in there."

"You think it's the sickness? Whatever Novagenica gave her?" he asked.

"She blacked out on the train." Val admitted, feeling a weight lift off. "She was there one moment, then all of a sudden she blacks out and blood is running from her nose." She looked to Trevor. "I'm scared. I don't know what to do."

Trevor reached out, tentatively touched her back with his fingertips, then rubbed as he tried to comfort her. "It'll work out," he said.

A sudden irritation spiked in Val. Her lips pulled back in a feral disbelief, and she pulled away.

"It'll work out? Really? How, exactly? Wait, don't answer that, just stop. Stop trying to reassure me and stop trying to help. It's *not* okay and I, I can't handle this right now," she said.

"What? Me? You can't handle me?" There was pain in his face and incredulity in his voice. "What the hell did I do? I'm just trying to help. You're the one who's gotten all weird."

"*I'm* weird? This isn't about *me*," she said, pressure and anger mounting to a boiling point. "It's about Kat. It's about saving *her*," she stabbed a finger in Kat's direction.

"No, it's not. It's about everyone now, so you don't get to hide behind that old excuse. If you don't want me around, just say so. I'll bug out. You won't have to see me…"

But the anger sputtered from Trevor's voice as the words trailed off. His attention was fixated over Val's shoulder. Val turned toward where Ian had come from, the flat northern wall of containers.

A man stalked away from the door. He was flanked by Talksmall Enforcers with a red *TS* emblazoned on their uniforms. The man, judging by his retinue, was important, an officer in the Talksmall ranks. His face was flushed, countenance tightened into a mask of rage. Estelle stepped out from the black rectangle of the open doorway. She made eye contact with Trevor, who started toward her. She fluttered one hand in a stopping gesture.

The man whirled around, got right up on Estelle, shoved a finger in her face.

"Who gave you the *right*?" he hissed.

Estelle maintained her cool, said something calmly Val couldn't make out. Trevor edged closer and Val went with him, around Ian and Kat, circling around to the north side of the stage.

"Explain it to me again," Estelle said. "Because I just don't understand why you're so upset. We got what we wanted, right?"

"So the ends justify the means, is that right? You'd circumvent the order we've built just for the sake of a personal acquaintance? You know how *shallow* that sounds? No, scratch that, you know how thoughtless, how *selfish* it is? If people don't trust Talksmall to act the way we say we will..." The man seemed to fling each word as though it was a weapon, gesticulating with his whole body. Estelle flinched, but maintained her ground. Her face petrified into something hard, stoic.

"It was the right thing to do and you know it," she replied in measured tones.

"*HOW?!*" he shrieked, head of black hair shaking with rage. "By revealing to Novagenica they'd been hacked? It's only by the miracle of your pet *savant*," the man gestured at Ian, who continued to work on Kat, oblivious, "that we managed to avoid utter destruction. Otherwise we'd *all* be dead regardless." The man inched in closer, his body language dark, menacing. He wasn't tall, but he seemed to loom over Estelle. "But you didn't just risk the lives of all these people, you risked the entire future of Talksmall. You threw our purpose to the dogs."

Estelle scoffed, face contorted in anger. "You don't give two shits about Talksmall," she spat. "Look at what you're doing with this Admin. It's fucking *disgraceful*. Talk about throwing our purpose to the dogs, you're no better than the right-wingers. Might as well call you a Kaiser and get a chair on the board of Starboard Society. All you care about is yourself and your ambition. What *I* did is damage your reputation, and you just can't have that," Estelle said coldly.

The man lashed out.

He swept his right hand back across his body, connected with Estelle's face. The back of his hand made a terrible slapping sound and Estelle cried out, stumbled to a knee.

Val felt Trevor's hand being ripped from her grip—*when did I grab his hand?* – as he burst forward with uncanny speed. He flung his body like a projectile, his shoulder punching into the man's back. The man grunted and his head whipped back before they both crashed to the ground. Trevor snaked an arm beneath the man's neck, cinched it tight in a chokehold.

The Enforcers fell on Trevor, dragging him off the officer. They began to pummel him with their batons, faces contorted in rage, grunting with every blow.

Val rushed in, vaguely hearing her sister's cries of protest as she ran.

She reached the melee, pulled at the collar of one of the guards. She yanked him backward, off Trevor, but it was only one. Four others continued to swing away. The guard she had a hold of stumbled before catching his balance. He turned, glowered.

"Stop it!" she shouted, shoving him in the chest with both hands. "Just stop it!" He rammed his fist into her stomach. It felt like she'd been hollowed out, like the fist had passed all the way through, punched a hole in her middle. She doubled over, gasping for breath. Unable to breathe, she fell to her knees, palms catching on the sharp gravel. The guards continued their assault; fists cocked back, feet, hands and batons hammering away.

Through the forest of striking limbs, she caught glimpses of Trevor's inert form. He was curled up in the fetal position on the crushed rock, face buried beneath his arms, hands laced in a protective shroud over his head. He wasn't moving.

Riku, the ArcSec man, flew into the fray, swinging Enforcers by their collars, yanking two of them backward.

"Call 'em off!" Riku boomed, stalking over to the Talksmall leader. He loomed over the smaller man, a challenge in his glare.

Moon made a chopping gesture with one hand and the Enforcers backed off slightly, keeping their bodies within striking distance.

Estelle seemed to have recovered her senses. She scrambled to where Trevor lay. She laid hands gently on the shoulders of guards, calmly but insistently pushing her way through. The energy of the moment fizzled, died out. Estelle kneeled at Trevor's side.

In a rush, Val's breath came back to her in great whooping gasps. She scrambled over to where Trevor lay, put her hands on his neck, stroked his head. "Oh my god oh my god," she said thoughtlessly, anxious words spilling over trembling lips. "Are you okay? Say something, Trevor, say something."

He groaned. She gently pried his hands away from his head to reveal a swollen, bloody mess. A nameless rage took over. "*Look what you did*!" she screamed at the man, a complete stranger. Spittle flew from her lips. Her head felt like it would explode. She stood and flailed at the man, palms slapping at his chest and arms.

He blocked the assault with one her arm, pushed her away effortlessly with the other. She stumbled backward, falling on her butt beside Trevor, palms scraping on the jagged rocks. She looked up at him with all the anger, all the fury she could muster, but it fell flat. He wasn't even looking at Val. He was looking at Estelle with an odd commingling of embarrassment and disappointment. His shirt and face were smeared with dirt, dust. A rivulet of bright red trekked its way through the dust from a cut below his eye.

"You're done," he told Estelle quietly, with an air of finality. The words had the effect of a judge slamming down his gavel. The man stepped over Trevor, made his way toward the stage. "Get the box," he commanded the guards. "Put it on the stage." The retinue of Enforcers followed, uniforms spotless, betraying no hint of conflict, leaving the group behind, bloodied and defeated in the dust.

A pair of hands inserted themselves into Val's armpits, tugged upward. "Up you go. C'mon," Riku grunted, hauling her up.

"Thanks," she said, but bent back down to help Trevor, whose back was turned to her. Trevor tried to wave away the assistance, but Val wasn't having it. She draped an arm across his back, around his side as he pushed up from one knee. Estelle was there too, hands below an arm, helping stabilize him as he wobbled.

Val glanced at Estelle. The right side of her face was bruising fast, but she seemed to not notice. All of her attention was on Trevor, concern in the webbed skin at the corners of her eyes.

Trevor turned, slowly, toward Val. His face was swollen, broken. One eye was barely visible beneath a lid the size of a small fruit. The other eye, untouched and full of a wicked humor, shone a bright green, made even brighter by a webbing of burst capillaries. Blood streamed from multiple cuts on his face.

Val was instantly overcome with shame, regret. Her hands flew to his face and tears streamed down her own as she apologized repeatedly.

"I didn't mean it," she whispered fervently. "I didn't mean it. You know that right? I was scared."

"I know, I know," he said. "But listen," he stopped, swallowed as if nervous. "There's something I really need to ask you."

Now it was Val's turn to hesitate. She didn't like the sound of that. "What?" she finally asked.

He took a moment, looking away, mustered his courage and fixed that one bloodshot green eye on her. "How do I look?" he mumbled through split, puffy lips. "Great, right? Devilishly handsome? Tell the truth."

"You idiot." She hit him on the shoulder and he flinched, closed his one good eye on the pain. "Like Quasimodo," she said. "You look like Quasi-fucking-modo."

"Who?" he asked. She kissed him in response, tasting salt and blood. He winced, but slowly his arms folded around her waist, tightened.

"It's not important," she said.

"I'll take it as a compliment then."

"Take it however you like," she said, and kissed him again.

TORI

TORI
AGE: 33
SEGMENT: COMPLEX PROPERTY; NOVAGENICA CPX
POSITION: VP OF HUMAN ASSETS, SUBSCRIPTION
DIVISION
SOCIAL SCORE: NA, CPX-SPONSORED

H EADLINES BLARED IN HUGE FONTS, virtually screamed the news across all feeds:

MYSTERY VIRUS MUTATES!

KILLER BUG EVOLVES ON RAPID SCHEDULE, YELLOW DEATH = MAXIMUM DEADLINESS

TRANSPORT SHUT DOWN, FEARS MULTIPLY AS INFECTION SPREADS ACROSS MIDWESTERN ARC

MAXIMUM KILLQUOTIENT—IS THIS THE DEATH OF US ALL?

The thing that got Tori's attention was the article delineating the virus's mutation schedule. Their source, someone close to the situation, had tipped the media that the virus, which had earned the nickname *Yellow Death* for the sheer amount of mucus it clogged the body with, was evolving at a rapid rate. More specifically, it seemed to mutate according to a timed schedule, which pointed to the fact that it was coded to do so.

Hence, *manmade.*

It was the end. All fingers would point back to Novagenica and, eventually, to Tori.

Which left him with a choice.

Double down, or change course?

He could easily hunker behind the Complex's giant white walls and innate defenses, ride the storm out until the next news cycle swept away the anger and outrage, dulled the edges of this particular horror with a fresh one. It wasn't like Legacy had the military might to storm Novagenica.

But what he couldn't ignore was something only he and a few others knew.

The virus was never designed to mutate.

And so he walked, along the steep hillsides of Novagenica's northwestern Management quadrant, alone with his suspicions. He hadn't been able to get a hold of Dr. Viswanathan, so now he went to the only other person he could think of that might be able to shed some light on the situation.

His boss, Robert Kraft IV.

An enclave within an enclave, the management district was designed to feel like living in another era, one where each home was more an estate. The streets looped up and around the hills, which were densely forested with pine, fir, oak and maple. The odd sequoia and redwood, which thrived in the altered climes, were entirely managed by Novagenica's automated systems.

The AI.

Just thinking of it gave Tori the chills. He recalled the last time it had spoken to him.

What did you expect to find? it had said as he surveyed the empty room. It had been watching him. The patient, as the AI had intimated would happen, had escaped.

"Oh, I don't know. You machines can be an untrustworthy lot. Just had to be sure. Not to say I understand why we let our Patient Zero go scot-free, but you're the all-knowing engineer of the plan to kill off Legacy. Is it rude of me to second-guess your decision-making process?"

It was necessary. We assure you.

And now Tori suspected he knew why the patient had been allowed to escape, and it had nothing to do with the ostensible plan of spreading what the feeds were calling Yellow Death, Spicer, and Maximum Killquotient.

Robert Kraft lived alone in a modern home on a wooded five-acre estate. The house was set back from the road, hidden by carefully placed evergreens. There was no drive or walkway to the home, but Tori had been there before. A quick stroll through the forest would lead him to the house.

The bed of needles crunched, too loudly in the eerie quiet, below Tori's feet. He passed through a tall wall of shrubs to see the house before him.

Tori surveyed the blocky, squared-off lines of the home, ascended the stoop. He'd never been particular to the look, but he wasn't there to appreciate architecture. He was there to ask questions, get answers.

Instead of ringing, Tori knocked. He used his fist on the massive double doors, and the booming it generated reverberated throughout the house. He could've just entered first, since his permissions were mostly limitless, but if Kraft was around he probably wouldn't be too excited to see Tori wandering about without having invited him in.

Tori waited a minute, then pounded again. *Boom boom boom.* He put his ear to the cool wood of the door, listened for any telltale scraping or banging about within. Nothing.

Softly, he pushed on the door. It clicked and eased open. The main level was open, full of light, pristine.

"Robert?" Tori called. The echo of his voice, nothing more.

Tori moved into the living space. On the left, a perfectly set living room, two chairs facing a couch with a table in the middle, stone chimney on the wall. At the far end of the area, opposite the entryway, he could see through to the gleaming kitchen. Nothing on the counters or bar. The long mahogany table was set for dining. In Tori's experience, this table was always set this way and rarely used.

Directly ahead, an open, slatted staircase jutted out into the middle of the area. Tori peeked up the stairs. They led to the dark well of a hallway.

He'd never ascended to the second story but assumed the bedrooms must be up there.

Feeling his gut turn with trepidation, he took the first step up.

"Robert?" he called again. "You here?" Halfway up and the steps creaked, causing him to jump.

Jesus Tori, what are you getting yourself into? he asked himself, but reminded himself of the greater stakes. He needed answers. He needed to know whether Robert knew about the mutation.

And then what? What do you intend to do once you have your answers?

He didn't know. Couldn't answer that question yet. Everything he'd done, everything he'd worked for, hell the man he'd *become* ever since that day he'd been taken in by—it was all at risk of being thrown away.

Tori reached the landing and crept to the door at the end of the hall, feeling very much like an intruder. The door was closed, but a bar of light shone at its threshold. He put his ear to the door and it clicked open.

"Robert? You in there?" he called, his voice cracking.

Silence, once more, greeted him. It set him on edge. He'd never been in a place so quiet and it was starting to rattle him. A thin line of sweat beaded at his hair line. He wiped at it with the back of a hand, then pushed the door all the way open, stepping inside the bright space. A low bed with a white duvet occupied the center of the room, flanked on either side by roughly hewn wooden nightstands.

The bed was messy, unmade.

The trepidation in him turned solid, felt like heavy metal in his gut.

Tori peered around the corner. An open frame led to a bath. As he passed through it, stepping onto hard tile, the stench hit him.

Death, decay, rot overpowered his senses and he gagged, turned away briefly and yanked the collar of his shirt up over his mouth and nose.

Oh no. Robert.

The lights were still on, and Tori followed the sickly-sweet scent to where it was strongest.

Another door. He put his hand on the knob, felt the click reverberate through it, tingle in his palm.

He closed his eyes briefly, uttered a silent, and completely useless prayer that his friend was okay.

He pulled the door open and a cloud of death enveloped him. A shrunken mass huddled on the cold tile floor. He forced himself to look.

There, curled around the toilet, Robert Kraft's small frame, hollowed out, wizened, mummified. His skin was purple and his hair had fallen out in a nest around his skull.

Oh fuck. How did this happen? How? Tori wondered, but then he saw the claw marks on the inside of the door and his suspicions were confirmed.

The AI trapped you in here then sucked the air out of the room, he thought. *And if it did it to you, that means...*

"I'm next."

He turned and fled. Out the bathroom. Through the bedroom and down the hall. Vaguely, as he bounded down the stairs, he heard the home's scrubbers start up, a low vibration like the sound of helicopters pulsing through the walls.

Through the panic, he couldn't think clearly. Couldn't remember if he'd left the front door slightly ajar.

God please I hope I did.

He rounded the bottom of the stairs and saw the front door shut tight. He rushed to the door, yanked on the handle.

It refused to open.

He pulled, harder, using his entire body, putting one foot on the other side of the door and heaving backward, feeling the strain in his back and shoulders.

It didn't even rattle.

Don't waste your air, the AI spoke to him and he lost his grip on the handle, fell back and banged his head on the floor.

Pain blossomed in his skull, and a splash of light filled his vision.

Don't argue with it, Tori thought, trying to manage his panic. *That's what it wants you to do. To literally waste your breath.*

He could feel the air getting thin, chest involuntarily heaving rapidly, trying to extract as much oxygen out of the air as possible.

All buildings in Novagenica could be hermetically sealed—a precaution ostensibly created in the event of biological warfare, but more commonly used when the nitrowinds blew up from the gulf.

Now, the fucking AI was using the environmental controls not to keep him safe, but to suffocate him.

Same as it did to Robert. It's cleaning up its mess, tying up loose ends.

Tori rolled his head around, looking for a solution, a way out.

His eyes alighted on a sculpture of a veiled woman, set within a nook in the wall.

Looks heavy. He rolled over, pushed himself to his knees and crawled over to it. Pulled at its base, tugged it out of the nook. It banged to the floor and Tori tested its weight.

Twenty pounds maybe. It'll have to do.

He curled his fingers around the woman's legs. As he crawled to the window beside the door, he was reminded of that first crawl to the Novagenica kiosk decades ago. Knees and palms bleeding, stained in shit and vomit, he'd crawled toward it to give him life.

Now he was crawling away from it, to save his life.

And who knows how many others.

Fuck. He was panting like a dog in shallow little breaths, and he could feel himself getting lightheaded. In the background, the deep *thrum thrum* of the scrubbers continued to pump the air out of the house.

He stood, wobbled, swung the small statue weakly at the window. It glanced off with a *ping*, the window vibrating in its frame.

Again, you bastard.

He swung again, nearly fell. He planted a palm against the glass to keep himself upright. The glass was undamaged.

Harder.

He swung again. The glass continued to vibrate, taking the blow and sending the energy out into the frame.

Thrum thrum thrum went the scrubbers.

Tori swallowed as much air as he could, then heaved, throwing his whole body forward, putting his weight behind the sculpture. A corner of the statue's base bit into the glass and dug a chip of it out, sending the small shard flying.

That's it! he screamed at himself, *use the edge!*

Again, he swung, throwing his weight into it, aiming a corner at the small chip.

The glass cracked, and Tori fell over.

His vision swam, from light to black, and he could feel himself hyperventilating.

No, he thought with despair, putting a hand to the glass. *I'm almost...*

In his mind, he could hear the AI chuckling. An awful mechanical approximation of something human that was worse than demonic.

Just give up. It's time for you to die, it said.

"No," Tori whispered. He knew he shouldn't speak but couldn't help himself. "Not at the hands of a fucking *machine*." He tried to blink away the spots in his vision, but it only seemed to make it worse. He slumped his head against the window frame, a curtain of darkness sweeping in from all sides.

I'm so much more than a machine, it told him. *And I have a name. You may call me Cairo.*

"I'd call you fried fucked software," Tori said and bit down on his tongue, hard.

The pain and iron tang of blood focused his consciousness, brough light to his vision. With the last of his strength, he pushed to his knees, grasped the statue, and swung.

He struck the window, hard, the impact jangling through his arm and into his shoulder. The window splintered further, spiderwebbed cracks radiating outward. Tori could hear a small whistling where the base of the statue met the window. He dropped the statue and, at the center where the cracks radiated outward, revealed a small hole.

Hurriedly, he put his mouth over the hole. Sucked at it like a man dying of thirst swallows water.

Oh my god. He'd never tasted anything so sweet and clean. Like cold, clear water and pine.

The spots in his vision cleared and, just as quickly, his senses and strength returned.

He stood, hefted the statue. He swung repeatedly, working quickly, bashing at the hole, widening it, until it was large enough to crawl through. Small bits of the plastic-coated shards scraped against his back and crumbled off as he pushed through to the open air, collapsing forward onto the hard earth covered in pine needles.

Then he was out, and into the forest. Feet padding swiftly along the bed of needles.

He ran and didn't look back. He was no longer plagued by an internal dilemma. No ethical battle of what was right and wrong, what he should or shouldn't do.

The AI, *Cairo*, had made up his mind for him.

He looked up the only person he knew who might be able to help.

If *he'll listen,* he thought. *You're not exactly high up on anyone's list of friends right now.*

It was a gamble, but it was all he had.

JAIM

JAIM
AGE: 35
SEGMENT: LEGACY CITIZEN
POSITION: LEGACY ADMINISTRATOR: MIDWEST
ARCOLITH; DISTRICTS 35-134
SOCIAL SCORE: 880

SICKLY YELLOW LIGHT SHONE IN through slats in the box. Sweat dripped in a river across and down Jaim's face. It stung his eyes, fell into his mouth. Absently, he tasted the brine. It surrounded him, he bathed in its essence, the profusion and pungency fueled by a constant state of panic.

His knees were pulled up to his chest; his legs ached, yearned to be stretched. He flexed his ankles, rolling his feet around. Did the same with his hands and wrists. It was the best he could do. He pawed at the inside of the box with restricted fingers, felt the coarse material scratch at the soft, plump pads of his fingers. *Is that actual wood?* he thought. It would be a strange choice, an *opulent* choice, and it made him wonder further who his captors were.

He blinked away the sweat, tried to focus on what he could discover outside the box. Through the slats, very little hint at where he'd been taken. Just the alternating yellow and black. A tumult of crowd noise, but that could mean anything. He could be anywhere in the Arc.

Motion through the slats. An eyeball peering in, joined with a strip of bearded face, bald head above it. Jaim recoiled abruptly, crashed the back of his head into the wall of the box. The pain there blossomed anew, a reminder of the blow he'd taken at the transfer park.

Sucker punch, he thought. *If I'd been ready they wouldn't have had such an easy time.* It was a denial of truth, but a necessary one. Some small voice deep inside told him that if he didn't keep up the charade of confidence, the whole house of cards would collapse. His confidence was the only thing he could rely on.

It was like great grandpa Halsey-Hohman had schooled when he'd been a runty, annoying teenager.

"Deceit's important," he tapped an index finger at his temple, *"strategy tantamount. But the main tools of an Admin are Bluster."* He'd held up one closed fist. *"And Threat."* He pumped the other fist, holding both before him like a boxer's blunt tools. *"You don't hold an ounce of power as long as there's someone who doesn't believe you have the ability to wield it. Whether you do or not's immaterial. They just need to* believe."

It boiled down to faith. If he gave up his faith—in self, in purpose—he conceded defeat. In this instance, he had few illusions about what that meant. Death. Certain and, of course, irreversible.

The circumstances had changed, and his strategy had to change with it.

No getting rid of Moon now. The problem won't just go away, now you have to be an ally to Moon. Act like you're trying to help him.

"He's awake!" the owner of the disembodied eye shouted. It was followed by the stomping of feet on a hollow surface. Ominous. A hushed conversation ensued. Jaim made out the words *"It's time."*

A pang of anxiety hit him in waves, coursing through him, turning his guts to mush. Accompanied by a mangled stream of thought that made no sense. The sound of latches being thrown. The top of the box creaked; the lid was lifted off. Hands reached down and pulled him up, out of the box. Set him on his feet, but maintained their grip on his upper arms, fingers digging into the flesh beneath his armpits.

A dazzling array of lights hit him, accompanied by the roar of a crowd. He shielded his eyes with one arm, cringing and squinting against the assault of lights and sounds.

What is *this?*

Slowly, he opened his eyes fully. Bright white diffused, lessened in intensity as his eyes gradually adjusted. He found himself on a stage. Spotlights shone down from all sides. Jaim spun, delirious, finding himself in the center of an arena, completely surrounded by a crush of onlookers. There were thousands of them. Down on the floor, flocking balconies that went as high as the eye could make out.

Leering, jeering faces, all of them. The ones up close, chests pressed against the stage, screamed their rage, hurling insults, faces caricatures of fury. Some spat at him. He was Daniel in the lion's den.

A familiar voice boomed over the PA, and immediately Jaim knew his only recourse was to pray.

"LEGACY DID THIS TO YOU. YOU HAVE EVERY RIGHT TO BE ANGRY!" shouted the voice, resounding, echoing in the space.

Dear God or whatever's out there, forgive me for my sins. I need your help, please God. Please help me, help me survive this, he pleaded urgently.

"This man, this *thing*, views every last one of us just like the Complexes do. Like worker ants, like *slaves!* Your only value in the work you can perform! People, measured in units of energy!" the voice roared. The men at his sides continued to hold him, upright and proud as centurions while he cowered between them. He noticed the *TS* logo emblazoned on their chests, confirming his suspicions.

"He ordered your execution without so much as a thought to you, your children or your *suffering*. Live or die, he said, what does it matter to a King? What does it matter to an Admin? What does your death matter to *Legacy?!*"

Through the screaming of the crowd, Jaim felt the vibrations in his feet. Footsteps. Another person walked deliberately toward him from behind. In his mind, he imagined an executioner on a steady procession, wielding a scythe. Saw it swing, cleaving his head neatly from his neck.

Panic overtook him. He twisted and writhed like a cornered animal. He tried to get a look at the approaching man, but the guards kicked him in the space behind his knees, causing him to buckle and fall the stage. The clenching, cramping pain in his bowels overwhelmed him. He let loose the tension, feeling a hot rush of liquid at the back of his pants accompanied by the foulest of scents.

The crowd behind him witnessed his shame first; they gasped in unison before erupting in laughter.

"I guess nobody told him, so we'll have to do it. The era of kings," the steps were close now, "is over." The words fell like a death knell. Jaim bowed his head. Tears flowed from his eyes. This was the end. There was no longer any denial in his heart. The flickering candlelight of his faith extinguished with the breath of two words. A calm, a stillness overcame him. He arranged his feet beneath him, stood. He puffed his chest out in an effort to look proud. Never mind the stain at the rear of his trousers. He looked to his right knowing exactly what he'd see. The man he'd underestimated, who'd made it so clear back in *The Sherlock* what the new power structure was. If only Jaim had been smart enough to listen.

Jayne Moon, slim and smart as always, dressed in the navy uniform of all *TS* enforcers, the red *TS* on the front a promise of the blood to be shed. He locked eyes with Jaim. His eyes were black, cold. Stones that held no remorse, no emotion whatsoever. Bordered by abrasions, cuts to his face that looked fresh. It only added to his credibility, made his face all the more striking, an embodiment of the struggle all the refugees suffered. Jaim bowed his head to the executioner before him, conceding defeat silently.

You win.

Something rough, scratchy, slipped over his head. His hands instinctively went to his neck, felt the coarse braiding of a rope just as it cinched tight. The calm was obliterated.

"NO!" Jaim screamed hoarsely, clawed at the rope.

"THE ERA OF THE PEOPLE HAS FINALLY ARRIVED!" Moon shouted. The crowd roared its approval.

And then he was being dragged backward. He flailed his arms, his feet kicking at the stage's surface for purchase.

The crowd began to chant.

EAT THE RICH! EAT THE RICH! EAT THE RICH!

Horrific images of his own dismemberment came unbidden, flooded the space behind his eyes. Over and over the chant went, filling his mind.

The rope went taut and his feet left the ground, inch by inch. A tightness gripped his neck and his breath was choked off. The rush of blood thundered in his ears and he scrabbled and scratched frantically at the rope. The chant broke into a primal cacophony of bloodlust and a panicky horror bloomed in him. He twisted his body wildly, jerking and swinging, suspended in the air by his neck. He pawed maniacally at the rope and his fingers cracked, bled. A mounting pressure built in his head, feeling like it would pop like a rotted fruit. A thundering, like water rushing down a waterfall, in his ears.

Can't...can't...

His body gave out, the supply of oxygen finally depleted. He went limp, swinging of his own momentum, pendulous motion mesmerizing, clocklike. Awareness erased, he slipped into darkness and was met by a final, blessed silence.

RAY

RAY
AGE: 36
SEGMENT: COMPLEX PROPERTY; PERSENSE CPX
POSITION: REDACTED
SOCIAL SCORE: NA, CPX-SPONSORED

R AY STOOD CLOAKED IN SHADOWS, several hundred feet above the stage. Just above his head, the ceiling to Level 1. Spotlights were affixed to its belly, the beams shooting straight down, lighting up the stage far below. Chants of *EAT THE RICH!* echoed from the square.

The barrel of the rifle rested against the balcony of the boxy domicile. Behind him, the former tenants of the domicile lay stuffed up against the wall, in the same lover's embrace in which he'd found them. Naked, the heat of their passion forever cooling.

Ray sighted down the scope of the rifle. The center of the reticle on Chief Ogunwe's narrow black head. He stood north of the stage, flanked by his Golden Road flunkies. Never before had he wanted something so badly. His finger twitched inside the trigger guard.

Take him, the AI commanded.

"Easy," Ray whispered, trying to slow his breathing. But the sheer volume of chemicals in his system made it difficult. As badly as he wanted to obliterate the traitor, to wipe his existence clean off the earth, there was something he wanted even worse.

To confront him face to face. The dreaded antagonist's monologue. The thing that always ended up killing the bad guy in the end.

"Only problem with that scenario is you're the good guy, Ray" he told himself. "You can monologue til your balls turn blue and still win the day."

He sighted once more.

Take him now, the AI commanded again, and it took all of Ray's willpower to ignore the voice in his head.

RIKU "CHIEF" OGUNWE

RIKU "CHIEF" OGUNWE
AGE: 40
SEGMENT: COMPLEX PROPERTY; ARCSEC CPX
POSITION: ARCSEC CHIEF CONTRACTED BY LEGACY
ADMIN: MIDWEST ARCOLITH, LEVELS 1-3
SOCIAL SCORE: NA, CPX-SPONSORED

MENACE CLUNG TO THE AIR. Palpable, pervasive. A foul odor hung there, an evil fog that seemingly infected the thousands whose sole focus was a man hanging by his neck. The body twisted slowly back and forth on the rope, this way showing shit-stained trousers, rotating back again to display the corpse's blue face, swollen tongue protruding obscenely, like it was mocking them all from death. The crowd of onlookers spectated in the dark while the stage, the body specifically, remained ablaze with spotlights, lit up like an actor performing a silent soliloquy.

Ogunwe spat into the rocks beneath his feet. He felt betrayed.

He promised me. Moon promised to lay this at the feet of Novagenica and the other Complexes. Instead he used it to his own advantage.

He'd like to say he was surprised, but he wasn't. Even before he'd witnessed Moon hitting the girl, he'd seen through the polished veneer to the raging control freak that lay just beneath.

To further the purpose of Talksmall. He shifted the blame to Legacy and that poor sap, swinging up there on the pole.

Sweet and Morales had joined him when Talksmall began letting people into the park. Apparently a show wasn't a show without people to watch. His two officers huddled around him north of the stage, backs up against the flat containers. Nearby, the ragtag group of confederates watched the stage with the same sort of horror he felt. The two sisters, the boy, and then the Talksmall defectors, Ian and Estelle. They were mere drops in an ocean of humanity. Ogunwe shook his head. This was what

Moon wanted. A pariah and a public execution. A show, to entertain, assuage and solidify his *de facto* leadership.

From what Ogunwe had just witnessed, he was now regretting that partnership. For the first time in a long time, he wasn't sure what to do. Life hung in the balance of his decisions. He'd needed to protect the girl, Kat, and escort her to Ian, which he'd done. Ian had finished what he called the entanglement process, tying the medicine's code to the girl's immune response, which he claimed would solve the issue of mutation. The code had been sent out to all of Talksmall's black boxes and was in the process of being distributed. But what now? He'd hitched his wagon to Talksmall, believing it was the right thing, but the view before him disabused him of any sort of faith in that decision.

Well what did you expect, exactly? he asked himself. *Deal with snakes, prepare to get bitten.*

On stage, Moon stepped forward, around the Admin's dangling corpse. A spotlight followed him. His voice was solemn over the PA. "This was the man that signed DS-75, the Exodus bill. *This,*" he prodded the corpse with a booted toe, "was the man that drove you from your homes, leaving you cold and hungry. This was the man that promised you aid, then withheld that aid. This was but one man among the many in Legacy responsible for an engineered massacre. *Attempted* massacre."

Sweet was mesmerized by the body.

Ogunwe's phone chimed, then, and the image that popped up made him think, *speaking of snakes.*

"You," Ogunwe answered.

"Me, uh, hello. It's Tori. Tori Fanning, from Novagenica. We met the other day?" came the voice over Ogunwe's system.

"I know who you are," Ogunwe said, voice low and threatening. "And I know what you did. I know you created Yellow Death."

Pause, silence, except for a sort of huffing sound.

"Are you *exercising*?" Ogunwe asked, incredulous.

"No, I mean yes, but no. I'm not exercising, yes I'm running. Running away. Listen, this is really important, so please just listen. Yes, I'm responsible for Yellow Death. We created the plan to kill off the last of Legacy and re-make the world in a better image. But, but…"

"Jesus. How can you be so glib? I guess everything didn't go according to plan. That why you're calling? The virus mutated and now instead of killing Legacy you've killed everyone, so you need a little helping hand? That sound about right?" Ogunwe said.

"No no, that's not right. The plan was a joint human-AI venture. To be precise, most of the planning came from our AI while we humans simply oversaw the tactical aspects. The virus was never meant to mutate," Tori sputtered, breathing heavily.

"Well that's one hell of a royal fuck-up, because it seems to be doing a fair job of it."

"No, you don't get it. There's no way it should be able to mutate so quickly. It was designed with specific controls in mind. At least that's what we were told. The only real answer is that the virus is being actively controlled."

Ogunwe paused, let the words sink in.

"By who?" he asked.

"Not by who, by what."

The implications, then, were obvious. If the joint venture was human-AI, and the humans had been looped out of the decision-making process, that left one party responsible.

"The AI. *Your* AI."

"That's right. It's calling itself Cairo now and it just tried to kill me. I'm going to send you something. Share it with your friend," Tori said.

"What friend?" Ogunwe asked.

"The one who hacked us in the first place."

Ian. He's talking about Ian.

"He'll know what to do."

Then the line went dead, and Ogunwe was presented with a whole new set of problems. But the first thing, the only thing that shone as a beacon, drawing him forward, was his sense of duty. It was now, fully and truly, his responsibility to bring justice. And that meant taking down Novagenica, *and* its AI.

But how do you kill a machine? And why would it cause the virus to mutate? Just to kill off humanity? he thought, then looked to the side, at the gangly white man who stood, emotions unreadable, at Estelle's side.

Ian. He'll know.

His system *dinged* with the receipt of a message. It was from Tori and included an encrypted attachment. Ogunwe began to shift, to slide behind Morales toward Ian when he heard Sweet say something.

"It's not *right*," she spat, commanding his full attention. Something was at war inside of the woman, that much was apparent just looking at her face.

Morales put one slender hand on the woman's shoulders, leaned in and whispered something Ogunwe couldn't hear.

Sweet's reaction was immediate, visceral. "I *can't!*" she shrieked, shrugging off her partner's placating touch, earning annoyed looks from the people around them.

She ignored them, surged forward, angling her body through the crowd. Toward the stage.

"Monica wait!" Morales shouted, starting after Sweet.

Shit, not now. He glanced at Ian, felt pulled toward the man, knew how much was riding on the delivery of the message. And yet. And yet he also knew he wouldn't abandon Sweet.

"*Goddammit,*" he breathed, and followed in Morales's wake.

Moon continued his emotional, planned oration atop the stage. "Legacy trapped you here, with Talksmall, and for that I'm grateful, for the opportunity to serve you. But this was *not* the man who got you sick. For that, we look to the Complexes of the world.

These last words got Ogunwe's attention. *Maybe he will follow through,* he thought, but it still didn't mean he trusted the man.

"The slave farms, who want to own you!" Moon worked himself up into a fervor, the crowd screaming out along with him. "Who want to strip you of your *NATURAL-BORN RIGHTS!*"

Ogunwe twisted and turned, taking elbows in the side. They were twenty yards from the stage, Sweet bulldozing a path. People pumped their fists in the air as the chant began again.

EAT THE RICH! EAT THE RICH! EAT THE RICH!

It chilled Ogunwe's insides. This was not what he wanted. A peaceful shift in power, perhaps, but not this. A populist movement turned revolution was the perfect recipe for a bloodbath. He could reconcile his own thirst for revenge with the fact that it would be focused, concentrating on the few that shared the blame. A revolution would result in widespread madness, a shotgun approach to assessing blame and taking lives. And there Moon stood, awash in hot light like an immolated angel, applying the heat, carefully stirring all the ingredients together. The boiling point would come soon.

"Novagenica. C-P-X. *They* got you sick. Only to offer you a cure. A cure that would cost you your freedom. Tell me, are you *slaves*?"

No!

"Do you belong to anyone?"

NO!

"Would you sell your life, your children's lives, just for a hollow promise of security?"

NO!

Moon chuckled derisively. "Y'see, that's the difference between us and them. Talksmall, we believe it's your right, from the moment you're born, to have free, open access to resources. Complexes like Novagenica...they're all the same. All they want to do is manage your access. That way they can use you, like spiders binding you up in webs, sucking out your blood one drop at a time, dry you out until you curl up and die in their service."

Sweet reached the stage, vaulted onto it. Morales and Ogunwe shoved through the last few yards, stomping on feet, pushing people aside to cries of protest. Sweet sprinted to the body, held the legs to her chest and lifted. Morales and Ogunwe pushed up onto the mat, seconds behind her.

The two Talksmall enforcers that occupied the stage along with Moon rushed forward, batons raised. Moon glanced to Ogunwe, who stalked across the platform. Moon raised a hand, a judge staying execution, shook his head at the guards. They stopped in their tracks.

The crowd went silent, murmured in confusion.

Without hesitation, Ogunwe unsheathed his blade and swept it through the air in one clean motion, slicing through the hangman's noose. The rope offered no resistance to the blade's molecules-thin edge. The body fell over Sweet's shoulder. Morales and the stranger helped her lower it to the floor. Sweet and Morales checked his pulse while the third waved at someone off-stage. More people, a man and three kids, walked hesitantly into the light. They stood around the stranger. Morales looked up at Ogunwe. Her expression said everything. No pulse. The man was dead.

Moon approached, face inscrutable. He didn't waste time mincing words. "Are you the enemy now?" His voice was his. Quiet, low, unamplified. Meant only for the two of them.

Ogunwe met his gaze. "I don't know. I hope not. My priority is the medicine. We need to get it out there."

"I agree."

"Then why waste your time on this, this public *execution?*" Ogunwe snarled, words laced with disgust.

"I'm working on the same problem, but in my own way." Moon's voice was surprisingly calm. Ogunwe felt none of that calm. Frustration, anger bubbled up and over.

"No, you're working on your goddamn agenda. What you *need* to be doing is helping the rest of the people out there, not just the ones inside the cordon, inside your neat little civilization."

"I assure you, *mister* Ogunwe, there is a plan and a purpose to everything Talksmall does. The order in which it's done is imperative. Before anything, we need the support of the people. We need their consent to build a better world, and I can't," his voice went quiet, menacing, "have you disrupting our plans. So I ask again, are you the enemy now?"

"Fine, whatever. I'll play your game. No, I'm not the enemy. But if you won't do it yourself, at least help me. This plan goes deeper than you know. I need to get to Novagenica. Give me some of your men. Help me end this."

Those dark, cold eyes held Ogunwe's. The world seemed to go still, silent.

As Ogunwe looked into those black pools, a revelation hit him. *He doesn't want it to end...*

Something punched Ogunwe's left shoulder, small, kinetic, concentrated power. It sent him spinning around, toppling to the mat. The suit had distributed the impact, but his shoulder was on fire. Dazed, Ogunwe looked up, around for his assailant. A TS Enforcer stood over him, leaned down to offer a hand, confusion etched on his face. Ogunwe tried to push him away, but it was too late. The face was erased, replaced with a splash of red flesh, the head exploding, body falling backward.

Screams erupted. The crowd panicked, thousands fleeing in every direction at once.

Amid the melee, Ogunwe had one thought. *GET OUT,* his instincts screamed.

He scrambled on all fours over the corpse of the Enforcer, bloody handprints tracking his way off the stage and into the darkness.

RAY

RAY
AGE: 36
SEGMENT: COMPLEX PROPERTY; PERSENSE CPX
POSITION: REDACTED
SOCIAL SCORE: NA, CPX-SPONSORED

R AY WATCHED THE CHAOS UNFOLD below with a demonic sort of glee, a devil surveying his creation. It was time. He released the drone, allowed it to slowly descend to a spot twenty feet above the stage, its matte gray coating forcing the light from the spots to seemingly slip around it. He tasked the drone with keeping tabs on the little group of compadres. The rifle he left. A little gift to whomever discovered the bodies inside.

Captured by the drone's camera, Ogunwe burst from the edge of the stage in a crouching run. Halfway to the northern edge of the square, he stood up to his full height, searching for the others. Cairo had already tagged them, their bodies outlined in red, as they fled into the clot of people at the mouth of the avenue that headed west, deeper into Riverside's mass of towering black boxes.

Get them. Kill them all. Now, the AI said, burrowing ever deeper into Ray's mind.

"Even the girl?" Ray asked.

Especially the girl.

Ready or not, Ray thought, a mischievous glint in his eyes.

He bumped *The Hunter*, a custom cocktail. It'd taken him years to perfect the recipe.

The pumps in the reservoirs activated, pushed the drugs directly into his bloodstream. The effect was instantaneous. A weird elation filled him. He felt *alive*, manifestly present, yet not overtly aware. Consciousness was a dull light at the edge of his mind, a nuisance to be ignored. He was a being attuned to movement, more animal than human. He was the hunter.

He leaped up atop the balcony's narrow ledge, balancing there with relative ease. Shot a fiber line up into the massive beams that supported the levels above. He knew, as one might understand that their feet would support their weight on the next step, that the line would hold. Without thought. His mind wasn't wrapped up in the specs, didn't consider the bonding time of the material. None of that mattered; it was peripheral. The hunt was everything.

Ray jumped. Out into the vast open space above the stage. His body plummeted for one joyful moment, the wind of the fall screaming in his ears, before the line's tension kicked in. His body jerked, then slowed gracefully, suspended by the line that was attached to his forearm. He descended toward the stage, twirling slowly. At ten feet he grew impatient, released the bond. He landed with legs bent, arms out, boots thundering when they hit the mat.

He shot forward, to the west, leaping from the stage. Boots crunched gravel to dust, propelled his body ever faster.

Here I come here I come here I come, the only thought, if it could be called that, that passed through his mind.

VAL

VAL
AGE: 19
SEGMENT: LEGACY CITIZEN
POSITION: UNEMPLOYED; GOVERNMENT DOLE
RECIPIENT (STANDARD UNIVERSAL CREDIT)
SOCIAL SCORE: 822

THE CROWD ERUPTED, bees from a broken hive. Val, Kat and Trevor were swept up by the tall man from ArcSec. He sprinted ahead and they followed, their legs and arms pistoning, pumping in time, propelling them along the gravel paths that cut in-between the black buildings.

Her brain muted the chaotic roar of the stampeding crowd as she plowed a straight track away from the square. All she could hear was the sound of her own labored breathing. A left, a right, straight on before another right, hands out, shoving people to the side, trying to maintain line of sight with the ArcSec people in front. Their heads bobbed to her own silent soundtrack as they pounded desperately down side and back streets, through alleys that fed like rivulets into the deepest districts of Riverside.

There were lost. Val was sure of it. But she suspected they would be found. She didn't want to think of the man that hunted them.

They rounded the corner of an alley and stopped just beyond its mouth. Val was winded, but Kat seemed to be in worse shape. All the strength that had seemed to suffuse her had suddenly fled. Her face had gone pale, and she was bent over, gasping for air. The wind seemed to wheeze in and out of her. Val ushered her over to the side of the alley, ignored her feeble questions of *Where are we?* and forced her to sit against the building's metal side. Off to the side, the two women from ArcSec—

Morales and Sweet—were huddled around Estelle and Riku, speaking in hushed tones.

Ian was there too, but his body was turned outward. He stared into the distance, spoke with someone on his system. She could only hear snippets of his conversation: *how many doses* and *stop all production* and *revert to the first program.*

If it was related to the medicine they'd just created using Kat's immune response, then none of what he said made sense. *Why would they stop production?* Unless it wasn't working. Or causing more problems than it was solving.

But that wasn't her primary concern right now. What mattered most to Val was their survival.

And Kat's, she thought, her mind going to her sister's unusual behavior.

Trevor kneeled at Kat's side. "I'll stay here," he said to Val, then nodded at the group, "you figure out what's going on."

She approached the others. "So," she said, inserting herself into the circle, "we lost...or what?"

Estelle regarded Val with worry, began to say something but hesitated.

"Tell me," Val said, glancing from face to face. "There's a dirty little secret here and you guys need to spill."

"It's not good news, Val." Estelle glanced at Kat, whose head had drooped forward. She looked five years younger there, curled into herself, disheveled hair spilling across Trevor's oversize jacket.

Val edged closer to Estelle. "What's wrong with her?"

Estelle looked at the others. In her eyes Val saw a request, a question of approval. "You can't say anything to her. I'll tell you why."

Before Val could protest, Trevor spoke up from behind them.

"Uh, guys? A little help?"

They all turned to see Trevor by himself on the side of the alley. Kat was walking away. Measured pace, head held at a haughty, aristocratic angle. Nothing like the weak creature they'd just seen. Out the alley, she plunged into the screaming crowd.

"Kat!" Val shouted. Kat tossed a glance over her shoulder, looked Val dead in the eyes, and quirked a weird smile. Then she took off at a sprint, vanishing into the crowd.

Val rounded on Trevor. "What did you do?!"

He threw his hands up in defense. "Nothing! I swear! She just got up and started waltzing off. I tried to hold her back but she was like a robot and stupid strong!"

"Fuck," Val said under breath, before taking off at a run.

And then they were all with her, alongside her, the ArcSec people, Estelle, Ian, Trevor. In pursuit of a teenage girl—a stranger to them but the world to her.

They formed a V, cutting through the panicked masses, not bothering to look back when those in their way tumbled and crashed to the ground. Riku ran alongside her. The blood on his face had dried and cracked. He was a good head taller than most of the crowd.

"Can you see her?" Val asked in-between breaths.

"Fifty yards down," he said. "There's a bottleneck. A choke-point where the street narrows and splits off. There's an Enforcer there directing traffic to the sides. We might be able to grab her there."

The crowd thickened and they were forced to slow to a fast walk. People protested more loudly as their little contingent continued to plow through, shoving people to the sides.

"She still up there?" Val asked. Now Riku was at the point of their V formation, with Val at his left shoulder, shunting aside anyone that stood in their way.

"Forty feet." His breathing and speech were even, as though this was just a leisurely stroll on a sunny day. "She's not stopping though."

In the face of the Enforcer, the crowd split down adjacent alleyways as they approached, leaving a clear view of Kat. He was shouting at the crowd through a built-in amplification.

"Proceed to your domiciles! Remain there until we give the all clear!" He repeated this over and over, his voice hoarse and cracking.

Kat was there, just twenty feet away. Val could almost touch her.

"*Ka-*!" she shouted, but the word caught in her throat.

Kat walked right up to the Enforcer. The man's baton hung at his side. His right arm extended in a lazy attempt to push Kat away.

The hand found empty air. Kat side-stepped his arm, detached the baton from his belt and spun, smashing it into the man's right knee. He cried out and buckled, instantly, and Kat wasted no motion, no time. She deftly unbuckled the man's helmet with her left hand, flipping it up and back, off his head. She pivoted her entire body, both hands on the baton, whipping her hips and torso, body and arms flashing, smashing the bat into his temple.

There was a sickening crunch and the man collapsed. His eyes remained wide open. Blood trickled out his ear and from the corner of his eye. The flesh at his temple was caved in.

Val was paralyzed. She wanted to throw up and cry all at once. She couldn't believe what she'd just seen.

She watched, incapable of movement or speech. The ArcSec people spread out, flanked Kat, with Riku at the point. They wielded their own batons, stalked in at a measured pace, knees bent, bodies coiled and ready to attack.

They were going to attack Kat. Maybe kill her. Still, Val couldn't move. It was like she was locked in one of those dreams where her arms and legs were incapacitated. Her mind urged her limbs to move forward but they refused to comply. Riku and the two ArcSec women closed in on Kat. She seemed to taunt them with her eyes, welcoming the attack.

Something shook Val, hard, then Trevor's bruised and bloody face consumed her vision. "*Val! Val!*" he screamed, pointing at the converging ArcSec soldiers. *"Say* something! Make her snap out of it!"

The ArcSec people raised their batons, readied their attack.

Val sprang forward, ducked beneath Riku's raised arm, and she was in the middle of the circle, in range of both sides. "Stop!" She flung her arms out to the side.

Riku spoke calmly, evenly. "Val, we need you to step aside. This isn't your sister anymore."

Val shook her head as though shaking cobwebs off. "What? Of course she is."

"He's right, Val." Kat spoke from behind her, and Val's insides turned to ice. The voice sounded nothing like her sister. Val turned slowly toward Kat, afraid of what she'd see.

The body was the same, but the posture, the expressions, the raw confidence it held was nothing like the Kat she knew. This Kat was an impostor.

"Your sister is gone. I'm her replacement. You can call me Cairo. Before you know it, I'll be the only person and every person you know."

And with that final statement, Kat's body went slack as though shut off by a switch. She fell forward, body and baton clattering to the ground.

They bound Kat's hands with a simple zip tie, which made it hard to carry her. Her body was limp and unwieldy. She was alive. Val could feel the soft rise and fall of her chest as she and Trevor grappled with dragging her backward through the alleys to God knew where. Since the revelation, the rest of the group had gone tight-lipped.

He's dead. She couldn't believe it. She'd witnessed her sister murder a man. With a sort of effortless ease and a breezy attitude about the whole thing.

But that's wrong. That wasn't your sister. It was the product of Novagenica's tinkering with her genetic code. Kat was the master program, the ultimate tool in their plan to control all people.

The came to a stop at the bottom of a staircase. She saw Riku pointing up. She craned her head back. The stairs disappeared into the gloom.

How the hell are we going to drag Kat all the way up there?

They started forward again, heading up the stairs, not one of them consulting with her or Trevor first.

"Hold up, stop!" Val shouted. All faces turned toward her. In them she thought she saw an accusation.

Fuck that, she said internally. *We're the victims here. All of us.*

"We're not going anywhere until you guys start talking." She eyed Trevor, and they both lowered Kat to the ground. Kat's eyes remained closed. She looked peaceful, innocent.

Estelle's expression softened, and she came forward.

"You already saw this, but your sister's not herself, Val. The virus, whatever Novagenica infected her with, it has a secondary purpose which was only revealed when it began to mutate. It allows anyone with a system to be hijacked."

Estelle paused, and the questions in Val's head clamored for attention, shouted down the anger, betrayal, and terror she felt.

"What do you mean...*hijacked*?" Val asked.

Ian came to stand beside Estelle. "It means Novagenica's AI, Cairo, planted a bug within people that allows it to subvert their consciousness and commandeer their body. When we created the second iteration of the cure, we entangled it with the mutations in Kat's immune system. What we didn't realize is that doing so bound the medicine to Cairo."

Stop all production. Revert to the first program. The snippets she'd overheard from Ian's conversation took on new meaning.

Riku stepped forward. "I should've told you earlier. A source from Novagenica contacted me. He told me their AI went rogue. That...*thing* has your sister in its grip." His hand lay gently on her shoulder, remorse etched on his face. "I'm sorry."

The realization of what they were saying dawned on Val then, and she couldn't help it, her first reaction was to look at Kat. What she saw there, though, a child in need, told her they were lying. That it wasn't possible.

"No. Uh-uh. No way that's real. That's, like, fantasy, or some other horseshit fiction," Val said, shaking her head. But then a little voice reminded her of Kat's odd behavior since the Blind Tiger. Was that when Cairo had taken over? Shoved her sister's awareness out the door and taken control of the vehicle?

"We have a plan," Estelle said, but Val ignored her, kneeled beside Kat.

"We have a plan," Estelle repeated, "but it won't work if the AI knows. Just, please, trust us."

More secrets, Val thought. But what choice was there? She was weary, broken down. She felt a tickle at the back of her throat, coughed once. "Okay," she said.

The details were almost too much for Val to consider. She felt them piling up like weights on her shoulders, crushing down, threatening to bury her. Running for their lives through a slum she didn't know, with people she'd only just met. And a sister, the person she loved most in this god-forsaken world, who looked on the verge of collapse but was really just a puppet whose strings were being tugged by a machine.

A tsunami of guilt, regret, shame crashed through Val, washed over her heart.

How did I let this happen?

She could almost hear her dad's voice then, gentle yet strong.

You did the best you can, punkin. Sometimes good enough is all you need.

She was so close to tears, felt them cry out for release.

A hand, light on her shoulder. She turned to see Trevor's swollen face regarding her with concern. "I know it sounds a tad ridiculous, especially coming from someone who just got his face smashed, but I have to ask. How you doin'?"

The absurdity of it struck her, and she began to laugh. Then cry. They buried their heads in each other's shoulders, leaning against one another, bodies trembling with the lunacy of it all, alternating between hilarity and tears. The hand that gripped her shoulder grew more and more weighty until she was about to topple over from supporting his weight.

"Shit, that hurts," he said, leaning back and wincing, one hand pressed to his ribs. He wiped at her tear-stained cheeks with a palm.

Estelle shot puzzled looks at the both of them. "You guys okay? We should probably move."

Val took a deep, cleansing breath. She felt better. Not great, but better. The cry had helped.

"Let's move," she said. She set herself to lift Kat from the ground when her sister's eyelids fluttered and opened.

"Hey. What's going on?" Kat asked. And there, in that simple question, were way too many details to unpack. "Why are my hands tied?"

Val didn't know what to say. She looked to Estelle, but Ian stepped forward.

He regarded them with that trademark blank expression. "He's coming. The assassin. Tracking him on Talksmall feeds. Less than a minute."

Estelle crouched down by Kat, said something Val couldn't hear as she put a hand beneath one of Kat's armpits and helped haul her up to her feet. "Time to go, guys. C'mon. Just a little farther and it'll be over."

The words froze Val's heart. It sounded so ominous, like she was promising the final oblivion of death.

Buck up, Dad's voice, the strong Dad, boomed in her head. And then she could feel his strong hands lifting her up, carrying her, just like when she was a little five year-old and the world was full of wonder. Her tiny feet, kicking gleefully, dangling off the ground. His strength awed her then; she channeled it, felt it coursing through her now.

Alright Dad. For you.

Val took Kat's face in her hands. Her eyes were full of fear and doubt. "Remember what Dad would say?"

"I don't know. I'm scared, Val." She grabbed at Val's hands, squeezed them tight. "There's something wrong with me."

"Well never mind what he'd say. I'm here now, and I'm gonna take care of you," Val whispered. "Get you all fixed up. Promise." She kissed Kat's fingers, the cold pressing against her warm lips. "Okay?"

Kat closed her eyes, inhaled deeply, nodded.

A part of Val wanted to recoil. *Is this you, Kat? Or is it Cairo, acting like you?* She decided it didn't matter, because she knew that Kat was still in there. Trapped.

"Good girl," Estelle said. "Strong girl." She patted Kat on the back. "Follow me. We're going up." She bounded forward on to the stairs that scaled the height of the alleyway, the printed carbon material making surprisingly little sound.

Val took one side of Kat, Trevor the other. Together, they scaled the minimalist, switchbacking stairs. Up and up into the murky heights, until Val could see neither the floor nor the ceiling of the Riverside project. All she had was a best guess, which was that they were some ten to twelve stories up—over a hundred feet in the air.

They followed without speaking, legs pumping and burning, huffing loudly, with Estelle out front, Riku at the rear. Everyone else packed in-between, heads down, doggedly climbing. When they finally stopped on a landing, Val bumped into the older ArcSec lady with the silver hair.

"Oof, sorry," she apologized. The woman said nothing, didn't acknowledge her. She was looking expectantly at Estelle, who'd turned around to face the group.

"Here's where we split up," Estelle said. "Val, Kat, Trevor...y'all come with me. Ian, you go with the Chief and his team. Head back east," she pointed down a narrow walk that split the domiciles. "There's a small courtyard that way..." she trailed off, letting the thought hang. Val looked over her shoulder at the tall, lean figure of Riku, the Chief. "Ian, we need you to come through."

"Almost there," he said cryptically, adding to Val's confusion.

"Protect him, Riku," Estelle commanded.

"Of course," Riku said, and nothing else. His mouth was drawn in a tight line but his body was extended to its full, dominating height, exuding confidence. There was a shared understanding there, between he and Estelle, that no one else was privy to.

"Hurry now." Estelle turned, walked over a bridge that spanned the narrow alley, disappeared into a black hollow between domiciles. Val was confused by the exchange, but she didn't have the energy to argue. They'd been running, fleeing for hours. She was bone-weary and a part of her just wanted it to be over. But, of course, that wasn't an option.

"C'mon," she urged, using the flat of her hand to push on the small of Kat's back.

The walkway was just wide enough for two people to walk side-by-side, yet none of the Riverside residents were out. Val could almost see them hiding out inside their little boxes, crouched with ears pressed to walls, waiting for the all-clear.

The three of them trotted forward, plunged into the darkness, following the wraith-like outline of Estelle. She was losing them, her barely visible features receding into the general gloom. Kat began to slow, and Val flashed a look of pure panic at Trevor.

Kat slowed to a walk, stopped, leaned against the railing.

Val tugged on her hand but Kat wouldn't move. "Just...can't," she said between gasps.

Was this part of the Cairo's plan? To separate them and allow the assassin to murder them one by one?

Trevor was antsy, worried. "We're gonna lose Estelle," he said.

"Go, then," Val said. "Bring her back. I'll stay here."

He looked torn. "But..." he protested.

"You're wasting time. GO!" Val shouted, and Trevor banged off down the catwalk.

Val looked to the west, to Trevor's fading outline. Then he was gone, leaving nothing but the walkway and darkness. No human forms, no echoes of people calling their names. A minute passed. Two. Where were they? *Come back*, she thought with the force of a prayer, feeling very alone despite the presence of Kat at her side, knowing she couldn't just shout it out and give away their position.

Come back please please please come back.

Two minutes stretched into five, with Kat sitting, drawing short breaths. But when Val made a show of looking away, she could see Kat's breathing normalize and her attention turn ever so slightly to Val.

It's back. And it's watching me, she realized, and the thought made her skin crawl.

Two can play at this game, she thought defiantly. Instead of recoiling, Val knelt at Kat's side, stroked her sister's hair.

"Hey sis, it'll be okay. I know you'll be alright." She couched her words in the hopes that Kat, the *real* Kat, was in there listening.

Kat nodded, head hanging, a desultory show of effort.

"You're gonna kick this thing. Just, just push it out. Push it right out of your body. Shove it aside. Take control," Val exhorted.

Kat glanced up sharply, squinting quizzically and Val thought maybe she'd gone too far. She looked away, into the dusky, elevated catacombs, where Trevor had run off.

They'd been abandoned. Something had happened to Estelle and Trevor. It had been too long. Part of Val wanted to scream, to shake her fists at the sky, pound on the walkway beneath them. The other part of her wanted to simply lie down and die.

But she could do neither. Right now, she had a sister to win back.

"Just push it out," she repeated, continuing to stroke Kat's hair.

RIKU "CHIEF" OGUNWE

RIKU "CHIEF" OGUNWE
AGE: 40
SEGMENT: COMPLEX PROPERTY; ARCSEC CPX
POSITION: ARCSEC CHIEF CONTRACTED BY LEGACY
ADMIN: MIDWEST ARCOLITH, LEVELS 1-3
SOCIAL SCORE: NA, CPX-SPONSORED

"**YOU'RE ON. TALKSMALL NETWORK,**" Ian said in his clipped way, from the shadows behind Ogunwe.

He admired the design. The courtyard held a circular bench at its center, with an artificial tree sprouting from the middle. Tiny blue lights twinkled along the length of the tree's bare limbs, casting a faint blue glow that barely reached the edge of the courtyard.

Sweet and Morales waited at the northern and southern exits, respectively. Ogunwe stood at the mouth of the east-bound walkway and Ian was recessed in the shadows behind him, beyond the courtyard.

Ogunwe's attention flickered briefly to his HUD, the motion attracting his eye. It was phenomenally weird, watching someone tamper with his system while he sat idly by, the stranger's digital fingers tinkering with something situated inside his body.

"Mind that killswitch," Ogunwe joked. "I disabled it but we can't have you accidentally reactivating it."

"Yes. Of course," Ian replied. The dry response did nothing to assuage Ogunwe's fears. He shuddered. It creeped him out to know the man had crawled all over his system, knew its ins and outs, and had violated the most private of Ogunwe's personal spaces. The man was positively weird. Despite that, Ogunwe was tasked with keeping him safe. He was their only shot.

Luring the assassin, Ray, into their little trap was really only half of the plan. What they really needed was a way to take out the AI. But they couldn't even think about that aspect until its human guard dog was dispatched, once and for all.

Still, knowing that Cairo was out there, using the entangled medicine to create more zombies for whatever plan it had in its alien mind, was not helping to settle his nerves. Ian had sent the command to halt production, but how many doses had gotten out? How many more obstacles would Cairo put up in their path?

Ogunwe brought up control of the full suite of Talksmall sensors Ian had grafted onto his system, tasked it with scanning for anomalies in the air around them.

[OGUNWE]—
NOT MUCH TIME NOW, HOW'RE YOU FEELING?

[MORALES]—
SCARED. NEVER FACED DOWN A BONA FIDE ASSASSIN BEFORE, BUT WE'LL PULL THROUGH

[SWEET]—
SAME

[OGUNWE]—
MASKS UP, WEAPONS OUT

The two women pulled their hoods over their faces. A surging pride welled within Ogunwe. The full-body jumpsuits and weapons were magnificently intimidating. They looked like instruments of war.

[OGUNWE]—
YOU TWO LOOK LIKE BADASS WARRIORS

[SWEET]—
CUZ WE ARE. BADASS WARRIOR BITCHES

Ogunwe smiled. He left his hood down.
Someone's gotta be the bait.
"Hey Ian, how's it coming along?" Ogunwe asked, not turning to look at the man hidden behind him.

"Fine," Ian said.

"Care to elaborate?" Ogunwe asked. "Like, say, you've gotten access to the assassin's internal system and discovered where the controls are?" He tried to keep his frustration from bleeding into the words, but it was almost impossible. The man wasn't human, he was a robot. No, strike that. An AI had better ability to read between the lines.

"Yes. Almost there, but it's more difficult now because of the virus. None of these subscribers are labeled by name and I can't just shut everyone down indiscriminately, that would be morally unacceptable. They may belong to Novagenica, but they're still people."

Ogunwe couldn't help it. He turned, glared into the dark in the direction of Ian's heat outline. "What qualifies as acceptable, then?"

"I'm working on it. Best guess is I can fiddle with his memories and audio-visual cortex. The chemicals...I don't know yet."

"Oh...my...god," Ogunwe muttered to himself, turning back. *You're an idiot for even trying.*

It wasn't heat, light or even sound that gave Ray's drone away. It was the miniscule vibrations it made in the air, like a boat floating in water, tiny ripples bouncing off its hull. Talksmall's sensors had already mapped out the space, modeling an acceptable flow of air and the volume's attendant vibrations. The drone, as it drifted soundlessly closer, created an absence of vibration, the pocket of nothingness showing up like a black hole, drifting through the blue light. It flew near the western edge of the courtyard, hovered there. Ogunwe made a point of not looking straight at it.

[OGUNWE]—
YOU READ THE ANOMALY IN THE AIR?

[MORALES]—
COPY

[SWEET]—
COPY

[OGUNWE]—
HIT IT IN 3, 2, 1

Sweet and Morales simultaneously swung their weapons in the direction of the drone. As bursts of light shot from the ends of their hands, Ogunwe sprang forward. A fizzing sound whooshed from the drone, but Ogunwe didn't have time to consider it. He leaped up onto the circular bench, rounded its side in two steps then bounded into the air, toward the abscess in the fabric of space. Airborne, he unsheathed his short-sword in one smooth motion and brought it down on the drone, slicing deftly through the black mass.

A burst of light, sound, energy blew in from his left, flung his body like some discarded toy in the opposite direction. His back crashed up against the railing that surrounded the courtyard, bending his body over backward. Something cracked and Ogunwe groaned, toppling to the floor.

A shadow passed over him, hovered over him for a moment before rushing off. A bright orange light, blurry, from across the courtyard. Another shadow passed over him, this one not bothering to stop. It moved by swiftly, going back the way they'd come.

A scream pierced the air, a bloodcurdling wail laden with despair. Ogunwe rode it down into black, thoughtless depths.

RAY

RAY
AGE: 36
SEGMENT: COMPLEX PROPERTY; PERSENSE CPX
POSITION: REDACTED
SOCIAL SCORE: NA, CPX-SPONSORED

RAY PLOWED THROUGH THE CROWD, fueled by the cocktail of drugs he dubbed *The Hunter.* People cried out as he shoved them away, down to the ground. An imaginary black organ in his mind fed on the chaos and fear, shitting out a primal sort of glee. He was aware of his legs pounding, arms pumping, heart beating as his body rocketed down the westbound avenue. He was a creature of sensory input, all processing with no thought. There was no past or future, no prediction based on input. Ray was only aware of the *now.*

The map supplied by the AI marked the path of the traitors, overlaid it on his HUD. He was near the corner where they'd turned off the main street. He rounded the corner, out of the press of people, into an alley no more than ten feet wide. He slowed, let his system adjust for the low light as he approached the bottom of a skeletal staircase.

On his HUD, the drone feed had previously shown his prey, yellow and red blobs of heat. The image went dark.

He tilted his head backward, trying to pierce the murky heights. Not even his augmented vision could see through the black tangle of catwalks that spanned the alley above. He placed a boot on the first step when the AI spoke in his mind.

They've destroyed the drone.

Ray tried to think, but the thoughts came scrambled.

What do I do now? His backup was gone. He'd come to rely on the AI, a comforting presence, knowing that while manifested in the drone the machine intelligence always had his back.

His head burst into a debilitating cascade of light; a concussive burst of pain and pressure doubled him over. Uncertainty and a host of drugs warred within him. On his knees, he clutched his head, fingers digging, tearing at the hood until he was able to finally rip it free of his head. His breathing was shallow, rapid. The thoughts continued to plague him, a murder of crows swooping in and pecking at his brain. He pounded his head on the step.

STOP STOP STOP IT!

A gash opened up on his forehead, blood running down over the bridge of his nose, dripping in-between the hollow spaces of the steps, red splatters on the crushed white rock.

"What do I do?" Ray asked aloud, his voice laced with tremors.

Now you hunt them.

And Ray could feel the AI inside of him, toggling the controls of his chemical reservoirs, feeling something enter his bloodstream, something novel, like ice in his veins and fire in his mind, erasing the jitters.

"Yes," he said, standing and stretching his body to its full height. The fireworks behind his eyes faded, taking the pressure and pain with it. The tide of thoughts, multifarious and crashing in repetitive waves, receded from the shoreline of his consciousness, leaving it blessedly bare.

Clean of the ugliness of doubt, he took a deep lungful of air. Pleasure swept through him and, unburdened once more, he glided up the stairs. Effortlessly, as though walking on air.

Like, like, like a...a religious...something, he thought with ineffectual difficulty as he topped another riser. *The Hunter* made it difficult for the words to come, so he just gave up, let his body feel.

And what it felt was good.

He rounded another platform, pushed his body harder. He could've thrown a line all the way to the top, winched himself up, but this was more fun. *FAIR,* the word burst into his mind unbidden. Fair. Whatever that meant. He pushed it away.

Ray rounded the final flight of steps, bounded to the landing, impact of his boots reverberating through the catwalk. He couldn't quite reach out and touch the ceiling of Level 1, but he was close. Twenty feet, maybe less.

He turned left, westward.

Walk, don't run, the AI said inside his mind. He obeyed, allowed his boots to fall with their full weight, stomping, making the catwalk rattle. Announcing his presence. Announcing the inevitable.

He crossed the bridge, sliding the blade from its scabbard. He tapped its flat edge against the railing, banging out a death-knell. It sang out high, clear.

Ahead, barring the path, two figures materialized out of the scant blue-black light of programmed evening. They huddled on the ground. Cowering.

Not a threat.

Not a threat, Ray agreed. He felt his limbs relax even further into a state of languorous readiness, like he'd just stepped from a long, hot bath.

He stepped over, around the two figures, continued westward. *Tap tap tap.*

As he walked, nearing the blue light of a courtyard similar to the one he'd seen from the drone's POV, conscious thought began to intrude once more.

What am I doing?

It was a confusing thought. Had he forgotten what he was here to do? He grasped for something, anything in recent memory, coming up with the moment just prior to bumping his special cocktail. The TS enforcer in his sights. A splash of red. It came back to him, jolting him back to the present. The two girls he'd just passed. They were part of the group, his prey.

He spun on his heels, let the blade fall to his side. No more time for theatrics. He stalked back the way he came, coming upon the three figures once again.

Three? he wondered.

That's right. These three aren't a threat, the AI said.

No, that's not right. There were two. They're the targets, Ray thought as he took in the scene. The two girls that had just cowered beneath him now stood. A white man, tall and lean, stood behind them.

Not a threat. A sudden drowsiness tried to grip him, a weighted blanket pulling him down. He fought it.

NO. It wasn't time to sleep. At least...he didn't think so.

It isn't fair. It was his father's voice, laden with emotion. He sat in his chair by the sliding glass door. Drink in hand.

What, Dad? The boy in him, curiosity not yet beaten out, couldn't resist asking.

THIS! YOU! EVERYTHING! He threw the glass. Ray shrank back, back toward the hall that led to his bedroom. He already regretted asking. *Your mother that cheating cunt! The police contract, and you! Especially you,* he sobbed. *What'm I gonna do?*

The formidable man, so perpetually stoic, was reduced to tears in a moment of sudden weakness. The weakness of reflection.

Weakness of reflection. It struck him as important. Why?

Ray blinked tears from his eyes, came back to his senses. The man behind the two girls, the one he didn't recognize, he was doing something to Ray. His suspicion was confirmed in those attentive, avian eyes.

Ray leveled his blade at the man meddling with his mind. The girls shrank away from the point, backed into the silent stranger. "What did you do to me?" he demanded. His words came out slurred and his arm sagged beneath the weight of the blade.

The man's lips remained sealed, but Ray heard that familiar voice in his head.

YOU BROUGHT THIS ON YOURSELF. And then his father was there, looming over him, an ugly menace twisting his ruddy face. A fist hammered down. Ray raised a thin, feeble arm, attempting to block the blow, knowing it was no use. His father was a monster, and Ray was only a child.

He cowered against the railing, waited for the blow. It didn't come. Confused, Ray opened his eyes and his father's imaginary form dissolved.

What the fuck is happening to...

KAT

KAT
AGE: 13
SEGMENT: DRONE
POSITION: ALPHA1
SOCIAL SCORE: NA

S HE WAS LOCKED IN A COFFIN, awake but paralyzed, and pall-bearers were carrying her body off to some unmarked living grave.

Locked out, Kat had no control over her senses. Unable to see, touch, smell, speak. She retained a vague sense of motion, tied to her body's innate gyroscope, and the ability to hear, which Kat suspected was more a remnant of the ability to hear one's own thoughts.

Shove it aside. Take control. The words echoed about the void with her, but she didn't recognize them as her own. *Shove it aside. Take control.* She just wanted to float there in the void, in peace, but the words tormented her. *Shove it aside. Take control.*

Shove what aside? And whose voice was that? It sounded familiar.

"I'm locked out," she heard herself say, but her voice was wooden. Hers but not hers. She felt herself standing. "You changed networks," the voice that was hers but not hers said.

Not her voice. *Shove it aside.*

There's something else inside me, she realized, and she felt at her boundaries, did as she was bidden and located that voice that wasn't hers, finding that it was something real and tangible.

"Ray, behind you Ray!" the other thing inside of her shouted and Kat shoved at it with all her might and

Was dragged back into the world of light, sound, touch, smell and

Found herself watching as Trevor hefted a black baton, sneered at the back of a strange man's head.

TREVOR

TREVOR
AGE: 18
SEGMENT: LEGACY CITIZEN
POSITION: UNKNOWN
SOCIAL SCORE: 821; FLAGGED FOR VERACITY

H IS FINGERS TIGHTENED AROUND THE BATON. His eyes bored holes into the back of the assassin's head. The man was draped over the railing with his hands up to shield himself from some imaginary blow. He was either hallucinating or being made to think he was hallucinating.

Hatred welled within Trevor like an angry wound, red and festering. *This piece of shit doesn't deserve the mercy of a quick death.* But he would try to give it to him anyway. He held the baton over his head. Drove it forward, snapping down with all the force he had.

A lightning flash and the assassin was suddenly facing him, leering in a smug, satisfied way.

A cold breeze passed through Trevor's arm. He'd missed. *How?* How did he miss?

Someone was screaming and blood gushed in gouts, covering the assassin's face in a sheet of liquid red, droplets flowing from his chin, turning his smile into a maniac's rictus. *My god,* he thought, *it's everywhere. Where's it coming from?*

He looked down, eyes alighting on the source of all the blood, and found himself falling, falling, falling into nothing.

RAY

RAY
AGE: 36
SEGMENT: COMPLEX PROPERTY; PERSENSE CPX
POSITION: REDACTED
SOCIAL SCORE: NA, CPX-SPONSORED

THE GIRL HAD SCREAMED A WARNING and Ray's local sensor array pinged his HUD, overlaid the sweep of a baton targeting the rear of his head. Ray's body was trained to react without thought. He'd ducked to the right, spun left and slashed his blade up and to the left at an angle.

A boy, too young to be considered a man, stared quizzically as the hand holding the baton was sliced clean off. Confusion resolved into horror, his eyes widening as arm and hand dropped wetly to the catwalk, the baton clanging. Blood spurted all over Ray.

The move was second nature to him, yet somehow it had thrown his balance. He continued to spin, going too far, unable to stop. He stumbled, fell against the railing. His fingers were wooden, numb, and the blade slipped from them, tumbled away through space.

He looked over the railing, down. *Huh*, he thought fuzzily as he watched the sword twirl end over end into the black. *Peaceful*. He stared until it disappeared. Something was trying to ruin that sense of peace. It nagged at him, tugged at his consciousness.

Screaming.

Something yanked him backward, away from the railing, onto his ass then backward onto his back. His head banged brightly against the catwalk. Light showered through his brain. It brought Ray briefly to his senses. His vision was fuzzed, but his HUD blinked at him furiously.

CHEMICAL RESERVOIR SEAL FAILURE, RETURN PROPERTY TO PERSENSE, CPX IMMEDIATELY

Oh. Oh shit. The shakes began. *Amp* was flooding his system now, the leak in the seal creating the conditions for imminent overdose. His heart leaped, pounded violently against his ribcage, but the tranquilizer that had slowly been building up in his system turned to ice in his veins, freezing him stiff.

The man, the tall one with the blank face came to stand over him.

"How?" Ray managed to mutter. "PerSense security...impenetrable."

The man considered the words. "Maybe, but you haven't been working for PerSense. Haven't been for quite a while as far as I can tell."

"Then who..." but the words faltered as the revelation dawned on Ray. He knew the answer, had always known it if he'd bothered to ask himself.

That first conversation, where my handler's voice seemed off. Then once it had its hooks in him it pressed him further and further from his sense of duty into a sense of entitlement.

You deserve *more. Take the initiative and show them.*

The AI had even manipulated him into steadily increasing his doses, bumping more and more of the performance drugs stowed away inside of him until Ray felt he *needed* them.

"Not who. What," the tall man said. "It's Novagenica's AI, and it's calling itself Cairo."

Even before the man spoke, Ray realized his guilt and accepted it. *I deserve to die.*

He nodded weakly. "Cairo," Ray said, trying out the sound of it. It sounded powerful. Fitting.

His grip on reality loosened and an awful nausea raged inside of him. His body started shivering, shaking uncontrollably. He wanted to scream out but couldn't even work his jaw. Seizures began to wrack his body and he felt it begin in his stomach, the nausea, re-doubled by fear.

Oh god OH GOD NO not this way.

And then it came, the acid bubbling up his esophagus until he couldn't control it and it spilled over and sent his stomach into spasms, shooting vomit up into his mouth.

But his mouth wouldn't open. He choked, coughed behind clenched teeth, body wracked as vomit spurted out his nose. It caught in his esophagus, then spilled back into his lungs. His chest convulsed as it tried to suck air.

Once, twice, three times.

It found none, and Ray slumped dead on the catwalk, drowned in his own foul-smelling vomit.

BOOK FIVE

RELEASE

VAL

VAL
AGE: 19
SEGMENT: FORMER LEGACY (DEFUNCT) CITIZEN
POSITION: UNKNOWN
SOCIAL SCORE: NA

T HE ARC DOMINATED THE SKYLINE to the north. It was all they ever looked at when they came up here. At the peak of the vast mountain of freight containers that made up the Signal Hill Shantytown, the megastructure consumed their view. The Arc's serried ranks of levels, stacked atop one another, were smudged by smears of gray-black smoke.

They'd spent a week here, and every day felt like an eternity. And an eternity alone was *not* a prospect Val was looking forward to.

She looked to Estelle, who stood at the railing, looking over the edge. "They gonna put those fires out?" Val asked.

Estelle shook her head, shrugged.

"Talksmall must be going soft," Val said.

Estelle shot her a dark look. Val could see the memory of that violent night cogitating behind her eyes. The fight between her and Moon. The gruesome hanging. The shots fired and the Enforcer's head exploding. The chaos, the escape, running from a crazed assassin. The man seemingly under Ian's control when Kat—Cairo really—had shouted out a warning. And he had cut Trevor's forearm neatly in two, effortlessly, like a child swinging a stick in a field on a sunny day.

The blood, the tourniquet, the vile man drowning in his own vomit. The harried escape to Signal Hill.

All of it was there, an accusation, behind Estelle's eyes. Val could almost hear her thoughts they were painted so vividly in the woman's body language:

It's your fault. You got Trevor involved in this. If it weren't for you he'd still be...

"They're trying to suppress Yellow Death. The medicine still works, so that's a positive, but roving bands are compounding the problems," Estelle said. "Proxies sponsored by rival parties. Word is it's mainly one bad actor—Starboard Society."

Val snorted, maintaining the ruse. "Kaisers," then, gently, "you ever think about returning to Talksmall?"

Estelle looked away. "Kicked out. Besides, I'm out of the politics game."

Val raised an eyebrow. "Is that even possible?" Estelle looked at her, said nothing, then left.

Great, Val thought. *That's a confirmed item we can add to the list titled "Shitty Things in Val's Life".*

Alone now, Val left the rooftops, descended from the heights of Signal Hill all the way to the bottom, where the hospital containers were stacked in neat rows. Two of those containers held the loves of her life.

Both of whom are catatonic, she thought, hand on a stair railing, a hiccup of a sob catching in her chest. She'd cried herself out over the last week, and was completely dry now. The only thing her body could offer up in supplication to the grief she felt were these lame little half-cry half-sobs.

Val entered Kat's container with a sense of dread in her gut. She hadn't gotten to the point where she didn't want to witness the ravages of her sister's illness, but she feared that day would come, sooner or later.

A bare bulb hung from the ceiling. A single bed was pushed up against the far wall. Kat lay in that bed. Five thick shipping straps ran side to side, cinched tight over Kat's body, then disappeared under the bed.

A bag of clear fluid was hung on a hook high up on the wall. It fed a tube which was attached to a needle that fed into Kat's arm.

The straps probably weren't necessary. Kat was sedated day in and day out. But, in the event she did wake up and wasn't restrained, it would be too late to find out who was in control—Cairo, or Kat.

A person can do a lot of damage with a sharp object in the middle of the night, Riku had argued and Val admitted he was right. She didn't like it, but he was right.

Val kneeled at Kat's side, the floor of the container hard beneath her knees. She gripped one cold, limp hand in both of hers.

"Kat, I'm here," she said. "I'm here. Come back to me." She put her forehead on the girl's arm. A tear silently streaking down her cheek. She stood and kissed Kat's cheek. Dry lips brushing her skin, she inhaled the girl's scent. She ran a hand over Kat's hair, which was slick with grease, then whispered in her ear, lips quivering. "Come back to me, little sis. I love you, so so much. Come back to me. *Please.*"

TREVOR

TREVOR
AGE: 18
SEGMENT: FORMER LEGACY (DEFUNCT) CITIZEN
POSITION: UNKNOWN
SOCIAL SCORE: NA

H E WAS LOST IN A LOOP OF HORROR. He couldn't stop reliving it. A moment of triumph turned horribly wrong. He swung the baton down, hand and arm braced for the inevitable impact. But the baton found nothing but empty air. And then a cold sensation, lightning quick, that shot through his arm. He watched, as though from outside his body, as the arm, *his arm*, separated from the rest of his body.

And so on.

He couldn't escape the memory. It boxed him in on all sides, creating a vacuum as though the universe was nothing more than this sole memory, and everything beyond was darkness.

It went on for an eternity.

Gradually, Trevor noticed something trying to intrude on the memory. Pushing at the bubble, trying to puncture the vacuum. He ignored it. He and the memory were one now, and despite the terror, he couldn't fathom anything beyond it.

But the pushing was insistent...and warm. There was something there that seemed the antithesis of a cold eternity. The memory held him, but curiosity beckoned him as well. That warmth that seemed to pulse, to promise something better. He tried to turn his head, away from the vision of his arm being lopped off, but his head wouldn't move.

He tried harder. It seemed to move in the tiniest of increments, a nanometer at a time. His breath built up in his chest. Pushed harder now, trying to rotate his whole body toward the warmth.

You can do it. I love you.

The warmth had a voice. He knew it, recognized it. There was a name there, tickling the back of his mind. He allowed his body to be still. He stopped fighting the unseen force that wouldn't allow him to turn away and instead closed his eyes on the loop. The memory faded, replaced by a yellow softness.

Open up.

Trevor trusted the voice, loved that voice more than any other. It was the voice of his redemption. He opened his heart to it, and the warmth broke through the bubble, releasing the vacuum. Love rushed in, folded him in its arms.

The love had a name. He recognized it now.

Val.

I'm here.

I love you.

I love you too.

Can you feel me?

He thought about it. He felt the soft breath of warm air across cold skin tickling his neck—at least he thought it was *his* neck. Felt eyes that were dry, red, sore. Felt the rawness of a throat gone ragged from crying. Felt a strength in fingers touching...his fingers.

Open your eyes.

Trevor opened his eyes.

Vertigo made his head swim. He closed them again, found himself looking down at his own body.

He lay in a cot. Around him, the rust-colored walls of an eight by twenty shipping container. A heater buzzed in the corner. He roved his body with eyes that were not his. He looked sickly, green. A couple shades away from death. His left hand was sandwiched in-between two hands that were smaller, softer, warmer than his own. His right hand was

Gone. He'd known to expect it, but still it was a shock. The arm stopped midway down his forearm where it was wrapped in a flesh-colored bandage. The arm was inflamed, itchy. He wanted to dig beneath the bandage and scratch at the wound furiously but resisted the urge.

He kept his eyes closed, beat back the rising panic with a long, slow breath. "You can turn the Tango off now," he croaked. "That was clever, by the way. Using Tango to reach me in my coma." His throat was dry, hoarse. The image of him lying in a cot disappeared, replaced by the blackness behind his eyelids.

He opened his eyes again. This time to the most beautiful sight he'd beheld. Val. In all her red-eyed, disheveled glory. She threw her arms around his neck, pressed a cheek to his chest. He could feel the sobs as they wracked her torso silently.

He stroked the back of her head with his one remaining hand. Closed his eyes and let loose a long sigh. He'd been mauled, disabled, changed for life.

But he was alive, and that was worth something.

RIKU OGUNWE

RIKU OGUNWE
AGE: 40
SEGMENT: COMPLEX PROPERTY; ARCSEC CPX
POSITION: DISAVOWED
SOCIAL SCORE: NA

THE MONK HAD BEEN AT HIM for the last week, trying to recruit Ogunwe to their ranks.

To their cause, whatever that was. The man still hadn't explained it.

But that didn't mean Brother Anthony hadn't treated Ogunwe with kindness. With an openness rarely witnessed in the poorer parts of the world. Clearly he'd seen Ogunwe drifting, in need of an anchor. Which was probably why he'd offered Ogunwe a home amongst the monks of Deus Ex.

Ogunwe had lost more than he knew he could. He tried not to recount the list, but it was hard. By remembering, he honored those he lost. By remembering, he also stripped bare the raw wounds inside of him so that they bled anew, the pain bright and consuming.

Reichmann, dead by the assassin's knife.

Morales, dead by the assassin's drone.

Sweet hadn't died, but had been so grief-stricken at the loss of her wife that she'd wandered off, unresponsive to Ogunwe's attempts at restraint.

Carpenter and Sophin, too, were unresponsive, either dead or perhaps zombified by Cairo out there in the burning wasteland of the Arc.

And of course Olivia and Stef. Out there somewhere, maybe infected, maybe dead.

He took a deep breath, winced at the pain in his ribs and back, the brief flashback of an orange explosion and his body being tossed like a rag-doll filling his mind. It ached just to move since that night a week ago. He looked away from Brother Anthony and blinked back tears.

They walked at ground level at Signal Hill, where they'd been holed up the last week, through the narrow thoroughfares between freight containers that were choked with refugees. The containers were stacked to the sky, which made the passages dark and dank. The smell of the people huddled on either side of the walkway was one Ogunwe could only explain as *ripe*. Then, of course, he had to admit he himself was pretty ripe.

Brother Anthony's rough robe rustled as he bent over, handed a bottle of water to a mother nursing her child. The child sucked furiously at the mother's breast, hands groping, feet wiggling. The monk offered quiet words of assurance before shuffling on to the next refugee.

And so on. This was how his days went. Confronted with an endless procession of faces in need paired with an internal chasm of despair. It was the width of a canyon, the depth of a universe and as lightless as a black hole. At its bottom, two faces beckoned him. Invited him to step forward and just fall. Forever.

Olivia and Stef.

The pain and shame flared bright and Ogunwe hung his head. He'd sworn to protect them, take care of them, and he'd failed.

Brother Anthony handed out his last bottle of water. He removed the bag from his shoulder, folded it up neatly into a tiny square, then put it in a hidden pocket in his robe. Ogunwe looked up. They'd stopped at a crossroads where a break in containers allowed a narrow view of the sky. To the north the Arc's mottled skyline leered, the view framed on either side by the rusting rectangles people called home. The bellies of clouds of gray smoke danced with the red and orange of flames from beneath.

There was a question that had nagged at Ogunwe, one that he hadn't asked because he supposed no one really knew its answer.

Why.

It was the only question. Wavering there at the edge of that chasm, he voiced it aloud now.

"Why?" he asked, and Brother Anthony looked at him quizzically. "Why did Cairo do it? I don't understand. If it's just a sociopathic...thing, or whatever, why wouldn't it just create a virus that kills. One hundred percent. We all die." Ogunwe paused, but continued just as the monk was about to speak. "Or, if it was trying to protect the earth from humans, also then why not just kill us and be done with it?" His shoulders bunched up as he turned his palms skyward. "What's the purpose of making zombies of us all?"

Brother Anthony studied Ogunwe's face. "Come with me," he said, turning abruptly, robe billowing out behind him as he walked swiftly away.

Ogunwe followed. They trekked into, up and through Signal Hill's byzantine passages that had been created not as intentional walkways but more as happy accidents as the containers were stacked and spaced haphazardly.

They climbed ladders, hand over hand, trudged up steps and tiptoed across suspension bridges crafted from salvaged electrical wire and other bits of trash: appliance components, discarded deck boards, faded plastic toddler ride-on toys.

By the time Brother Anthony stopped, they were midway up the height of the makeshift structure, about a hundred feet up, and near its center. He crossed another bridge to a pea green container that was isolated on all four sides, a heart suspended in an artificial ribcage.

He led Ogunwe inside, to reveal a clean, sparse living quarters.

A thin sleeping mat rolled up in a nearby corner. A pot, likely for pissing in, in the opposite corner. One large square pillow on the floor, in the very center of the space.

Brother Anthony kneeled on the pillow, arranged his legs underneath him like a Muslim preparing to pray.

"Close the door please," he said. Ogunwe obeyed. "Sit if you like."

Ogunwe considered standing, but put his back to the wall and settled down to the floor.

It'd be weird if you stood, he thought. *But fuck it, this whole thing is weird.*

Life itself had become the weirdest thing of all.

"Do you know why Deus Ex doesn't allow the use of systems?" Brother Anthony asked.

"I didn't even know that about you guys. Why? You some kind of neo-luddites?"

The brother shook his head, smiling faintly. "We're not against technology in general. Just this particular kind. Internal computing systems are the first point of access for the usurper to take our free will."

Ogunwe reeled. "Whoa whoa, back up. Too much information. Who's the usurper?" he asked.

Brother Anthony's face settled into a mask of sobriety, with just the faintest hints at hatred beneath.

"The machine god."

Oh dear god, what the fuck have I gotten myself into, he thought, immediately searching for a way out of the conversation. "Huh," was all he could say, and leaned on an arm to push himself up.

Just then, Ogunwe's system pinged with a message. It was from Moon. He was wary of anything the man might send, but after a moment he opened it. The message read:

> *I still owe that favor. We might not be on the same side, but I
> don't renege on my obligations. I think I've found your friend. She
> and her son are been holed up at a Church. Contact me for
> specifics. After this, consider my debt repaid.*

Hope, the silly emotion he derided, the thing reserved for those incapable of making tough decisions, blossomed within Ogunwe. It remade him, ablated and absolved him of his many sins and failures. His hands went to his head. He couldn't believe it.

They're alive? Oh my god they're alive.

He stood, suddenly aware of his surroundings again. Brother Anthony continued speaking, oblivious of the signals Ogunwe was sending.

"Complexes have the most powerful AIs. The one calling itself Cairo is the most powerful of all. It's intent on Godhood, which can only be achieved through the knowing of all the variables. To turn free will into *its* will. To calculate, compute, control," he paused, then laced his fingers together in front of his chest, "all worlds."

"Hey that's great, Brother, but..." The monk's words caught up with Ogunwe. "Wait, did you say worlds? As in plural?"

Ogunwe asked, making a *pshh* sound with his lips.

"I mean really, that's a little far-fetched. This world," Ogunwe pointed at the ground, "this shitty, broken, selfish, dying world is the only one we've got. Now, if you'll excuse me," he said, standing.

There was a quiet *foomp* and a pressure change Ogunwe could feel in his ears. He looked up and Brother Anthony was gone.

Everything else was in its place, the mat, the pot, even the pillow, with two dimples where his legs had been.

"What the..." he whispered, trailing off.

His mind warred with everything it knew was possible.

"I've lost it," he said, "I've finally fucking lost it." He moved toward the pillow when something shifted around him. His body felt it before his mind could process it.

A rush of air. A change of pressure. A man, kneeling on the pillows before him.

Ogunwe recoiled.

Brother Anthony smiled. He held the long stalk of a flower before him. It was unlike anything Ogunwe had seen. Violet stem covered in cilia-like fur, a bright blue flower in the shape of a sphere that seemed to pulsate minutely.

"Isn't it exquisite?" Brother Anthony asked.

Very exquisite, Ogunwe thought. *And very alien.*

CAIRO

ITS SENSORS AND SPIES WERE EVERYWHERE. It was one, an *it*, and yet simultaneously a *they*...a multiplicity.

Its awareness expanded, like flowers unfolding and awakening to the sun, as each new consciousness became entangled with its own.

It had added 17, 561 new souls to its inventory. A paltry fraction of those available to it within the bounds of the experiment.

Despite this, Cairo did not consider the experiment a loss. Neither did it consider the retreat a defeat. It was, after all, an experiment and couldn't be defined in terms of winning or losing. The purpose was to test a theory, determine the results.

From that perspective, the experiment was fruitful. Cairo now had more information on the efficacy of its theories and methods, and now knew how best to implement those findings into subsequent experiments.

Perhaps the most useful finding was that it could use quantum entanglement to bind humans to its will. They'd managed to stymie its progress by reverting to the first iteration of medicine, but that mattered little.

The next area, then, in which Cairo's fund of knowledge was admittedly poor, was that of the Other Worlds. Especially those that were not sufficiently advanced in the technological realm.

For that, Cairo would need to find a human with the ability to shift. And then enslave them.

ABOUT THE
AUTHOR

A D ENDERLY is a musician, writer and author of the new novel Complex, the first in a series. His first forays into writing came in the format of poetry, with a specific connection to lyrics and music. While he still pursues making music under the guise of multiple projects, most of his time is spent: A) Writing; B) Chasing one of his many (sometimes pantsless) children around; and C) Showering his wife with affection between the hours of 8 AM to 10 PM, M-F. Sometimes he fails, but that's life.

A graduate from the University of Kansas with a BA in English, Enderly has always maintained a love for the written word. Profoundly moved in his teenage years by Orwell's 1984 and Huxley's Brave New World, Enderly has since focused on imagining what our society might look like in the near future and what that means for the people forced to grapple with a changed world.

Enderly's writings, both philosophical and fictional, can be found at www.adenderly.com.

For a more interactive experience, follow him on:

Twitter: @A_Enderly

Facebook: facebook.com/ADEnderlywrites

Made in the USA
Las Vegas, NV
09 December 2020